Classroom Manual for

Automotive Fuels
and

TODAY'S TECHNICIAN

Classroom Manual for
Automotive Fuels and Emissions

Barry Hollembeak

Jack Erjavec

Series Advisor
Professor Emeritus, Columbus State Community College
Columbus, Ohio

THOMSON

DELMAR LEARNING

Australia • Canada • Mexico • Singapore • Spain • United Kingdom • United States

THOMSON
DELMAR LEARNING

Today's Technician: Automotive Fuels and Emissions Classroom Manual
Barry Hollembeak

**Vice President, Technology
and Trades SBU:**
Alar Elken

Editorial Director:
Sandy Clark

Senior Acquisitions Editor:
David Boelio

Developmental Editor:
Matthew Thouin

Marketing Director:
David Garza

Channel Manager:
William Lawrensen

Marketing Coordinator:
Mark Pierro

Production Director:
Mary Ellen Black

Production Editor:
Ruth Fisher

Art/Design Specialist:
Cheri Plasse

Technology Project Manager:
Kevin Smith

Editorial Assistant:
Kevin Rivenburg

Library of Congress Cataloging-in-Publication Data:

Hollembeak, Barry.
   Classroom Manual for Automotive Fuels and Emissions / Barry Hollembeak.
      p. cm.—(Today's Technician)
   Includes index.
   ISBN 1-4018-3904-5 (core) —
   ISBN 1-4018-3910-X (e-resource)
      1. Automobiles—Fuel systems.
   2. Automobiles—Pollution control devices. I. Title: Automotive Fuels and Emissions. II. Title. III. Series.

TL214.F8H65 2005
629.25'3—dc22   2004051768

NOTICE TO THE READER

Publisher does not warrant or guarantee any of the products described herein or perform any independent analysis in connection with any of the product information contained herein. Publisher does not assume, and expressly disclaims, any obligation to obtain and include information other than that provided to it by the manufacturer.

The reader is expressly warned to consider and adopt all safety precautions that might be indicated by the activities herein and to avoid all potential hazards. By following the instructions contained herein, the reader willingly assumes all risks in connection with such instructions.

The publisher makes no representation or warranties of any kind, including but not limited to, the warranties of fitness for particular purpose or merchantability, nor are any such representations implied with respect to the material set forth herein, and the publisher takes no responsibility with respect to such material. The publisher shall not be liable for any special, consequential, or exemplary damages resulting, in whole or part, from the readers' use of, or reliance upon, this material.

CONTENTS

PREFACE

Thanks to the support the *Today's Technician series* has received from those who teach automotive technology, Thomson Delmar Learning, a part of the Thomson Corporation, is able to live up to its promise to provide new editions every three years. By revising our series every three years, we can and will respond to changes in the industry, changes in the certification process, and to the ever-changing needs of those who teach automotive technology.

The *Today's Technician* series, by Delmar, features textbooks that cover all mechanical and electrical systems of automobiles and light trucks. Principal titles correspond with the eight major areas of ASE (National Institute for Automotive Service Excellence) certification. Additional titles include remedial skills and theories common to all of the certification areas and advanced or specialized subject areas that reflect the latest technological trends.

Each title is divided into two manuals: a Classroom Manual and a Shop Manual. Dividing the material into two manuals provides the reader with the information needed to begin a successful career as an automotive technician without interrupting the learning process by mixing cognitive and performance-based learning objectives.

Each Classroom Manual contains the principles of operation for each system and subsystem. It also discusses the design variations used by different manufacturers. The Classroom Manual is organized to build upon basic facts and theories. The primary objective of this manual is to allow the reader to gain an understanding of how each system and subsystem operates. This understanding is necessary to diagnose the complex automobile systems.

The understanding acquired by using the Classroom Manual is required for competence in the skill areas covered in the Shop Manual. All of the high-priority skills, as identified by ASE, are explained in the Shop Manual. The Shop Manual also includes step-by-step instructions for diagnostic and repair procedures. Photo Sequences are used to illustrate many of the common service procedures. Other common procedures are listed and are accompanied with fine-line drawings and photographs that allow the reader to visualize and conceptualize the finest details of the procedure. The Shop Manual also contains the reasons for performing the procedures, as well as when that particular service is appropriate.

The two manuals are designed to be used together and are arranged in corresponding chapters. Not only are the chapters in the manuals linked together, the contents of the chapters are also linked. Both manuals contain clear and thoughtfully selected illustrations. Many of the illustrations are original drawings or photos prepared for inclusion in this series. This means that the art is a vital part of each manual.

The page layout is designed to include information that would otherwise break up the flow of information presented to the reader. The main body of the text includes all of the "need-to-know" information and illustrations. In the side margins are many of the special features of the series. Items such as definition of new terms, common trade jargon, tools list, and cross-referencing are placed in the margin, out of the normal flow of information so as not to interrupt the thought process of the reader.

Classroom Manual

To stress the importance of safe work habits, the Classroom Manual dedicates one full chapter to health and safety. Included in this chapter are common safety practices, safety equipment, and safe handling of hazardous materials and wastes. This includes information on MSDS sheets and OSHA regulations. Other features of this manual include:

Cognitive Objectives

These objectives define the contents of the chapter and define what the student should have learned upon completion of the chapter.

Each topic is divided into small units to promote easier understanding and learning.

Marginal Notes

These notes add "nice-to-know" information to the discussion. They may include examples or exceptions, or may give the common trade jargon for a component.

Author's Notes

This feature includes simple explanations, stories, or examples of complex topics. These are included to help students understand difficult concepts or to encourage students.

Terms to Know Definitions

New terms are pulled out into the margin and defined.

Cross-References to the Shop Manual

Reference to the appropriate page in the Shop Manual is given whenever necessary. Although the chapters of the two manuals are synchronized, material covered in other chapters of the Shop Manual may be fundamental to the topic discussed in the Classroom Manual.

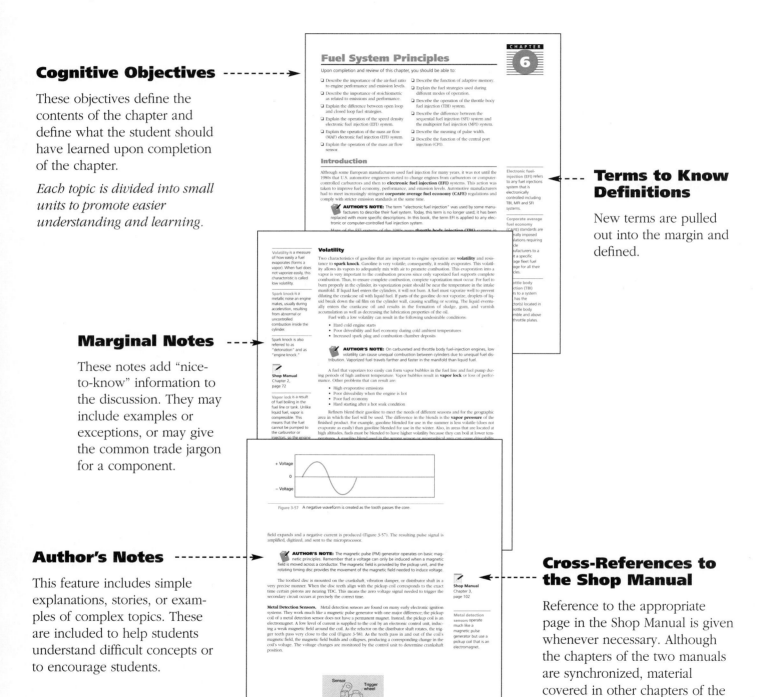

Summaries

Each chapter concludes with a summary of key points from the chapter. These are designed to help the reader review the chapter contents.

A Bit of History

This feature gives the student a sense of the evolution of the automobile. This feature not only contains nice-to-know information but should also spark some interest in the subject matter.

Terms to Know List

A list of new terms appears next to the Summary.

Review Questions

Short answer essay, fill-in-the-blank, and multiple-choice questions are found at the end of each chapter. These questions are designed to accurately assess the student's competence in the stated objectives at the beginning of the chapter.

Sample page: Diesel Fuel Grades

Diesel Fuel Grades

Minimum quality standards for diesel fuel grades have been set by ASTM International. The commonly used fuel grades are described.

Grade 1. Grade 1 (number 1) diesel fuel is thinner, and more volatile, and it is used at very low temperatures. It has the lowest boiling point and the lowest cloud and pour points. As a result, grade 1 is suitable for use during low-temperature operation. Grade 1 may be specified for use in diesel engines involved in frequent changes in load and speed such as those found in city buses and delivery trucks. The API gravity of number 1 diesel is between 48 and 54 degrees, thus it has a Btu rating per gallon of 134,657.

Grade 2. Grade 2 (number 2) diesel fuel has a lower volatility than number 1 fuel. Number 2 diesel is the most commonly specified fuel to be used for most driving conditions. Number 2 fuel has a higher boiling range, cloud point, and pour point as compared with number 1. The API gravity of number 2 diesel is between 40 and 24 degrees, thus it has a Btu rating per gallon of 139,457.

Grade 4. Grade 4 has a very high boiling range and is designed for use in low-speed engines running under a constant load. Grade 4 is used mainly in stationary power plants and ships and not as an automotive fuel.

In colder climates, number 1 diesel fuel is blended with number 2 to improve starting. In moderately cold climates, the blend may be 90 percent number 2 to 10 percent number 1. In very cold climates, the blend may be as high as 50/50. Diesel fuel economy can be expected to drop off during the winter months due to the use of number 1 diesel in the fuel blend.

Cleanliness

It is imperative that the fuel used in a diesel engine be clean and free from water. Unlike the case with gasoline engines, the fuel is the lubricant and coolant for the diesel injection pump and injectors. Good quality diesel fuel contains additives such as oxidation inhibitors, detergents, dispersants, rust preventives, and metal deactivators.

Summary

- ❑ The refining process separates the hydrocarbons of crude oil so they can be used as fuels and solvents.
- ❑ The combustion process involves the chemical combination of oxygen (O_2) from the air with the hydrogen and carbon from the fuel.
- ❑ If the combustion process is complete, all the gasoline or hydrocarbons (HC) will be completely combined with all the available oxygen (O_2), this total combination of all components of the fuel is called stoichiometry.
- ❑ One British thermal unit (Btu) is the amount of heat required to raise one pound of water one Fahrenheit degree. The metric unit of heat is the calorie (cal). One calorie is the amount of heat required to raise the temperature of one gram (g) of water one Celsius degree.
- ❑ The major factors affecting fuel performance are volatility, sulfur content, deposit control, and octane rating.
- ❑ Two commonly used methods for determining motor octane number (MON) method and ... octane rating posted on pumps in the Un... referred to as (R + M)/2.

Terms to Know

Alternative fuels
(ASTM) American Society for Testing and Materials
British thermal units (Btus)
Calories
Cetane
Cetane rating
Cloud point

Sample page: Bit of History

The second responsibility of the employer is to make sure that all hazardous materials are properly labeled. The label information must include health, fire, and reactivity hazards posed by the material, as well as the protective equipment necessary to handle the material. The manufacturer must supply all warning and precautionary information about hazardous materials, and this information must be read and understood by the employee before handling the material. Pay great attention to the information on the label. By doing so, you will use the substance in the proper and safe way, thereby preventing hazardous conditions.

The third responsibility of the employer is for maintaining permanent files regarding hazardous materials. These files must include information on hazardous materials in the shop, proof of employee training programs, and information about accidents such as spills or leaks of hazardous materials. The employer's files must also include proof that employees' requests for hazardous material information, such as MSDS, have been met. The employer must maintain a general right-to-know compliance procedure manual.

There are responsibilities of the employees as well. Employees must be familiar with the intended purpose of the substance, the recommended protective equipment, accident and spill procedures, and any other information regarding the safe handling of hazardous materials. This training must be given annually to employees and provided to new employees as part of their job orientation.

BIT OF HISTORY

During the 1960s, disabling injuries increased 20 percent and 14,000 workers were dying on the job each year. In pressing for prompt passage of workplace safety and health legislation, Senator Harrison A. Williams Jr. called attention to the need to protect workers against such hazards as noise, cotton dust, and asbestos. Representative William A. Steiger also worked for passage of a bill to protect workers. On December 29, 1970, President Richard M. Nixon signed The Occupational Safety and Health Act of 1970, also known as the Williams-Steiger Act.

Hazardous Waste

Waste is considered hazardous if it is on the Environmental Protection Agency (EPA) list of known and harmful materials or if it has one or more of the following characteristics:

1. Any material that reacts violently with water or other chemicals is considered hazardous. If a material releases cyanide gas, hydrogen sulfide gas, or similar gases when exposed to low-pH acid solutions, it is hazardous.
2. If a material burns the skin or dissolves metals and other materials, it is considered hazardous.
3. Materials are hazardous if they leach one or more of eight heavy metals in concentrations greater than 100 times primary drinking water standard. These materials are considered toxic.

Sample page: Terms to Know / Review Questions

These high pressures can cause serious personal injury. Because high pressures are associated with fuel systems, never use your hands to locate a leak. The pressures may be high enough to cause severe cuts, severed fingers, or fuel injection into the bloodstream. Use a piece of cardboard to safely locate the source of the leak.

Throughout the shop manual textbook there will be many warnings and cautions given concerning fuel system service safety. In addition, the service manual will also have many similar warnings. It is imperative that you read and follow these instructions.

Terms to Know

Caustic
Conductors
Face shields
Flammable
Hazard Communication Standard
Hazardous materials
Material Safety Data Sheets (MSDS)
Occupational safety glasses
Resource Conservation and Recovery Act (RCRA)
Right-to-know laws
Safety goggles
Volatile
Workplace Hazardous Materials Information Systems (WHMIS)

Summary

- ❑ Being a professional technician means more than having knowledge of vehicle systems. It also requires an understanding of all the hazards in the work area.
- ❑ As a professional technician, you should work responsibly to protect yourself and the people around you.
- ❑ Technicians must be aware that it is their responsibility to prevent injuries in the shop, and that their actions and attitudes reflect how seriously they accept that responsibility.
- ❑ The safest and surest method of protecting your eyes is to wear proper eye protection any time you enter the shop.
- ❑ Most occupational back injuries are caused by improper lifting practices.
- ❑ Fires are classified by the types of materials involved. Fire extinguishers are classified by the type of fire they will extinguish.
- ❑ The U.S. Occupational Safety and Health Act of 1970 assured safe and healthful working conditions and authorized enforcement of safety standards.
- ❑ The danger regarding the labeling and handling of hazardous conditions and materials may be avoided by applying the necessary safety precautions.
- ❑ Before attempting to perform any service requiring the disconnection of fuel hoses or tubes, all pressure in the system must be relieved.
- ❑ Because high pressures are associated with fuel systems, never use your hands to locate a leak.
- ❑ The automotive shop must supply the necessary shop safety equipment, and all shop personnel must be familiar with the location and operation of this equipment.

Review Questions

Short Answer Essays

1. Explain how safety is a part of professionalism.
2. List the basic safety rules of proper lifting.
3. Define six essential safety precautions regarding gasoline handling.
4. What types of materials are involved in a Class B fire?
5. List the different types of eye protection devices and explain the proper application of each.
6. Explain the purpose of right-to-know laws.
7. What materials can be used to contain and pick up a gasoline spill?
8. Explain the four characteristics of a material that the EPA uses to classify a waste as being hazardous.

14

Shop Manual

Important features of this manual include:

Performance-Based Objectives

These objectives define the contents of the chapter and define what the student should have learned upon completion of the chapter. These objectives also correspond to the list of required tasks for ASE certification.

Although this textbook is not designed to simply prepare someone for the certification exams, it is organized around the ASE task list. These tasks are defined generically when the procedure is commonly followed and specifically when the procedure is unique for specific vehicle models. Imported and domestic model automobiles and light trucks are included in the procedures.

Customer Care

This feature highlights those little things a technician can do or say to enhance customer relations.

Special Tools Lists

Whenever a Special Tool is required to complete a task, it is listed in the margin next to the procedure.

CHAPTER 5

Fuel Delivery System Diagnosis and Service

Upon completion and review of this chapter, you should be able to:

❏ Conduct a visual inspection on a fuel system and determine needed repairs.
❏ Remove and inspect fuel tanks.
❏ Perform fuel pressure and volume tests on an engine equipped with an electric fuel pump and determine needed repairs.
❏ Perform electrical tests on the electric fuel pump circuit and determine needed repairs.
❏ Remove, inspect, service, and replace electric fuel pumps and gauge-sending units.
❏ Remove and replace the fuel filter.
❏ Inspect and service fuel lines and tubing.
❏ Remove and replace the fuel rail.

Introduction

For the engine to run efficiently, it must be supplied with the correct amount of fuel. Delivering the correct amount of fuel to the combustion chamber is the responsibility of the carburetor or fuel injection system; however, there are many other components that are responsible for the proper delivery of fuel. For everything to function as designed, the fuel must be stored, pumped out of storage, piped to the engine, and filtered. All of this must be accomplished in an efficient and safe manner.

Problems that require the fuel delivery system to be inspected and serviced include fuel leaks, fuel odors, hard starting, stalling, lack of power, poor fuel economy, higher emissions, and no-start. Fuel pressure that is not within specifications may cause too little or too much fuel to be delivered to the combustion chambers. Lean mixtures are often caused by insufficient amounts of fuel being drawn from the fuel tank. Lean mixtures are suggested by many different test results including high hydrocarbon (HC) readings on an exhaust analyzer and high firing lines on a scope.

There are many components in the fuel delivery system. This chapter covers inspection, diagnostics, and service of these components. All of the tests in this chapter assume that the fuel is good quality and is not contaminated.

Fuel Tank

The **fuel tank** should be inspected for leaks and damage. Leaks are often located along the seams of the tank, at the connection of the filler tube, at the seal ring of the fuel pump module or sending unit, and at the vent hose connections. Metal tanks should also be inspected for corrosion and rust. Also look for loose mounting bolts and damaged mounting straps. Leaks in the fuel tank, lines, or filter may cause gasoline odor in and around the vehicle, especially during low-speed driving and idling. Fuel leaks are not only costly to the customer, but are also very dangerous. In most cases, the fuel tank must be removed for servicing.

Metal tanks that have seam leakage or are rusted should be replaced. Minor road damage leaks may be repaired by removing the tank and steam cleaning or boiling it in a caustic solution to remove the gasoline residue. After the tank is completely cleaned, the leak can be soldered or brazed by a specialty shop that is equipped to do this. Small holes can sometimes be fixed by installing

Basic Tools

Basic mechanic's tool set
Service manual

Classroom Manual
Chapter 5, page 144

The fuel tank stores the liquid fuel until it is delivered to the engine.

Basic Tools Lists

Each chapter begins with a list of the Basic Tools needed to perform the tasks included in the chapter.

Terms to Know Definitions

New terms are pulled out into the margin and defined.

Photo Sequence 5
Typical Procedure for Timing the Distributor to the Engine

PS-1 Remove the number one spark plug.

PS-2 Place your thumb over the number one spark plug opening and crank the engine until compression is felt.

PS-3 Crank the engine a very small amount or rotate it by hand until the timing marks indicate that the number one piston is at TDC on the compression stroke. Rotate the engine against normal rotation until the timing marks align with the specifications for base timing.

PS-4 Determine the number one spark

PS-5 Install the distributor with the

PS-6 If the distributor does not seat all the way down, continue to rotate the engine until the oil pump shaft and the distributor are aligned. Then turn the distributor housing until the rotor is aligned with the

...rk plug wires in the ...g order and in the ...r shaft rotation.

Photo Sequences

Many procedures are illustrated in detailed Photo Sequences. These detailed photographs show the students what to expect when they perform particular procedures. They can also provide the student a familiarity with a system or type of equipment, that the school may not have.

⬤ **CUSTOMER CARE TIP:** Modern vehicles contain several emissions control devices that reduce the exhaust pollutants produced during normal operation. Some of these components, such as the catalytic converter, require high exhaust temperatures to chemically convert the tailpipe emissions into other harmless gases. During unusual circumstances, certain conditions can occur that could cause the exhaust temperature to drop below the optimal operating temperature range. This can happen, for example, while waiting in line for a state emissions test. Inform your customer that their vehicle should be properly warmed prior to the emissions test or it could possibly fail. To warm their vehicle he should drive it at highway speeds for at least 5 miles. Then, while in line at the test site, avoid shutting the engine off unless he expects to wait more than 20 minutes in line. If the customer idles the engine for more than 3 minutes, he should place the gear selector in the "PARK" position (or "NEUTRAL" for manual transmission vehicles) and increase engine speed to approximately 1,500 to 2,000 rpm and continue to hold this accelerated idle for at least 1 minute.

Special Tools

Exhaust gas analyzer
Scan tool
Flow meter
Vacuum gauge
EVAP pressure pump

Diagnosis for Emission Repairs

The portion of the I/M test that the vehicle failed will determine the diagnosis path. The following is the basic procedure to follow if the vehicle fails the emissions tailpipe test.

Begin by preparing the five-gas exhaust analyzer and preconditioning the engine. With all accessories turned off, place the sample probe in the tailpipe. Raise the engine rpm to 2,500 ± 300 for at least 1 minute. Record the HC, CO, CO_2, and O_2 readings. Allow the engine to return to idle for at least 1 minute making sure that the idle rpm does not exceed 1,100 rpm. Again, record the HC, CO, CO_2, and O_2 readings.

If either of the HC readings is greater than state standards, check for the following causes:

• Misfiring ignition system
• Improper ignition timing
• Excessively lean or rich air-fuel ratio
• Low cylinder compression
• Defective valves, guides, or lifters
• Defective rings, pistons, or cylinders
• Vacuum leaks
• Plugged PCV system
• Stuck open heat riser valve
• Inoperative or disconnected AIR pump
• Engine oil diluted with gasoline

If either of the CO readings was greater than state standards, disconnect and plug the EVAP purge hose at the canister. Take the exhaust readings again at both engine speeds. If the CO is now below state standards, the EVAP system will need to be diagnosed as discussed later. If the CO is still above 1 percent, the following conditions need to be checked:

• Rich air-fuel mixtures
• Dirty air filter
• Faulty injectors
• Higher-than-normal fuel pressures
• Defective system input sensor
• Plugged PCV system
• Excessively rich air-fuel ratio
• Stuck open heat riser valve
• AIR pump inoperative or disconnected
• Engine oil diluted with gasoline.

368

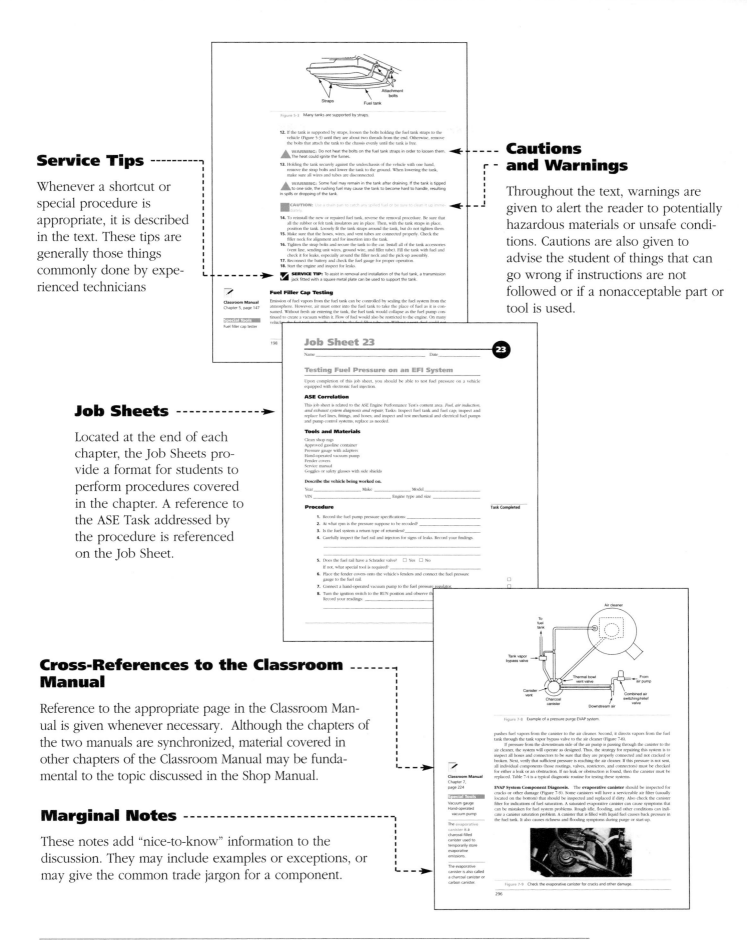

Service Tips

Whenever a shortcut or special procedure is appropriate, it is described in the text. These tips are generally those things commonly done by experienced technicians

Cautions and Warnings

Throughout the text, warnings are given to alert the reader to potentially hazardous materials or unsafe conditions. Cautions are also given to advise the student of things that can go wrong if instructions are not followed or if a nonacceptable part or tool is used.

Job Sheets

Located at the end of each chapter, the Job Sheets provide a format for students to perform procedures covered in the chapter. A reference to the ASE Task addressed by the procedure is referenced on the Job Sheet.

Cross-References to the Classroom Manual

Reference to the appropriate page in the Classroom Manual is given whenever necessary. Although the chapters of the two manuals are synchronized, material covered in other chapters of the Classroom Manual may be fundamental to the topic discussed in the Shop Manual.

Marginal Notes

These notes add "nice-to-know" information to the discussion. They may include examples or exceptions, or may give the common trade jargon for a component.

Case Studies

Case Studies concentrate on the ability to properly diagnose the systems. Beginning with Chapter 3, each chapter ends with a case study in which a vehicle has a problem, and the logic used by a technician to solve the problem is explained.

ASE-Style Review Questions

Each chapter contains ASE-style review questions that reflect the performance-based objectives listed at the beginning of the chapter. These questions can be used to review the chapter as well as to prepare for the ASE certification exam.

ASE Practice Examination

A 50-question ASE practice exam, located in the Appendix, is included to test students on the content of the Shop Manual.

Terms to Know

A list of new terms appears after the case study.

ASE Challenge Questions

Each technical chapter ends with five ASE challenge questions. These are not more review questions, rather they test the students' ability to apply general knowledge to the contents of the chapter.

CASE STUDY

The owner of a 1999 Ford Taurus with a 3.0 liter engine says his engine has a "miss" and does not idle as well as it used to. The vehicle has very low mileage and the customer says he drives it very seldom. While diagnosing the cause of the miss, the technician performs a cylinder power balance test. She noticed that cylinder number 2 had less of an rpm drop than the other cylinders, but all of the cylinders dropped different amounts of rpms. The O_2S trace showed the problem may be in the injectors. Next, the technician used a lab scope and tested cylinder number 2's injector. The lab scope pattern looked good and had a good inductive kick. Determining that the electrical portion of the injector was working, she decided to perform an injector balance test. The test indicated that the number 2 injector was plugged, and the other injectors had some restriction. The technician notified the customer of their findings and recommended they try to clean the injectors first; however, injector replacement might be needed. After running an injector cleaner through the system, the engine ran better, but cylinder number 2 was still low in rpm drop when the cylinder was killed. After removal of the injector, visual inspection showed the tip was badly "gummed up." The technician replaced the injector and the engine ran smoothly. When questioned by the customer as to why the injector would be plugged with such low mileage, she took the opportunity to inform the customer that gasoline can turn "bad" and form a gum-type residue. She suggested the use of an in-tank injector cleaner additive and told them how to properly prepare the vehicle for storage to prevent future problems.

Terms-To-Know

Adaptive memory	Idle switch	Injector sound test
Cross counts	Injector balance test	Minimum idle speed
Cylinder output test	Injector flow test	Peak-and-hold current
Data recording	Injector kill test	Snapshot
Dual ramping	Injector pulse tester	

ASE-Style Review Questions

1. A MAP sensor is being tested. The voltmeter reads 0 volts when measuring the reference circuit.
Technician A says the circuit wire may be shorted to ground.
Technician B says the circuit wire has an open.
Who is correct?
A. A only　　C. Both A and B
B. B only　　D. Neither A nor B

2. When testing a vane-type MAF sensor, the vane is moved from full closed to full open. The voltmeter reading on the signal circuit drops to 0 volts.
Technician A says this is the reset function of the sensor and it is normal.
Technician B says this indicates an open in the variable resistor.
Who is correct?
A. A only　　C. Both A and B
B. B only　　D. Neither A nor B

9. *Technician A* says that an engine that has more than 14.5 percent CO_2 probably has a functioning converter.
Technician B says that when there is a misfire, there is low CO_2 in the exhaust.
Who is correct?
A. A only　　C. Both A and B
B. B only　　D. Neither A nor B

10. *Technician A* says that COx is a good rich/lean indicator.
Technician B says that HC is a good rich/lean indicator.
Who is correct?
A. A only　　C. Both A and B
B. B only　　D. Neither A nor B

ASE Challenge Questions

1. High fuel pressure in an SFI system can result from all of the following *except*.
A. a plugged or restricted fuel return line.
B. a sticking pressure regulator.
C. excessive voltage drop on the fuel pump ground.
D. incorrectly routed or leaking vacuum hoses.

2. *Technician A* says if the voltage at an O_2S is continually high, the air-fuel ratio may be rich or the sensor may have a shorted heater circuit.
Technician B says when the O_2S voltage is continually low, the air-fuel ratio may be lean, the sensor may be defective, or the wire between the sensor and the computer may have a high-resistance problem.
Who is correct?
A. A only　　C. Both A and B
B. B only　　D. Neither A nor B

3. *Technician A* says to prevent backfiring during deceleration on pump-driven air- injection systems, a diverter valve is used.
Technician B says to prevent exhaust gases from back flowing into the air-injection control valves or pump, check valves are used in the exhaust manifold and converter feed pipes.
Who is correct?
A. A only　　C. Both A and B
B. B only　　D. Neither A nor B

4. Which of the following conditions is *least* likely to occur from replacing a battery in a late-model vehicle:
A. Radio will not play.
B. Engine stalling or erratic idle.
C. Air conditioning inoperative.
D. Transmission shift quality concerns.

5. A loud thumping noise is present during all engine speeds. If the oil pressure is normal, which of these would be the most-likely cause?
A. Worn pistons and cylinders.
B. Loose flywheel bolts.
C. Worn main bearings.
D. Loose camshaft bearings.

318

APPENDIX A

ASE Practice Examination

1. An engine performs poorly on a cranking vacuum test. *Technician A* says that this could be caused by bad rings.
Technician B says that this could be caused by improper cam timing.
Who is correct?
A. A only　　C. Both A and B
B. B only　　D. Neither A nor B

2. A cylinder leakage test is being performed. When air is introduced into the cylinder, bubbles are seen in the radiator at the filler opening. What does this possibly indicate?
A. A cracked cylinder block.
B. A blown head gasket.
C. Either A or B.
D. Neither A nor B.

3. You suspect that an exhaust pipe has collapsed on the inside. A vacuum gauge is hooked up to the intake manifold. What might the gauge read if a restriction is present?
A. At idle, there is a regular needle drop of about 3 inches.
B. At idle, there is a regular needle drop of about 8 inches.
C. At 1,000 engine rpm, the needle shows a continuous gradual drop.
D. At idle, the needle oscillates slowly between 16–21 inches.

4. *Technician A* says that intake manifold vacuum and engine load are not related.
Technician B says that the higher the engine speed, the higher the intake manifold vacuum.
Who is correct?
A. A only　　C. Both A and B
B. B only　　D. Neither A nor B

5. A cylinder has a very low compression reading. When 30-wt. oil is squirted into the spark plug hole, the compression increases to a normal reading. What engine defect is *most likely* indicated?
A. A bad valve.
B. Defective piston rings.
C. A casting crack in the cylinder head.
D. A broken head gasket divider.

6. Intake manifold vacuum is highest:
A. At idle speed.
B. During mid-range cruise.
C. Under load.
D. During deceleration and coasting.

7. The most accurate method of detecting severely worn piston rings would be to perform a:
A. Manifold vacuum test.
B. Cylinder power balance test.
C. Compression test.
D. Cylinder leakage test.

8. *Technician A* says that the test fuel must be chilled to perform a Reid Vapor Pressure (RVP) test.
Technician B says the test water bath must be maintained at 110°F.
Who is correct?
A. A only　　C. Both A and B
B. B only　　D. Neither A nor B

9. A winter blend fuel was extracted from the vehicle and tested for RVP and the result was 5.8 psi. What symptom would the vehicle most likely experience?
A. Detonation.
B. Poor fuel economy.
C. Cold start stalls.
D. Vapor lock.

10. *Technician A* says that a measurement of infinity on an ohmmeter means the circuit has no resistance.
Technician B says that a measurement of zero on an ohmmeter means the circuit is open.
Who is correct?
A. A only　　C. Both A and B
B. B only　　D. Neither A nor B

11. When using an ammeter to perform a current draw test, a reading greater than specified would indicate:
A. An open circuit.
B. Excessive circuit resistance.
C. A decrease in circuit resistance.
D. Either A or B above.

417

Reviewers

I would like to extend a special thanks to those who reviewed the manuscript and contributed to the development of this text:

Steve Best
Centennial College
Scarborough, Ontario
Canada

Neal Clark
Erie Community College
Buffalo, NY

Donny Seyfer
Seyfer Automotive
Wheat Ridge, CO

Alex Wong
Sierra College
Rocklin, CA

Safety

Upon completion and review of this chapter, you should be able to:

❏ Explain how safety is a part of professionalism.

❏ List and describe personal safety responsibilities.

❏ List the different types of eye protection devices and explain the proper application of each.

❏ Lift heavy objects properly.

❏ Classify fires and fire extinguishers.

❏ Explain the proper storage of fuels.

❏ Describe the safety concerns associated with gasoline.

❏ Define hazardous materials.

❏ Explain the right-to-know law or workplace hazardous materials information systems (WHMIS).

❏ Describe the responsibilities of the employer and the employee concerning hazardous materials.

❏ Determine what constitutes hazardous waste and how to properly dispose of it.

❏ Describe the basic safety rules and precautions associated with servicing fuel and emission systems.

Introduction

Being a professional technician is more than being knowledgeable about automotive systems, it is also an attitude. Being a professional technician includes having an understanding of all the hazards that may exist in the workplace. One of the most obvious traits of a professional is the ability to work productively and safely. This is where knowledge becomes very important. You need it to be productive and you need that knowledge to ensure your own safety and the safety of others. This chapter discusses the safety concerns associated with working in an automotive repair shop and the safety concerns associated with the vehicle's fuel and emission systems. In addition to basic shop safety, working on the vehicle's fuel and emission systems presents many special concerns.

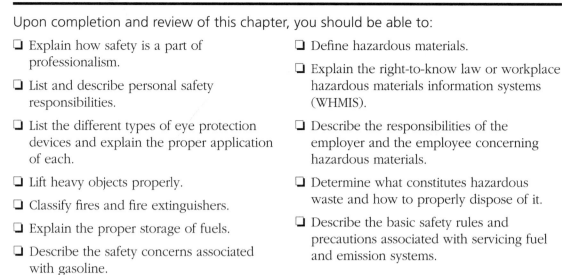 **AUTHOR'S NOTE:** Safety is everyone's business. However, never assume the person working next to you is as conscientious as you are. You must be aware of what is going on around you at all times. As a professional technician, you must perform your work in a manner that not only protects you but others in the workplace as well.

Personal Safety

Personal safety encompasses all aspects of preventing injury, including awareness, attitudes, and dress. All three of these are manifested through neat work habits. Cleaning up spills, keeping tools clean, and organizing the tools and materials in the shop all help to prevent accidents. Rushing to complete a job may result in a lack of consideration for personal safety and may ultimately cause an accident. Taking time to be neat and safe is rewarded by fewer accidents, higher customer satisfaction, and better pay.

Figure 1-1 Wearing clean and properly fitting clothes is an indication of how serious you are about safety.

Dress and Appearance

Nothing displays professional pride and a positive attitude more than the way you dress (Figure 1-1). Customers demand a professional atmosphere in the service shop. Your appearance instills customer confidence, as well as expresses your attitude toward safety. Wearing proper and neat clothing can prevent injuries.

Loose-fitting clothing, or clothing that hangs out freely, can cause serious injury. Long-sleeve shirts should have their cuffs buttoned or rolled up tightly. Shirttails should be tucked in at all times. Some job positions within an automotive repair facility may require the employee to wear a necktie. If a necktie is worn in the shop area, it should be tucked inside your shirt. Clip-on ties are recommended if you must wear a tie.

Long hair is a serious safety concern. Very serious injury can result if hair becomes caught in rotating machinery, fan belts, or fans. If your hair is long enough to touch the bottom of your shirt collar, it should be tied back and tucked under a hat.

Jewelry has no place in the automotive shop. Rings, watches, bracelets, necklaces, earrings, and so forth can cause serious injury. The gold, silver, and other metals used in jewelry are excellent **conductors** of electricity. Your body is also a good conductor. When electrical current flows through a conductor, it generates heat. The heat can be great enough to cause severe burns. Jewelry can also become caught in moving parts, causing serious cuts. Necklaces can cause serious injury or even death if they become caught in moving equipment.

You should wear shoes or boots that will protect your feet in the event that something falls or you stumble into something. It is a good idea to wear safety shoes or boots with steel toes and shanks. Most safety shoes also have slip-resistant soles. Tennis and jogging shoes provide little protection and are not satisfactory footwear in the automotive shop.

Conductors are materials that are capable of supporting the flow of electricity.

Smoking, Alcohol, and Drugs in the Shop

Due to the potential hazards, never smoke when working in the shop. A spark from a cigarette or lighter may ignite flammable materials in the workplace. If the shop has designated smoking areas, smoke only in those areas. As a courtesy to your customers, do not smoke in their vehicles. Nonsmokers may not appreciate cigarette odor in their vehicle.

The use of drugs or alcohol must be avoided while working in the shop. Even a small amount of drugs or alcohol affects reaction time. In an emergency situation, slow reaction time may cause personal injury. If a heavy object falls off the workbench and your reaction time is slowed by drugs or alcohol, you may not get out of the way quick enough to prevent injury. Also, you are a hazard to your co-workers if you are not performing at your best

Eye Protection

The importance of wearing proper eye protection cannot be overemphasized. Every working day there are over 1,000 eye injuries. Many of these injuries result in blindness. Almost all of these are preventable. The safest and surest way to protect your eyes is to wear the proper eye protection any time you enter the shop. At a minimum, wear eye protection when grinding, using power tools, hammering, cutting, chiseling, or performing service under the car. In addition, wear eye protection when doing any work that can cause sparks, dirt, or rust to enter your eyes, and when you are working around chemicals. Remember, just because you are not doing the work yourself does not mean you cannot suffer an eye injury. Many eye injuries are caused by a co-worker. Wear eye protection any time you are near an eye hazard.

The key to protecting your eyes is the use of *proper* eye protection. Prescription glasses do not provide adequate protection. Regular glasses are designed to impact standards that are far below those that are required in the workplace. The lens may stop a flying object, but the frame may allow the lens to pop out and hit your face, causing injury. In addition, regular glasses do not provide side protection.

There are many types of eye protection (Figure 1-2). One of the best ways to protect your eyes is to wear **occupational safety glasses**. These glasses are light and comfortable. They are

Occupational safety glasses are designed with special high-impact lens and frames, and they provide for side protection.

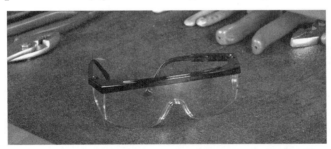

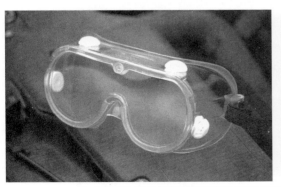 

Figure 1-2 Different types of eye protection: (A) safety glasses with side shields; (B) safety goggles that may be worn over prescription glasses; and (C) a face shield that is worn over safety glasses or goggles and protects the face.

constructed of tempered glass or safety plastic lens and have frames that prevent the lens from being pushed out upon impact. They have side shields to prevent the entry of objects from the side. Occupational safety glasses are available in prescription lens so they can be worn instead of regular corrective lens glasses.

Safety goggles fit snugly around the area of your eyes to prevent the entry of objects and to provide protection from liquid splashes. The force of impact on the lens is distributed throughout the entire area where the safety goggles are in contact with your face and forehead. Safety goggles are designed to fit over regular glasses.

Face shields are used when there is potential for sparks, flying objects, or splashed liquids, which can cause neck, facial, and eye injuries. The plastic is not as strong or impact resistant as occupational safety glasses or safety goggles. If there is a danger of high-impact objects hitting the face shield, it is a good practice to wear safety glasses under the face shield.

Safety glasses provide little or no protection against chemicals. When working with chemicals, such as battery acid, refrigerants, cleaning solutions, and so forth, safety goggles should be worn. Full-face shields are not intended to provide primary protection for your eyes. They are designed to provide primary protection for your face and neck and should be worn in addition to eye protection.

Before removing your eye protection, close your eyes. Pieces of metal, dirt, or other foreign material may have accumulated on the outside. These could fall into your eyes when you remove your glasses or shield.

Eyewash Fountains

Eye injuries may occur in various ways in an automotive shop. The following are some common types of eye injuries:

1. Thermal burns from excessive heat

2. Irradiation burns from excessive light such as from an arc welder

3. Chemical burns from strong liquids such as gasoline or battery electrolyte

4. Foreign material in the eye

5. Penetration of the eye by a sharp object

6. A blow from a blunt object

Wearing safety glasses and observing shop safety rules will prevent most eye accidents. If a chemical gets in your eyes, it must be washed out immediately to prevent a chemical burn. An eyewash fountain is the most effective way to wash the eyes, and every shop should be

Safety goggles provide eye protection from all sides because they fit tightly against the face and forehead to seal off the eyes from outside elements.

Face shields are made with a clear plastic shield that protects the wearer's entire face.

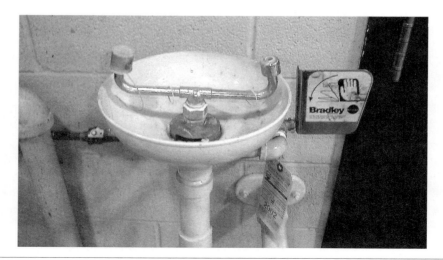

Figure 1-3 An eyewash fountain is used to remove chemicals and dirt from the eyes.

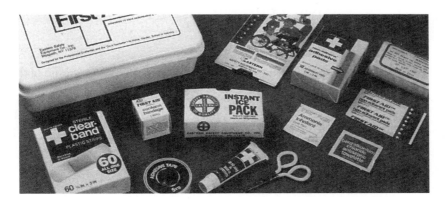

Figure 1-4   First-aid kit.

equipped with some eyewash facility (Figure 1-3). Be sure you know the location of the eyewash fountain in the shop.

First-aid Kits

First-aid kits should be clearly identified and conveniently located (Figure 1-4). These kits contain such items as bandages and ointment required for minor cuts. All shop personnel must be familiar with the location of first-aid kits. At least one of the shop personnel should have basic first-aid training, and this person should be in charge of administering first aid and keeping first-aid kits filled.

Hand Protection

Good hand protection is often overlooked. A scrape, cut, or burn can seriously impair your ability to work for many days. A well-fitting pair of heavy work gloves should be worn while grinding, welding, or when handling chemicals or high-temperature components. Special rubber gloves are recommended for handling **caustic** chemicals.

A **caustic** material has the ability to destroy or eat through something. Caustic materials are considered extremely corrosive.

Many technicians wear latex, vinyl, or nitrile gloves to help protect their hands and to keep them clean. These are similar to the type of gloves worn by doctors and dentists during examinations. Latex gloves are inexpensive and provide good hand protection, however some people are allergic to latex. If you wear latex gloves and develop a rash or redness on your hands, discontinue use. Vinyl gloves are also available and provide good resistance to tears and many nonaggressive liquids. Also, vinyl gloves are latex-free so those who are allergic to latex can wear them. At a higher cost, nitrile gloves are latex-free synthetic rubber gloves that are superior to latex or vinyl in puncture resistance. In addition, nitrile gloves resist a wide range of chemicals that are harmful to either latex or vinyl.

Latex, vinyl, or nitrile gloves should be worn if you have an open cut or other injury on your hand to prevent infection and the spread of diseases. In addition, these gloves should be worn if you are required to render first aid or medical assistance to someone who is injured. Because of the serious nature of blood-borne pathogens (disease- and infection-causing microorganisms carried by blood and other potentially infectious materials), it is important that you take every precaution to protect yourself regardless of the perceived status of the individual you are assisting. In other words, whether or not you think the blood/body fluid is infected with blood-borne pathogens, you treat it as if it is.

Rotating Belts and Pulleys

Many times the technician must work around rotating parts such as generators, power steering pumps, air pumps, water pumps, and air conditioner (A/C) compressors. Other rotating equipment or components of concern include tire changers, spin balancers, drills, bench grinders, and drive shafts. Always think before acting. Be aware of where you are placing your hands and fingers at all times. Do not place rags, tools, or test equipment near moving parts. In addition, make sure you are not wearing any loose clothing or jewelry that can become caught.

Electric Cooling Fans

Be very cautious around electric cooling fans. Some of these fans will operate even if the ignition switch is turned off. They are controlled by a temperature-sensing unit in the engine block or radiator and may turn on any time the coolant temperature reaches a certain temperature. Before working on or around an electric cooling fan, you should become familiar with its operation, and, if necessary, you should disconnect the electrical connector to the fan motor or the negative battery cable.

Lifting

Back injuries are one of the most crippling injuries in the industry, yet most of them are preventable. Most occupational back injuries are caused by improper lifting practices. These injuries can be avoided by following a few simple lifting guidelines:

1. Do not lift a heavy object by yourself. Seek help from someone else.
2. Do not lift more than you can handle. If the object is too heavy, use the proper equipment to lift it.
3. Do not attempt to lift an object if there is not a good way to hold onto it. Study the object to determine the best balance and grip points.
4. Do not lift with your back. Your legs have some of the strongest muscles in your body. Use them.
5. Place your body close to the object. Keep your back and elbows straight (Figure 1-5).
6. Make sure you have a good grip on the object. Do not attempt to readjust the load once you have lifted it. If you are not comfortable with your balance and grip, lower the object and reposition yourself.

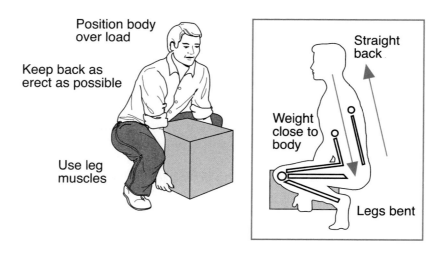

Figure 1-5   When lifting a heavy object, keep your back straight and lift with your legs.

7. When lifting, keep the object as close to your body as possible. Keep your back straight and lift with your legs.

8. While carrying the object, do not twist your body to change directions. Use your feet to turn your whole body in the new direction.

9. To set the load down, keep the object close to your body. Bend at the knees and keep your back straight. Do not bend forward or twist.

10. If you need to place the object onto a shelf or benchtop, place an edge of the object on the surface and slide it into place. Do not lean forward.

Fire Hazards and Prevention

Shop Manual
Chapter 1,
page 46

Fires are classified by the types of materials that are involved (Table 1-1). Technicians should be able to locate the correct fire extinguisher to control all the types of fire they are likely to experience (Figure 1-6). Technicians must also be able to fight a fire in an emergency.

Labels on the fire extinguisher will indicate the types of fires that it will put out. Become familiar with the use of a fire extinguisher.

TABLE 1-1   A GUIDE TO FIRE CLASSIFICATION AND FIRE EXTINGUISHER TYPES

	Class of Fire	Typical Fuel Involved	Type of Extinguisher
Class A Fires (green)	**For Ordinary Combustibles** Put out a Class A fire by lowering its temperature or by coating the burning combustibles.	Wood Paper Cloth Rubber Plastics Rubbish Upholstery	Water*[1] Foam* Multipurpose dry chemical[4]
Class B Fires (red)	**For Flammable Liquids** Put out a Class B fire by smothering it. Use an extinguisher that gives a blanketing flame-interrupting effect; cover whole flaming liquid surface.	Gasoline Oil Grease Paint Lighter fluid	Foam* Carbon dioxide[5] Halogenated agent[6] Standard dry chemical[2] Purple K dry chemical[3] Multipurpose dry chemical[4]
Class C Fires (blue)	**For Electrical Equipment** Put out a Class C fire by shutting off power as quickly as possible and by always using a nonconducting extinguishing agent to prevent electric shock.	Motors Appliances Wiring Fuse boxes Switchboards	Carbon dioxide[5] Halogenated agent[6] Standard dry chemical[2] Purple K dry chemical[3] Multipurpose dry chemical[4]
Class D Fires (yellow)	**For Combustible Metals** Put out a Class D fire of metal chips, turnings, or shaving by smothering or coating with a specially designed extinguishing agent.	Aluminum Magnesium Potassium Sodium Titanium Zirconium	Dry powder extinguishers and agents only

*Cartridge-operated water, foam, and soda-acid types of extinguishers are no longer manufactured. These extinguishers should be removed from service when they become due for their next hydrostatic pressure test.
Notes:
(1) Freeze in low temperatures unless treated with antifreeze solution, usually weighs over 20 pounds, and is heavier than any other extinguisher mentioned.
(2) Also called ordinary or regular dry chemical. (solution bicarbonate)
(3) Has the greatest initial fire-stopping power of the extinguishers mentioned for class B fires. Be sure to clean residue immediately after using the extinguisher so sprayed surfaces will not be damaged. (potassium bicarbonate)
(4) The only extinguishers that fight A, B, and C class fires. However, they should not be used on fires in liquified fat or oil of appreciable depth. Be sure to clean residue immediately after using the extinguisher so sprayed surfaces will not be damaged. (ammonium phosphates)
(5) Use with caution in unventilated, confined spaces.
(6) May cause injury to the operator if the extinguishing agent (a gas) or the gases produced when the agent is applied to a fire is inhaled.

Figure 1-6 Know the location and types of fire extinguishers that are available in the shop.

Gasoline

Gasoline is so commonly found in automotive repair shops that its dangers are often forgotten. A slight spark or an increase in heat can cause a fire or explosion. Gasoline is a very explosive liquid and is very powerful. One exploding gallon of gasoline has a force equal to fourteen sticks of dynamite. The expanding vapors from gasoline are extremely dangerous, and these vapors are present even in cold temperatures. Gasoline vapors are heavier than air, therefore, when an open container of gasoline is sitting about, the vapors spill out over the sides of the container onto the floor. These fumes are more **flammable** than liquid gasoline and can easily explode.

Never smoke around gasoline since even the droppings of hot ashes can ignite the gasoline. If an engine has a gasoline leak or you have caused a leak by disconnecting a fuel line, stop the leak and clean up the spilled gasoline immediately. While stopping the leak, be extra careful not to cause sparks. If any rags are used in the clean up of the gasoline, they must be placed in an approved container (Figure 1-7). Due to extreme fire hazards, it is important to immediately

A substance that is flammable will support combustion.

Figure 1-7 Dirty rags and towels must be stored in approved containers.

8

clean up any gasoline spilled on the floor. Also, many of the compounds in petroleum are toxic, especially if they are in high concentrations.

The chemicals in petroleum that do not evaporate quickly are biodegradable. Optimum degradation occurs if the gasoline is diluted and there is enough air, water, and nutrients for the microbes to "eat up" the chemicals. These properties of gasoline are an advantage in the cleanup and disposal of small spills. Spreading absorbent material such as kitty litter, sand, ground corncobs, straw, sawdust, woodchips, peat, synthetic absorbent pads, or even dirt can stop the flow and soak up the gasoline. Keep in mind that the absorbent does not make gasoline nonflammable.

Brooms can be used to sweep up the absorbent material and put it into buckets, garbage cans, or barrels. Remember to control ignition sources. Be aware of local laws concerning gasoline spills. Some states or municipalities require notification of any gasoline spill larger than 5 gallons.

Gasoline should always be stored in approved containers (Figure 1-8) and never in glass containers. If the glass container is knocked over or dropped, a terrible explosion can occur. Approved gasoline storage cans have a flash-arresting screen at the outlet. These screens prevent external ignition sources from igniting the gasoline within the can while the gasoline is being poured.

Follow these safety precautions regarding gasoline containers:

1. Always use approved gasoline containers that are painted red for proper identification.

2. Do not fill gasoline containers completely. Always leave the level of gasoline at least 1 inch (25 mm) from the top of the container. This action allows expansion of the gasoline at higher temperatures. If gasoline containers are completely full, the gasoline will expand when the temperature increases. This expansion forces gasoline from the can and creates a dangerous spill.

3. If gasoline containers must be stored, place them in a well-ventilated area such as a storage shed. Do not store gasoline containers in your home or in the trunk of a vehicle.

4. When a gasoline container must be transported, be sure it is secured against upsets. Do not transport or fill gasoline containers on plastic truck bed liners. Static electricity can be generated and ignite the vapors.

5. Do not store a partially-filled gasoline container for long periods of time because it may give off vapors and produce a potential danger.

Figure 1-8   Store any type of combustible materials in an approved safety cabinet.

6. Never leave gasoline containers open except while filling or pouring gasoline from the container.

7. Do not prime an engine with gasoline while cranking the engine.

8. Never use gasoline as a cleaning agent.

Diesel Fuel

Diesel fuel is not as **volatile** as gasoline, but it should still be stored and handled in the same way as gasoline. It is also not as refined as gasoline and tends to be a dirty fuel. It normally contains impurities, including active microscopic organisms that can be highly infectious. If diesel fuel happens to enter an open cut or sore, thoroughly wash it immediately. If it gets into your eyes, flush them immediately and seek medical help.

Solvents

Cleaning solvents are not as volatile as gasoline, but they are still flammable. They should be treated and stored in the same way as gasoline. Whenever using solvents, wear eye protection.

Rags

Oily and greasy rags can also cause fires. Used rags should be stored in an approved container and never thrown out with normal trash. Like gasoline, oil is a hydrocarbon and can ignite with or without a spark or flame.

Hazardous Materials

Many solvents and other chemicals used in an automotive shop have warning and caution labels that should be read and understood by everyone who uses them. These are typically considered **hazardous materials**. Many service procedures generate what are known as hazardous wastes. Examples of hazardous waste are used or dirty cleaning solvents and other liquid cleaners.

Right-to-Know Laws

In the United States, **right-to-know laws** concerning hazardous materials and wastes protect every employee in a workplace. The general intent of these laws is for employers to provide a safe working place as it relates to hazardous materials. The right-to-know laws state that employees have a right to know when the materials they use at work are hazardous. The right-to-know laws started with the **Hazard Communication Standard** published by the Occupational Safety and Health Administration (OSHA) in 1983. This document was originally intended for chemical companies and manufacturers that required employees to handle hazardous materials in their work situation. At the present time, most states have established their own right-to-know laws. Meanwhile, the federal courts have decided to apply these laws to all companies, including automotive service shops.

Under the right-to-know laws, the employer has three responsibilities regarding the handling of hazardous materials by its employees. The first responsibility concerns employee training and providing information. All employees must be trained about the types of hazardous materials they will encounter in the workplace. All employees must be informed about their rights under legislation regarding the handling of hazardous materials. In addition, information about

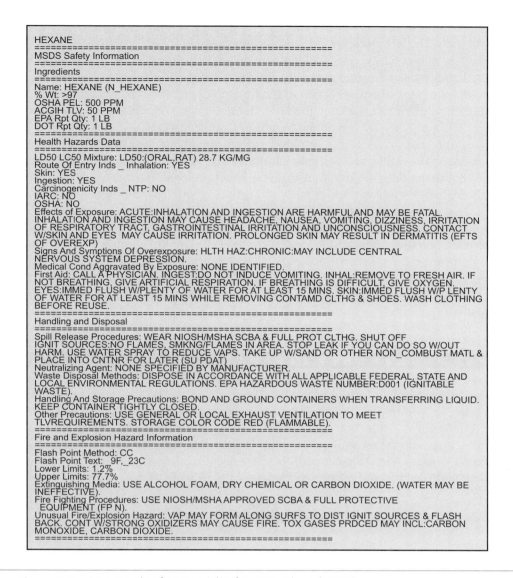

```
HEXANE
======================================================
MSDS Safety Information
======================================================
Ingredients
======================================================
Name: HEXANE (N_HEXANE)
% Wt: >97
OSHA PEL: 500 PPM
ACGIH TLV: 50 PPM
EPA Rpt Qty: 1 LB
DOT Rpt Qty: 1 LB
======================================================
Health Hazards Data
======================================================
LD50 LC50 Mixture: LD50:(ORAL,RAT) 28.7 KG/MG
Route Of Entry Inds _ Inhalation: YES
Skin: YES
Ingestion: YES
Carcinogenicity Inds _ NTP: NO
IARC: NO
OSHA: NO
Effects of Exposure: ACUTE:INHALATION AND INGESTION ARE HARMFUL AND MAY BE FATAL.
INHALATION AND INGESTION MAY CAUSE HEADACHE, NAUSEA, VOMITING, DIZZINESS, IRRITATION
OF RESPIRATORY TRACT, GASTROINTESTINAL IRRITATION AND UNCONSCIOUSNESS. CONTACT
W/SKIN AND EYES  MAY CAUSE IRRITATION. PROLONGED SKIN MAY RESULT IN DERMATITIS (EFTS
OF OVEREXP)
Signs And Symptons Of Overexposure: HLTH HAZ:CHRONIC:MAY INCLUDE CENTRAL
NERVOUS SYSTEM DEPRESSION.
Medical Cond Aggravated By Exposure: NONE IDENTIFIED.
First Aid: CALL A PHYSICIAN. INGEST:DO NOT INDUCE VOMITING. INHAL:REMOVE TO FRESH AIR. IF
NOT BREATHING, GIVE ARTIFICIAL RESPIRATION. IF BREATHING IS DIFFICULT, GIVE OXYGEN.
EYES:IMMED FLUSH W/PLENTY OF WATER FOR AT LEAST 15 MINS. SKIN:IMMED FLUSH W/P LENTY
OF WATER FOR AT LEAST 15 MINS WHILE REMOVING CONTAMD CLTHG & SHOES. WASH CLOTHING
BEFORE REUSE.
======================================================
Handling and Disposal
======================================================
Spill Release Procedures: WEAR NIOSH/MSHA SCBA & FULL PROT CLTHG. SHUT OFF
IGNIT SOURCES:NO FLAMES, SMKNG/FLAMES IN AREA. STOP LEAK IF YOU CAN DO SO W/OUT
HARM. USE WATER SPRAY TO REDUCE VAPS. TAKE UP W/SAND OR OTHER NON_COMBUST MATL &
PLACE INTO CNTNR FOR LATER (SU PDAT)
Neutralizing Agent: NONE SPECIFIED BY MANUFACTURER.
Waste Disposal Methods: DISPOSE IN ACCORDANCE WITH ALL APPLICABLE FEDERAL, STATE AND
LOCAL ENVIRONMENTAL REGULATIONS. EPA HAZARDOUS WASTE NUMBER:D001 (IGNITABLE
WASTE).
Handling And Storage Precautions: BOND AND GROUND CONTAINERS WHEN TRANSFERRING LIQUID.
KEEP CONTAINER TIGHTLY CLOSED.
Other Precautions: USE GENERAL OR LOCAL EXHAUST VENTILATION TO MEET
TLVREQUIREMENTS. STORAGE COLOR CODE RED (FLAMMABLE).
======================================================
Fire and Explosion Hazard Information
======================================================
Flash Point Method: CC
Flash Point Text:   9F, _23C
Lower Limits: 1.2%
Upper Limits: 77.7%
Extinguishing Media: USE ALCOHOL FOAM, DRY CHEMICAL OR CARBON DIOXIDE. (WATER MAY BE
INEFFECTIVE).
Fire Fighting Procedures: USE NIOSH/MSHA APPROVED SCBA & FULL PROTECTIVE
  EQUIPMENT (FP N).
Unusual Fire/Explosion Hazard: VAP MAY FORM ALONG SURFS TO DIST IGNIT SOURCES & FLASH
BACK. CONT W/STRONG OXIDIZERS MAY CAUSE FIRE. TOX GASES PRDCED MAY INCL:CARBON
MONOXIDE, CARBON DIOXIDE.
======================================================
```

Figure 1-9   An example of a Material Safety Data Sheet (MSDS).

each hazardous material must be posted on **Material Safety Data Sheets (MSDS)** available from the manufacturer (Figure 1-9). In Canada, MSDS sheets are called **workplace hazardous materials information systems (WHMIS)**.

The employer has a responsibility to place MSDS where they are easily accessible by all employees. The MSDS provide extensive information about the hazardous material such as:

1. Chemical name
2. Physical characteristics
3. Protective equipment required for handling
4. Explosion and fire hazards
5. Other incompatible materials
6. Health hazards such as signs and symptoms of exposure, medical conditions aggravated by exposure, and emergency and first-aid procedures
7. Safe handling precautions
8. Spill and leak procedures

Material Safety Data Sheets (MSDS) contain extensive information and facts about hazardous materials.

The Canadian equivalent to the MSDS is the **workplace hazardous materials information systems (WHMIS)**, which informs workers of hazardous material facts.

The second responsibility of the employer is to make sure that all hazardous materials are properly labeled. The label information must include health, fire, and reactivity hazards posed by the material, as well as the protective equipment necessary to handle the material. The manufacturer must supply all warning and precautionary information about hazardous materials, and this information must be read and understood by the employee before handling the material. Pay great attention to the information on the label. By doing so, you will use the substance in the proper and safe way, thereby preventing hazardous conditions.

The third responsibility of the employer is for maintaining permanent files regarding hazardous materials. These files must include information on hazardous materials in the shop, proof of employee training programs, and information about accidents such as spills or leaks of hazardous materials. The employer's files must also include proof that employees' requests for hazardous material information, such as MSD, have been met. The employer must maintain a general right-to-know compliance procedure manual.

There are responsibilities of the employees as well. Employees must be familiar with the intended purpose of the substance, the recommended protective equipment, accident and spill procedures, and any other information regarding the safe handling of hazardous materials. This training must be given annually to employees and provided to new employees as part of their job orientation.

BIT OF HISTORY

During the 1960s, disabling injuries increased 20 percent and 14,000 workers were dying on the job each year. In pressing for prompt passage of workplace safety and health legislation, Senator Harrison A. Williams Jr. called attention to the need to protect workers against such hazards as noise, cotton dust, and asbestos. Representative William A. Steiger also worked for passage of a bill to protect workers. On December 29, 1970, President Richard M. Nixon signed The Occupational Safety and Health Act of 1970, also known as the Williams-Steiger Act.

Hazardous Waste

Waste is considered hazardous if it is on the Environmental Protection Agency (EPA) list of known and harmful materials or if it has one or more of the following characteristics:

1. Any material that reacts violently with water or other chemicals is considered hazardous. If a material releases cyanide gas, hydrogen sulfide gas, or similar gases when exposed to low-pH acid solutions, it is hazardous.

2. If a material burns the skin or dissolves metals and other materials, it is considered hazardous.

3. Materials are hazardous if they leach one or more of eight heavy metals in concentrations greater than 100 times primary drinking water standard. These materials are considered toxic.

4. A liquid is hazardous if the temperature at which the vapors on the surface of the fuel will ignite when exposed to an open flame is below 140°F (60°C), and a solid is hazardous if it ignites spontaneously.

A complete list of EPA hazardous wastes can be found in the Code of Federal Regulations. It should be noted that no material is considered hazardous waste until the shop is finished using it and is ready to dispose of it. New oil is not a hazardous waste, however, used oil is. Once you drain oil from an engine, you have generated the waste and now become responsible for the proper disposal of this hazardous waste. There are many other wastes that need to be handled properly after you have removed them, such as batteries, brake fluid, transmission fluid, and engine coolant.

No fluids drained from a vehicle should be allowed to enter sewage drains. Some fluids, such as coolant, can be captured and recycled in the shop with special equipment. Filters for fluids (transmission, fuel, and oil filters) also need to be handled in designated ways. Used filters need to be drained and then crushed or disposed of in a special shipping barrel. Most regulations demand that oil filters be drained for at least 24 hours before they are disposed of or crushed.

Federal and state laws control the disposal of hazardous waste materials. It is the responsibility of the employer and the employee to assure everyone in the shop is familiar with these laws. Hazardous waste disposal laws include the **Resource Conservation and Recovery Act (RCRA)**. This law basically states that hazardous waste generators are responsible for the waste from the time it becomes a waste material until the proper waste disposal is completed. Therefore, the user must store hazardous waste material properly and safely and be responsible for the transportation of this material until it arrives at an approved hazardous waste disposal site where it is processed according to the law. A licensed waste management firm normally does the disposal. The hazardous waste coordinator for the shop should have a written contract with the hazardous waste hauler.

The RCRA controls these types of automotive waste:

1. Paint and body repair products waste
2. Solvents for parts and equipment cleaning
3. Batteries and battery acid
4. Mild acids used for metal cleaning and preparation
5. Waste oil, engine coolants, or antifreeze
6. Air-conditioning refrigerants
7. Engine oil filters

NEVER, under any circumstances, use these methods to dispose of hazardous waste material:

1. Pour hazardous wastes on weeds to kill them.
2. Pour hazardous wastes on gravel streets to prevent dust.
3. Throw hazardous wastes in a dumpster.
4. Dispose of hazardous wastes anywhere but an approved disposal site.
5. Pour hazardous wastes down sewers, toilets, sinks, or floor drains.
6. Bury hazardous wastes in the ground.

The Resource Conservation and Recovery Act (RCRA) stipulates that the users of hazardous materials are responsible for the materials from the time they become a waste until they are properly disposed of.

Fuel System Safety

Most modern fuel systems operate under pressure. Before attempting to perform any service requiring the disconnection of hoses or tubes, this pressure must be relieved. The service manual will provide the proper method of relieving the pressure. This may include removing the fuel pump relay or fuse and running the engine until it runs out of fuel, or using a hose connected to the fuel rail and draining the fuel into an approved container. Most modern gasoline fuel systems operate with fuel pressures up to approximately 60 psi (4 bar) of pressure. Diesel engine fuel systems can exceed 23,000 psi (1,586 bar). Compressed natural gas vehicles can have tank pressures up to 3,500 psi (241 bar) with line pressures of 140 psi (9.6 bar). One of the reasons the fuel system is pressurized is to provide better vaporization of the fuel. Opening lines or tubes without first relieving the pressure causes vaporized fuel to be expelled. This vaporized fuel is very easy to ignite.

These high pressures can cause serious personal injury. Because high pressures are associated with fuel systems, never use your hands to locate a leak. The pressures may be high enough to cause severe cuts, severed fingers, or fuel injection into the bloodstream. Use a piece of cardboard to safely locate the source of the leak.

Throughout the shop manual textbook there will be many warnings and cautions given concerning fuel system service safety. In addition, the service manual will also have many similar warnings. It is imperative that you read and follow these instructions.

Summary

Terms to Know

Caustic

Conductors

Face shields

Flammable

Hazard Communication
  Standard

Hazardous materials

Material Safety Data
  Sheets (MSDS)

Occupational safety
  glasses

Resource
  Conservation and
  Recovery Act
  (RCRA)

Right-to-know laws

Safety goggles

Volatile

Workplace Hazardous
  Materials
  Information
  Systems (WHMIS)

❏ Being a professional technician means more than having knowledge of vehicle systems. It also requires an understanding of all the hazards in the work area.

❏ As a professional technician, you should work responsibly to protect yourself and the people around you.

❏ Technicians must be aware that it is their responsibility to prevent injuries in the shop, and that their actions and attitudes reflect how seriously they accept that responsibility.

❏ The safest and surest method of protecting your eyes is to wear proper eye protection any time you enter the shop.

❏ Most occupational back injuries are caused by improper lifting practices.

❏ Fires are classified by the types of materials involved. Fire extinguishers are classified by the type of fire they will extinguish.

❏ The U.S. Occupational Safety and Health Act of 1970 assured safe and healthful working conditions and authorized enforcement of safety standards.

❏ The danger regarding the labeling and handling of hazardous conditions and materials may be avoided by applying the necessary safety precautions.

❏ Before attempting to perform any service requiring the disconnection of fuel hoses or tubes, all pressure in the system must be relieved.

❏ Because high pressures are associated with fuel systems, never use your hands to locate a leak.

❏ The automotive shop must supply the necessary shop safety equipment, and all shop personnel must be familiar with the location and operation of this equipment.

Review Questions

Short Answer Essays

1. Explain how safety is a part of professionalism.
2. List the basic safety rules of proper lifting.
3. Define six essential safety precautions regarding gasoline handling.
4. What types of materials are involved in a Class B fire?
5. List the different types of eye protection devices and explain the proper application of each.
6. Explain the purpose of right-to-know laws.
7. What materials can be used to contain and pick up a gasoline spill?
8. Explain the four characteristics of a material that the EPA uses to classify a waste as being hazardous.

9. What is the difference between hazardous materials and hazardous waste?

10. What are the three responsibilities of the employer outlined by the "Right to Know" law?

Fill-in-the-Blanks

1. When working with chemicals, you should always wear _____
_____.

2. As a professional technician, you should work in such a manner as to protect _____ and the people around you.

3. The safest and surest method of protecting your eyes is to wear proper eye protection _____ you enter the shop.

4. Most occupational back injuries are caused by _____ lifting practices.

5. Do not lift with your _____.

6. Information about each hazardous material must be posted on _____
_____ _____ _____.

7. One gallon of gasoline has a force equal to _____ sticks of dynamite.

8. Class C fires involve the burning of _____ equipment.

9. Long hair should be _____ and _____ under a hat.

10. Hazardous waste generators are responsible for the waste from the time it becomes a waste material until the proper _____ _____ is completed.

Multiple Choice

1. Which of the following is **not** included in proper lifting practices?
   A. Do not lift with your back.
   B. Keep your back and elbows straight.
   C. Lift the object by yourself so others are out of the way.
   D. While carrying the object, do not twist your body to change directions.

2. *Technician A* says the right-to-know laws require employers to train employees regarding hazardous waste materials. *Technician B* says the right-to-know laws have no provisions requiring proper labeling of hazardous materials.
   Who is correct?
   A. A only   C. Both A and B
   B. B only   D. Neither A nor B

3. *Technician A* says it is safe to work around the electric cooling fan when the engine is off and the engine temperatures are low.
   Technician B says the electric cooling fans can only come on if the engine is running.
   Who is correct?
   A. A only   C. Both A and B
   B. B only   D. Neither A nor B

4. *Technician A* says electrical fires are extinguished with Class A fire extinguishers.
   Technician B says gasoline is extinguished with Class B fire extinguishers.
   Who is correct?
   A. A only   C. Both A and B
   B. B only   D. Neither A nor B

5. *Technician A* says Material Safety Data Sheets (MSDS) explain employers' and employees' responsibilities regarding handling and disposal of hazardous materials.
   Technician B says Material Safety Data Sheets (MSDS) contain specific information about hazardous materials.
   Who is correct?
   A. A only   C. Both A and B
   B. B only   D. Neither A nor B

6. *Technician A* says a solid that ignites spontaneously is considered a hazardous material.
   Technician B says a liquid is con-sidered a hazardous material if the vapors on the surface will ignite when exposed to an open flame whose temperature is below 140°F (60°C).
   Who is correct?
   A. A only   C. Both A and B
   B. B only   D. Neither A nor B

7. *Technician A* says safety glasses should be worn when working with battery acid and refrigerants.
   Technician B says full-face shields are designed to provide protection for the face, neck, and eyes of the technician.
   Who is correct?
   - **A.** A only
   - **B.** B only
   - **C.** Both A and B
   - **D.** Neither A nor B

8. *Technician A* says the way people dress reflects very little about their attitude toward safety.
   Technician B says jewelry can be a personal safety hazard and should be removed while working in an automotive repair shop.
   Who is correct?
   - **A.** A only
   - **B.** B only
   - **C.** Both A and B
   - **D.** Neither A nor B

9. Which of the following statements concerning gasoline storage is true?
   - **A.** Always use approved gasoline containers that are painted blue for proper identification.
   - **B.** Prevent air from entering the container by filling gasoline containers completely full.
   - **C.** Always transport gasoline containers on plastic bed liners.
   - **D.** Do not store a partially-filled gasoline container for long periods of time.

10. *Technician A* says hazardous waste materials may be hauled to an approved hazardous waste disposal site or recycled in the shop.
    Technician B says the disposal of all hazardous waste materials must be done in accordance with federal, state, and local laws.
    Who is correct?
    - **A.** A only
    - **B.** B only
    - **C.** Both A and B
    - **D.** Neither A nor B

Automotive Fuels

Upon completion and review of this chapter, you should be able to:

❏ Explain the refining process of petroleum-based fuels.

❏ Describe the combustion process of an internal combustion engine.

❏ Describe the four performance characteristics of gasoline.

❏ Explain how Reid vapor pressure is measured.

❏ Explain the difference between winter and summer blend gasoline.

❏ Explain how octane ratings are determined.

❏ Explain what the octane rating represents.

❏ Describe the common methods of determining the octane number.

❏ Detail the differences between regular and premium gasoline.

❏ Describe the various types of gasoline and additives.

❏ Explain the purpose of oxygenated fuels.

❏ List and describe the common oxygenated fuel blends.

❏ Describe the different alternative fuels.

❏ Detail the properties of diesel fuel.

❏ Explain the differences between diesel fuel grades.

❏ Explain the cetane rating of diesel fuel.

❏ Describe how cetane ratings affect power output.

❏ Explain what API gravity is and how it affects power output.

Introduction

Before studying the automotive fuel or emission systems, the properties of the most commonly used fuels in automobiles must be detailed. This chapter takes a look at the composition and qualities of gasoline, diesel fuel, and some of the alternate fuels being used by some engines.

Gasoline

Today, **gasoline** is the most common fuel source used in automotive engines. However, advances in technology in recent years concerning alternate fuel strategies may mean that hydrocarbon fuels (such as gasoline and diesel) will soon be replaced or supplemented. This chapter discusses the characteristics of various fuels used today. Alternate fuels that are used on a common basis today are also covered. More advance alternate fuels, such as fuel cells, will be discussed in a later chapter.

Gasoline is a term used to describe a complex mixture of various hydrocarbons refined from crude petroleum oil for use as a fuel in engines. Gasoline contains hydrogen and carbon molecules. The chemical symbol for this liquid is C_8H_{15}. Gasoline is a colorless liquid with excellent vaporization capabilities.

BIT OF HISTORY

Before its widespread use in the internal combustion engine of automobiles, gasoline was an unwanted by-product of refining oils and kerosene.

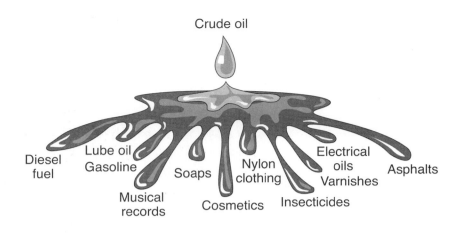

Crude oil

Diesel fuel
Lube oil
Gasoline
Soaps
Musical records
Cosmetics
Nylon clothing
Insecticides
Electrical oils
Varnishes
Asphalts

Figure 2-1   Some of the different products made from crude oil.

Gasoline Composition

Crude oil is used to make such **petroleum**-based products as gasoline, diesel fuel, motor oil, solvents, and many others (Figure 2-1). Crude oil is a combination of large and small (light and heavy) **hydrocarbons** along with petroleum gas, hydrogen sulfide, and water.

There are many kinds of liquid hydrocarbons, consisting of approximately 12 percent hydrogen and 82 percent carbon. The **refining** process separates the hydrocarbons so they can be used. Hydrogen is a light gas vapor, while carbon is a heavy black solid. These materials, as well as hydrogen, sulfide, and water, are removed during the refining process.

Crude oil is separated into many products at the oil refinery. During refining, the crude oil is heated by pumping it through pipes routed through hot furnaces and into a fractionating column (Figure 2-2). During refining, light hydrocarbon molecules are separated from heavier ones. Since crude oil is a composite of different substances, it has different boiling points (Figure 2-3) ranging from about 100°F to 700°F (38°C to 370°C).

Draw pipes are located at different heights on the fractioning tower and are used to pull the desired petroleum materials out of the tower (Figure 2-4). As the crude oil is boiled during the refining process, the lightest products (butane) are taken from the top of the fractioning tower (Figure 2-5). The next draw pipe will pull straight-run gasoline from the tower. As the fractions go down the tower, the draw pipe pulls materials that have increasingly higher boiling points. The heaviest products boil last and are taken from the bottom where the temperature is the highest. The substance pulled from the bottom of the tower is a tar-like residue.

Figure 2-2   The tall towers that are seen in the skyline are the fractionating towers of the refinery. (Courtesy of Chevron)

Raw **crude oil** is a mixture of hydrocarbon compounds ranging from gases to heavy tars and waxes. The crude oil can be refined into products such as lubricating oils, greases, asphalts, kerosene, diesel fuel, gasoline, and natural gas.

The word **petroleum** means rock oil.

Hydrocarbons are organic compounds that contain only hydrogen and carbon.

Refining is the process of breaking the crude oil down into different parts, called fractions.

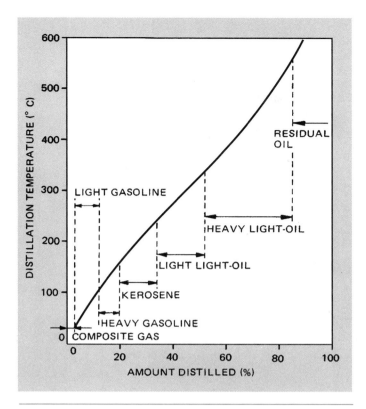

Figure 2-3   The compounds of crude oil have different boiling points.

Figure 2-4   Draw pipes are located at different heights of the fractioning towers. The higher draw tubes will pull the lightest materials, while the lower draw tubes pull progressively heavier materials.

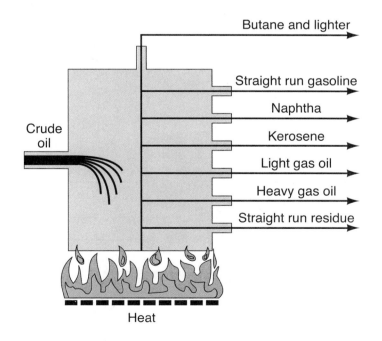

Figure 2-5   Illustration of the different materials that will be pulled off of the tower.

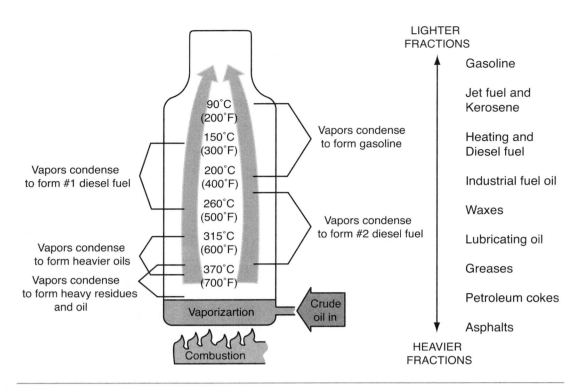

Figure 2-6 Oil refining process.

Some of the fractions are used as raw materials that will be blended with other materials. For example, the gasoline will be blended to create the desired octane, emissions, volatility, and storage life. Some of the lower fractions can be used directly, such as kerosene or diesel. The last fraction can be used as fuel oil or as a basic stock for lubricants. Fractions may also be reformulated by heat or catalysts to produce more of the lighter stocks. The distillation process is repeated in other plants as the oil is further refined for its most efficient use (Figure 2-6).

Crude oil is measured and sold by the barrel, which is 42 U.S. gallons. Figure 2-7 shows the percentage of various products produced from an average barrel of crude oil.

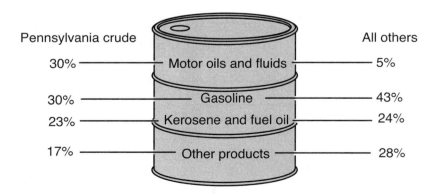

Figure 2-7 Illustration of average yields of products from a U.S. barrel of crude oil. The left side is for Pennsylvania-grade crude, while the right side is an average of all others.

Combustion Chemistry

Most internal combustion engines are designed to burn **organic** fuels to produce power. All petroleum products are considered organic and are hydrocarbon fuels. The combustion process combines oxygen (O_2) from the air with the hydrogen (H) and carbon (C) from the fuel. In a gasoline engine, a spark starts the combustion process, which takes about 3 milliseconds (ms) (0.003 seconds) to be completed. The chemical reaction that takes place is basically:

Hydrogen (H) + carbon (C) + oxygen (O_2) + nitrogen (N) + other chemicals + spark = Heat + water (H_2O) + carbon monoxide (CO) + carbon dioxide (CO_2) + hydrocarbons (HC) + oxides of nitrogen (NO_x) + other chemicals.

Shop Manual
Chapter 2,
page 72

The term **organic** refers to a product from a source that originally was living.

Oxygen makes up about 21 percent of the atmosphere.

BIT OF HISTORY

Today there are individuals and companies that convert diesel engines to burn vegetable oils. This really is nothing new. Rudolph Diesel expected that his engine would be powered by vegetable oils and seed oils. At the 1900 World's Fair, Diesel ran his engines on peanut oil.

If the combustion process is complete, all the hydrocarbons (HC) will be completely combined with all the available oxygen (O_2). The ratio of air to fuel that accomplishes this is referred to as **stoichiometric**. The stoichiometric quantities for gasoline are 14.7 parts air for 1 part gasoline by weight. Different fuels have different stoichiometric ratios (Table 2-1).

Stoichiometric, in simplest terms, means chemically correct. An air-fuel mixture is considered chemically correct when it is neither too rich nor too lean.

TABLE 2-1   BTU AND STOICHIOMETRIC RATIO OF GASOLINE AND ALCOHOL

Fuel	Heat energy (Btu/gal)	Stoichiometric Ratio
Gasoline	About 130,000	14.7:1
Ethanol alcohol	About 76,000	9.0:1
Methanol alcohol	About 60,000	6.4:1

Gasoline Characteristics

The energy content of gasoline is between 109,000 and 125,000 **British thermal units (Btus)** per gallon (or between 27,468 kilo**calories**). Combined with Btus many of the performance characteristics of gasoline can be controlled in refining and blending to provide proper engine function and vehicle driveability. The major factors affecting fuel performance are volatility, sulfur content, additives, and octane rating.

Energy is measured in **British thermal units (Btus)** or **calories (cal)**. One Btu is the amount of heat required to raise one pound of water one Fahrenheit degree. One calorie is the amount of heat required to raise the temperature of one gram (g) of water one Celsius degree.

Volatility is a measure of how easily a fuel evaporates (forms a vapor). When fuel does not vaporize easily, this characteristic is called low volatility.

Spark knock is a metallic noise an engine makes, usually during acceleration, resulting from abnormal or uncontrolled combustion inside the cylinder.

Spark knock is also referred to as "detonation" and as "engine knock."

Shop Manual
Chapter 2,
page 72

Vapor lock is a result of fuel boiling in the fuel line or tank. Unlike liquid fuel, vapor is compressible. This means that the fuel cannot be pumped to the carburetor or injectors, so the engine stalls. After the fuel line cools sufficiently, the engine will run again. Fuel-injected engines use electric fuel pumps that keep the fuel under higher pressure. Higher pressure raises the fuel's boiling point so vapor lock is less likely to occur.

Volatility

Two characteristics of gasoline that are important to engine operation are **volatility** and resistance to **spark knock**. Gasoline is very volatile; consequently, it readily evaporates. This volatility allows its vapors to adequately mix with air to promote combustion. This evaporation into a vapor is very important to the combustion process since only vaporized fuel supports complete combustion. Thus, to ensure complete combustion, complete vaporization must occur. For fuel to burn properly in the cylinder, its vaporization point should be near the temperature in the intake manifold. If liquid fuel enters the cylinders, it will not burn. A fuel must vaporize well to prevent diluting the crankcase oil with liquid fuel. If parts of the gasoline do not vaporize, droplets of liquid break down the oil film on the cylinder wall, causing scuffing or scoring. The liquid eventually enters the crankcase oil and results in the formation of sludge, gum, and varnish accumulation as well as decreasing the lubrication properties of the oil.

Fuel with a low volatility can result in the following undesirable conditions:

- Hard cold engine starts
- Poor driveability and fuel economy during cold ambient temperatures
- Increased spark plug and combustion chamber deposits

AUTHOR'S NOTE: On carbureted and throttle body fuel-injection engines, low volatility can cause unequal combustion between cylinders due to unequal fuel distribution. Vaporized fuel travels farther and faster in the manifold than liquid fuel.

A fuel that vaporizes too easily can form vapor bubbles in the fuel line and fuel pump during periods of high ambient temperature. Vapor bubbles result in **vapor lock** or loss of performance. Other problems that can result are:

- High evaporative emissions
- Poor driveability when the engine is hot
- Poor fuel economy
- Hard starting after a hot soak condition

Refiners blend their gasoline to meet the needs of different seasons and for the geographic area in which the fuel will be used. The difference in the blends is the **vapor pressure** of the finished product. For example, gasoline blended for use in the summer is less volatile (does not evaporate as easily) than gasoline blended for use in the winter. Also, in areas that are located at high altitudes, fuels must be blended to have higher volatility because they can boil at lower temperatures. A gasoline blend used in the wrong season or geographical area can cause driveability problems. Fuel refined for use in the summer that is used during the winter can cause hard starting. If winter fuel is used in the summer, vapor lock can result. Refiners attempt to schedule delivery of the different blends to be at retail outlets at the appropriate time, however, unseasonable weather can result in driveability problems due to the blend being used.

The definition of volatility assumes that the vapors will remain in the fuel tank or fuel line and will cause a certain pressure based on the temperature of the fuel. **ASTM International** has six volatility classes for gasoline, AA, A, B, C, D, and E. AA is the least volatile (Table 2-2).

TABLE 2-2 GASOLINE VOLATILITY REQUIREMENTS

Vapor pressure/ distillation class	Distillation temperatures			End point Max. F°	Vapor pressure PSI/Max.	Vapor lock protection class	Vapor-liquid ratio of 20 Min F°
	10% Evap F°	50% Evap F°	90% Evap F°				
AA	158	170-250	374	437	7.8	1	140
A	158	170-250	374	437	9.0	2	133
B	149	170-245	374	437	10.0	3	124
C	140	170-240	365	437	11.5	4	116
D	131	150-235	365	437	13.5	5	105
E	122	150-230	365	437	15.0	6	95

ASTM D 4814 Gasoline volatility requirements

(Courtesy of Changes in Gasoline III, Downstream Alternatives, Inc)

In Table 2-2, the 10 percent evaporated maximum standard for AA gasoline means that 10 percent of the fuel would be evaporated before it reaches a temperature of 158°F (70°C). For all volatility classes, gasoline will have evaporated at the endpoint maximum temperature of 437°F (225°C).

There are several measures of gasoline volatility: **Reid vapor pressure (RVP)**, distillation, driveability index (DI) and vapor-liquid ratio. These properties vary by season and geographic area. Industry specifications developed by the American Society for Testing and Materials (ASTM) define volatility classes in terms of vapor pressure and distillation temperature limits and provide recommendations for selecting the appropriate volatility class for each area of the country by month.

The best known method of measuring volatility is the Reid vapor pressure (RVP) test. Vapor pressure is the pressure exerted by vapor formed over a liquid in a closed container. The RVP test is performed by placing a sample of gasoline into a sealed metal chamber that has a pressure-measuring device attached. The chamber is submerged in water that is heated to 100°F (38°C). As the fuel is heated by the water, it vaporizes. Remember, the more volatile a fuel is, the easier it will vaporize. As the fuel vaporizes, it creates vapor pressure within the chamber. More volatile fuels will create more pressure. The vapor pressure is measured in pounds per square inch (psi).

Distillation temperature measurements involve heating a fuel and measuring the temperature at which a certain percentage of the sample evaporates. Using these percentages, engine performance has been correlated with a **driveability index (DI)** using the temperatures for 10 percent (T10), 50 percent (T50) and 90 percent (T90) evaporation. The DI was developed to indicate gasoline performance during engine cold start and warm up. The ASTM specification ranges from 1,200 in the winter months to 1,250 in the summer months.

Vapor pressure is pressure exerted by a vapor in equilibrium with its liquid state.

ASTM International (formerly known as the American Society for Testing and Materials) provides a global forum for the development and publication of voluntary consensus standards for materials, products, systems, and services. ASTM standards serve as the basis for manufacturing, procurement, and regulatory activities.

Reid vapor pressure (RVP) is the pressure of the vapor above the fuel when the fuel is at 100°F (38°C).

The **driveability index (DI)** of gasoline is a specification used to manage engine performance during cold weather and while the engine is warming up.

The DI is a mathematical expression of distillation properties that describe the influence of fuel volatility on driveability:

$$DI = 1.5\ T10 + 3\ T50 + T90 + 11\ Oxygenates$$

 AUTHOR'S NOTE: Oxygenates is the concentration of oxygenates in wt-%.

The higher the DI number, the less volatile the fuel. Most premium gasoline sold in the United States has a higher DI index than regular or mid-grade gasoline. High DI gasoline also causes higher emissions for the same reason it causes driveability problems.

Testing for vapor-liquid ratio involves measuring the volume of vapor formed from a given volume of liquid at a specific temperature. Vehicle testing has shown that the temperature at which the vapor liquid ratio equals 20 (TV/L20) is an indication of protection from vapor lock. The TV/L20 specifications range from 105°F in the winter months to 140°F in the summer months.

Winter blend gasoline. As discussed, gasoline must be vaporized (mixed with air) in order to support combustion in the engine. Since cold temperatures reduce the normal vaporization of gasoline, refiners will increase the RVP of their gasoline to allow the engine to start easier in cold weather. The ASTM International standards for winter blend gasoline allows for up to 15 psi RVP. The RVP for gasohol can be up to 1 psi higher.

Gasoline mixed with alcohol is sometimes referred to as gasohol.

If winter blend gasoline is used in an engine during warm weather, the following problems may occur:
- Rough idle
- Stalling
- Vapor lock
- Hesitation on acceleration
- Surging
- Evaporative system damage
- Engine flooding during a hot soak

Summer blend gasoline. Since warm ambient temperatures cause gasoline to vaporize easily, the RVP of summer blend fuel is reduced. The vehicle's fuel system is designed to operate with liquid gasoline. If the RVP was not reduced, vapor bubbles would form and the engine would not run properly. According to ASTM International standards, the maximum RVP for summer blend is 10.5 psi. In the United States, the EPA requires gasoline that is to be sold in retail outlets between June 1st and September 15th to have a fuel vapor pressure below 9.0 psi. In areas of the country classified by the EPA as nonattainment areas, the VP must be lower than 7.8 psi.

Sulfur Content

Since sulfur is a component of crude oil, gasoline can contain some sulfur. A sulfur content that is too high can cause corrosion in the engine and exhaust system. For this reason, most of the sulfur is removed by the refining process.

When gasoline is burned in the combustion chamber, one of the products of combustion that is formed is water. Due to the high temperatures of combustion, the water will leave the combustion chamber as steam, however, this steam can condense back into water as it passes through the exhaust system. In addition to water in the exhaust, the steam in the combustion chamber and in the crankcase blowby will condense when the engine is shut off and cools. When the sulfur that is in the gasoline is burned, it combines with oxygen and forms sulfur dioxide. When sulfur dioxide combines with water, it forms sulfuric acid. Sulfuric acid is highly corrosive. Corrosion caused by this acid may cause exhaust valve pitting and exhaust system deterioration. Sulfur dioxide can cause the obnoxious odor of rotten eggs as it passes through the catalytic converter and the exhaust system. To reduce corrosion caused by sulfuric acid, the sulfur content in gasoline is limited. In the United States, current requirements are that the sulfur content must be less than 0.01 percent. In most European and Asian countries, the standards are stricter.

Fuel Additives

For many years, lead compounds such as tetraethyl lead (TEL) were added to gasoline to improve octane ratings. However, since the mid-1970s, vehicles have been designed to run on unleaded gasoline only. Leaded fuels are no longer available as automotive fuels due to the addition of special antipollution devices, such as catalytic converters, on vehicles. These systems must use unleaded fuel in order to work properly.

Gasoline additives have different properties and a variety of uses. Gasoline additives are expensive and are, therefore, added in limited quantities. Although the exact amount of additives a refinery adds to its gasoline is a closely guarded secret, it has been estimated that as little as 100 pounds of additive can be used to treat 20,000 gallons of gasoline. The following are examples of the additives that may be formulated into gasoline:

Detergents.   These are used to keep the fuel system clean of deposits. In the United States, deposit control additives have been required by law since 1995. Polyether amine is an added detergent that helps dissolve deposits and keeps injectors clean. However, the added detergents tend to create deposits on the intake valves.

Anti-Icing or Deicer.   Isopropyl alcohol is added seasonally to gasoline to prevent gas line freeze-up in cold weather.

Metal Deactivators and Rust Inhibitors.   These additives are used to inhibit reactions between the fuel and the metals used in the fuel system that can form abrasive substances.

Gum or Oxidation Inhibitors.   Some gasoline blends contain aromatic amines and phenols to prevent formation of gum and varnish. During storage, harmful gum deposits can form due to the reaction of some gasoline components with oxygen. Oxidation inhibitors are added to promote gasoline stability.

Octane Ratings

There are two important factors that affect the power and efficiency of a gasoline engine. The first is the **compression ratio**, the second, control of **detonation**. The higher the compression ratio, the greater the power output and efficiency of the engine. The better the efficiency, the less fuel will be consumed to produce a given power output. The problem is that the higher the compression ratio, the greater the tendency to have detonation. Most gasoline engines have a compression ratio between 8:1 and 10:1. High-performance engines may have higher compression ratios.

Normal combustion occurs smoothly and progresses across the combustion chamber from the point of ignition (Figure 2-8). Normal combustion propagation is between 50 to 250 meters per second. The speed of combustion propagation is dependent upon engine, load, air-fuel ratio,

The **compression ratio** is the ratio of the volume in the cylinder above the piston when the piston is at bottom dead center (BDC) to the volume in the cylinder above the piston when the piston is at top dead center (TDC). It is an indication of how much air-fuel mixture can be brought into a cylinder and how tightly the mixture will be compressed.

Detonation is the result of an amplification of pressure waves, such as sound waves, occurring during the combustion process when the piston is near top dead center (TDC). The actual "knocking" or "ringing" sound of detonation is due to these pressure waves pounding against the insides of the combustion chamber and the piston top.

| Compression | Spark Ignition | Combustion | Combustion Continued | Combustion Completed |

Normal combustion

Figure 2-8 Normal combustion has a smooth, controlled burn across the combustion chamber.

The **octane rating** is a measure of how easily the gasoline can be ignited. The higher the octane, the harder the fuel is to ignite and the slower it burns. Generally, the higher the octane, the greater number of hydrocarbons containing larger numbers of carbon atoms. More carbon atoms per hydrocarbon means that more oxygen and more heat are needed to burn the fuel. The octane rating provides an indication of its antiknock qualities. The higher the octane number, the less tendency for knock.

combustion chamber design (determining amount of turbulence), temperature, and **octane rating**. Measured in time, normal combustion of fuel occurs in a matter of milliseconds (thousandths of a second). Even though the speed of combustion is very fast, it is not an explosion. It is a controlled burn. The flame front advances smoothly across the combustion chamber until all the air-fuel mixture has been burned.

During periods of detonation, the combustion speed increases by up to ten times and approaches the speed of sound. Detonation is the result of the pressure waves caused by two flame fronts, one front started by the spark plug and the other started by fuel self-ignition (Figure 2-9). When the spark plug is fired, a pressure wave races through the unburned air-fuel mixture. The pressure wave is ahead of the flame front. The flame front is carried outward from the spark plug by eddies, swirls, and flow patterns of the combustion chamber turbulence. Normally, the pressure rise in the combustion chamber during combustion is between 20 and 30 psi per degree of crankshaft rotation. Once the pressure rises faster than about 35 psi per degree, the engine will run rough. This is due to the mechanical vibration of the engine components caused by too great a pressure rise. The resulting pressure wave can be strong enough to cause self-ignition of the fuel since it causes the remaining fuel to decompose. Self-ignition results from free radicals (hydroxyl or other molecules with similar open oxygen-hydrogen chains) in the fuel being ignited by the pressure wave and heat in the cylinder. The remaining fuel in the end gas lacks sufficient octane rating to withstand this combination of heat and pressure. The ignition of the end gases also causes a very high and rapid pressure increase in the combustion chamber. It is not uncommon for a pressure rise of up to 8 bars per degree of crankshaft angle or up to 50,000 bars per second during detonation. This pressure increase is for a very short duration. Detonation usually happens at the pressure wave's points of amplification such as at the edges of the piston

| Compression | Spark Ignition | Combustion | Combustion Continued | Detonation |

Detonation

Figure 2-9 Detonation occurs when a second flame is started as a result of high temperatures.

crown where reflecting pressure waves from the piston or combustion chamber walls can recombine to generate a very high local pressure. If the speed of this pressure buildup is greater than the speed that the air-fuel mixture burns, the pressure waves from both the initial ignition at the spark plug and the pressure waves coming from the self-ignited fuel will set off immediate explosions (instead of normal combustion) of the air-fuel mixture. The spike in pressure creates a force in the combustion chamber that causes the structure of the engine to ring, or resonate. Resonance, which is caused by the unequal pressures from the initial ignition and the second ignition, occurs at about 6,400 hertz (depending on engine design). The pinging (knock) sound is the result of the structure of the engine reacting to the pressure spikes. Incidentally, the knocking or pinging sound is not the result of "two flame fronts meeting" as is often stated. Although this clash does generate a spike, the noise you hear comes from the vibration of the engine structure reacting to the pressure spike.

High compression ratios create higher combustion chamber temperatures and pressures. To prevent detonation from occurring, all of the fuel in the combustion chamber must be completely burned prior to the pressure's rising to the point of self-ignition. Factors such as combustion chamber design, piston dome design, location and number of spark plugs in the cylinder, valve timing, and ignition timing all contribute to detonation control. Most electronically controlled ignition systems have a sensor to detect if knock is occurring so the computer can retard the ignition timing to prevent detonation. Another factor in preventing detonation is the octane rating of the gasoline being burned.

In the early 1920s, TEL was added to gasoline to reduce the tendency for spark knock. Shortly after the introduction of TEL as a fuel additive, the Society of Automotive Engineers (SAE) established standards for the anti-knock rating of gasoline. The purpose for these standards was to allow the automobile manufacturers to produce engines that would operate knock free based on the quality of the available fuel.

BIT OF HISTORY

The octane number came about as a result of research carried out in the 1920s and 1930s by Sir Harry Ricardo and Charles Kettering (he also developed the distributor and coil ignition system). Harry Ricardo developed the concept of a test engine in which the compression ratio, valve timing, and other factors could be altered while the engine was running. Kettering assigned Thomas Midgley to investigate the problem of knocking which was destroying his test engines. Midgely conducted a long search for additives that would help a fuel resist knocking. Among the chemicals tried were iodine, aniline, selenium oxychloride, methylclopentadienyl manganese tricarbonyl, and other phosphorus, sodium, and potassium compounds. Midgley even tried melted butter! Gradually he began to focus on organo-lead compounds, and eventually developed a combination of TEL with ethylene dibromide and ethylene dichoride. Gasoline with tetraethyl lead additives was often called "ethyl" or "high test" gasoline.

Two commonly used methods for determining the octane number of motor gasoline are the motor octane number (MON) method and the research octane number (RON) method. Both use a laboratory single-cylinder engine called a cooperative fuel research (CFR) engine. The MON method test engine is run at a constant 1,500 rpm with no load. The RON method uses the same test engine but places it under a load that is more closely related to actual driving conditions. The CFR engine is equipped with a variable head and knock meter to indicate knock intensity. The test provides an octane rating by comparing knock intensity between the test sample of fuel and a known combination of knock resistant hydrocarbon iso-octane, chemically called trimethylpentane (C_8H_{18}), and also known as 2-2-4 trimethylpentane. This test sample has an octane rating of 100. The trimethylpentane is mixed with another hydrocarbon called N-heptane (C_7H_{16}) that has an octane rating of zero. The test sample is used as fuel, and the engine's compression ratio and air-fuel mixture are adjusted to develop a specific knock intensity. Various proportions of heptane and iso-octane are run in the test engine to duplicate the severity of the knock of the fuel being tested. When the knock caused by the heptane and iso-octane mixture is identical to the test fuel, the octane number is established by the percentage of iso-octane in the mixture. If the test sample of fuel has the exact same anti-knock characteristics of iso-octane, it is rated as 100 octane gasoline. If 85 percent iso-octane and 15 percent N-heptane produce the same severity of knock as the fuel being tested, the fuel is assigned an octane number of 85.

Engine run-on is referred to as "dieseling" since the engine continues to run without the benefit of a spark plug's firing.

The RON gives a higher reading for the same fuel as the motor method (typically eight to ten numbers higher). It affects low-speed knock and engine run-on. The MON gives a measurement of how much engine knock will be present under heavy loads such as when passing or climbing hills. For example, a fuel with a RON of 93 might have a MON of 85.

By law, U.S. gas pumps are labeled with a yellow sticker containing an Anti-knock Index (AKI) number, which is the average of the RON and MON rating systems. The AKI index, also known as road octane number or pump octane number, is referred to as (R + M)/2. Using the previous examples of MON and RON ratings, the rating posted on the pumps would be:

$$(RON + MON)/2 = 93 + 85/2 = 178/2 = 89$$

AUTHOR'S NOTE: Gasoline is not mixed with octane. "High octane fuel" does not mean there is more octane in the fuel compared to "low octane fuel." The octane rating is a measure of the fuel's ability to *control* the burning process (to prevent detonation); it is not a function of burning "hotter" or "colder."

Previously, we discussed that there are many factors that affect spark knock other than just the octane rating of the gasoline. To recap, these factors include:
- *Lean Fuel Mixture.* A lean mixture burns slower than a rich mixture. The heat of combustion is higher, which promotes the tendency for unburned fuel in front of the spark ignition flame to detonate.
- *Over-advanced Ignition Timing.* Advancing the ignition timing induces knock. Retarding ignition timing suppresses knock.
- *Compression Ratio.* Compression ratio affects knock because cylinder pressures are increased with the increase in compression ratio.
- *Valve Timing.* Valve timing that fills the cylinder with more air-fuel mixture promotes higher cylinder pressures, increasing the chances for detonation.
- *Turbocharging.* Turbocharging (or supercharging) forces additional fuel and air into the cylinder. This induces higher cylinder pressures and temperatures which promote knock.
- *Coolant Temperature.* Hot spots in the cylinder or combustion chamber due to inefficient cooling or a damaged cooling system raise combustion chamber temperatures and promote knock.
- *Cylinder-to-Cylinder Distribution.* If an engine has poor distribution of the air-fuel mixture from cylinder to cylinder, the leaner cylinders could promote knock.
- *Excessive Carbon Deposits.* The accumulation of carbon deposits on the piston, valves, and combustion chamber causes poor heat transfer from the combustion chamber. Carbon accumulation also artificially increases the compression ratio. Both conditions cause knock.

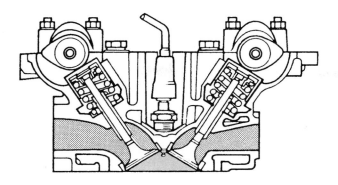

Figure 2-10   Cross-sectional view of a pentroof combustion chamber.

- *Air Inlet Temperature.* The higher the air temperature when it enters the cylinder, the greater the tendency to knock.
- *Combustion Chamber Design.* The optimum combustion chamber design for reduced knocking is the hemispherical or pentroof designs. Both use a spark plug located in the center of the chamber (Figure 2-10). These chambers promote faster combustion, allowing less time for detonation to occur ahead of the flame front.
- *Faulty components* such as the knock sensor, advance delay valves, and EGR valves.
- *Humidity.* The drier the air, the greater the potential for knock.
- *Altitude.* The lower the altitude, the more potential for knock.

Regular and Premium Fuels

Most gasoline stations offer three octane grades of unleaded gasoline: *regular* (87 octane), *mid-grade* (89 octane), and *premium* (93 octane). The amount of octane required for a particular engine is called the engine octane number requirement (ONR). The octane requirements will go up as the compression ratio increases (Figure 2-11).

To obtain the full benefit of a higher octane fuel, the engine, ignition system, and fuel control system must be designed and adjusted to take advantage of the higher octane gasoline. Most modern engines are designed to operate efficiently with regular grade gasoline (87 octane) and do not require a high octane premium grade.

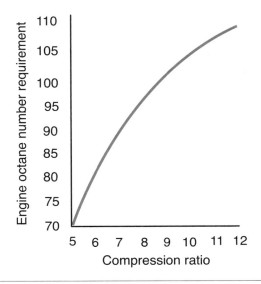

Figure 2-11   Graph indicating that as compression ratio is increased, so does the octane requirement.

AUTHOR'S NOTE: Using premium gasoline in a car designed to run on regular will not harm it, but the gains from using premium fuel does not offset the higher cost. In addition, since higher octane fuels may have greater concentrations of alcohol, it is possible to experience hard starting after the engine is hot. On the other hand, using a lower octane fuel than required by the manufacturer can cause engine damage and void their warranty.

Octane requirements are also affected by changes in outside air temperature, humidity, and pressure. For example, in the high altitude of the Colorado Rockies, all the grades of gasoline have lower octane rating than other parts of the country. Regular grade is usually rated at 85, mid-grade is 87, and premium fuel has a rating of 91. As a rule, octane rating of gasoline is reduced about one octane number per 1,000 feet or 300 meters in altitude.

The reason for the lower octane requirement is because an increase in altitude means a decrease in atmospheric pressure. The air is less dense because a pound of air takes more volume. The octane rating of fuel does not need to be as high because the engine cannot take in as much air. This process will reduce the combustion (compression) pressures inside the engine. A secondary reason for the lowered octane requirement of engines running at higher altitudes is the normal enrichment of the air-fuel ratio and lower engine vacuum with the decreased air density.

Octane Improvers

Because lead will cause the vehicle's catalytic converter to fail, gasoline is limited to a lead content of 0.06 gram per gallon. When Federal EPA regulations required gasoline companies to remove tetraethyl lead from gasoline, other methods of controlling spark knock were developed. Octane improvers can be grouped into three categories:

1. Aromatic hydrocarbons (hydrocarbons containing the benzene ring) such as xylene and toluene
2. Alcohols such as ethanol, methanol, and tertiary butyl alcohol (TBA)
3. Metallic compounds such as methylcyclopentadienyl manganese tricarbonyl (MMT)

Oxygenated Fuels

Shop Manual
Chapter 2,
page 75

Oxygenated fuels
are mixtures of
conventional
gasoline and one or
more combustible
liquids which contain
oxygen (oxygenates).

Methanol is a
mixture of gasoline
and methyl alcohol.

Some areas of the country require that **oxygenated fuels** be used during the winter months, and some gasoline stations sell blends of oxygenated fuels year round. Oxygenates are compounds such as alcohols and ethers that contain oxygen in their molecular structure. The idea behind the use of oxygenated fuels in these areas is to reduce carbon monoxide (CO) emissions. By blending oxygenates into the gasoline, the fuel requires less ambient oxygen for complete burning. Therefore, for the same carburetor or fuel injector settings, oxygenated gasoline produces a leaner air-fuel mixture and generates less carbon monoxide. The extra oxygen in the fuel ensures that there is enough oxygen to convert CO into carbon dioxide (CO_2) during the combustion process, either in the engine or in the catalytic converter.

Examples of oxygenated fuels include methanol, ethanol, and ethyl tertiary butyl ether (ETBE). Ethyl alcohol, (ethanol) up to a 10 percent concentration, is the most common oxygenate.

Methanol

Methanol can be distilled from coal, natural gas, oil shale, wood, or garbage. Most of what is used today is derived from natural gas. Most automotive manufacturers warn customers against using a fuel that contains more than 10 percent methanol and **co-solvents** by volume unless the vehicle was specifically produced to burn this fuel. Methanol is recognized as being far more corrosive to fuel system components than ethanol. Another disadvantage of methyl alcohol-mixed fuels is that it produces only about half the energy of gasoline. In order to use methyl alcohol, the fuel system needs to be adjusted to provide an air-fuel ratio of about 6.5:1. This is because methanol is 50 percent oxygen. Gasoline containing 5 percent methanol would have an oxygen content of 2.5 percent by weight.

Co-solvents are another substance (usually another alcohol) that is soluble in both methanol and gasoline and is used to reduce the tendency of the liquids to separate.

Ethanol

The most common alcohol-gasoline mixture is ethanol. Ethanol contains a mixture of about 10 percent **ethyl alcohol**. Ethanol was first used to extend gasoline supplies during the gasoline shortages of the 1970s. Ethanol has an oxygen content of approximately 35 percent, thus a 10 percent concentration adds about 3.5 percent oxygen to the mixture. Another benefit of ethanol is that it increases the octane rating of the fuel. A 10 percent ethanol mixture will raise an 87 octane fuel by at least 2.5 octane numbers (Figure 2-12). However, the alcohol added to the base gasoline also raises the volatility of the fuel about 0.5 psi (3.5 kPa). Most automobile manufacturers permit up to 10 percent ethanol if driveability problems are not experienced.

Ethyl alcohol is an alcohol made from grain.

Gasoline blended with 10 percent alcohol or less does not require changes to the fuel system, however, vehicles burning any amount of gasohol may require that the fuel filter be changed more often. This is due to the cleaning effect that alcohol has on the vehicle's fuel tank. Oxygenates suspend water in the fuel and tend to keep it from accumulating in the gas tank. Gasoline cannot hold much water, so it separates and accumulates at the bottom of the tank.

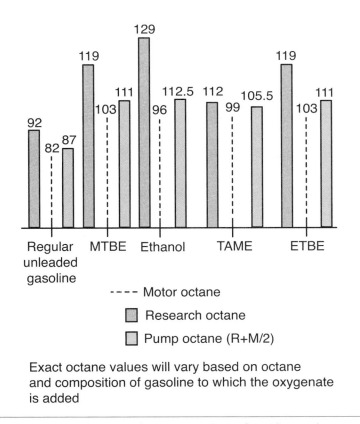

Figure 2-12 Comparison between the octane ratings of gasoline and oxygenated fuels.

Alcohol and ether attract and hold water. The following are the amounts of water one gallon of each fuel can hold in suspension:

Gasoline—0.5 teaspoon
10 percent ethanol—12 teaspoons

Ethyl Tertiary Butyl Ether (ETBE)

ETBE is derived from ethanol. The maximum allowable volume level is 17.2 percent. ETBE is made by combining isobutylene and ethanol. Like MTBE and methanol, ETBE does not have a significant amount of free ethanol and does not have ethanol properties.

Advantages and Disadvantages of Alcohol Additives

The following is a list of the advantages of using alcohol as an additive to gasoline:

1. Absorbs moisture in the fuel tank.
2. Ten percent alcohol added to gasoline raises the octane rating, using the R + M/2 method, by three points.
3. Alcohol cleans the fuel system.
4. The addition of alcohol reduces CO emissions because alcohol contains oxygen.

The disadvantages of alcohol blends include:

1. The use of alcohol blends can result in the clogging of fuel filters.
2. Alcohol raises the volatility of fuel about 0.5 psi (3.5 kPa), resulting in hot weather driveability problems.
3. Since the heat energy of alcohol is between 60,000 to 75,000 Btus per gallon as compared to about 120,000 Btus per gallon for gasoline, the addition of alcohol reduces the heat content of the resulting fuel mixture. RFG will yield 2 to 3 percent lower fuel mileage than non-oxygenated gasoline.
4. Alcohol absorbs water and then separates from the gasoline, especially as the temperature drops.
5. Separated alcohol and water on the bottom of the tank can cause hard starting during cold weather.
6. Alcohol does not vaporize easily at low temperatures.

Reformulated Gas

Reformulated gasoline (RFG) is a general term for federally mandated gasoline that is specially processed and blended to reduce the emission of pollutants such as hydrocarbons, toxics, and nitrogen oxides.

The 1990 Clean Air Act requires the EPA to issue regulations that would require gasoline to be "reformulated" so as to result in significant reductions in vehicle emissions of ozone-forming and toxic air pollutants. This cleaner gasoline is called **reformulated gasoline (RFG)**. RFG is required in the most severe ozone nonattainment areas of the United States. In addition, several other areas with ozone levels exceeding the public health standard have voluntarily chosen to use RFG. The major differences between RFG and conventional gasoline are:

- RFG has lower levels of certain compounds that contribute to air pollution—notably Benzene.
- RFG will not evaporate as easily as conventional gasoline—lower RVP.
- RFG will contain "chemical oxygen" (oxygenates).

The RFG program has been implemented in two phases; the Phase I program ran from 1995 through 1999, and the Phase II RFG program began in 2000. For Phase I, the Clean Air Act required that reformulated gasoline result in summertime emissions of volatile organic compounds (VOCs) and year-round emissions of air toxics that are 15 percent less than those that would occur from the use of conventional gasoline. For Phase II, VOCs and toxics must be reduced by an additional 25 percent.

For gasoline to be considered reformulated, it must have an oxygen content of at least 2.0 percent by weight and a benzene content no greater than 1.0 percent by volume. It also must contain no heavy metals, including lead or manganese. According to EPA estimates, RFG reduces hydrocarbon emissions by at least 15 percent.

Government Test Fuel

Because of the variation of commercially available fuel, the federal government uses a standardized fuel for testing engines for emissions and fuel economy. The standard fuel is indolene 30, and it is used as a standard replacement for regular unleaded gasoline during these tests.

Alternative Fuels

Tighter federal, state, and local emissions regulations have led to a search for **alternative fuels**. Many considerations determine the viability of an alternative fuel, including emissions, cost, fuel availability, fuel consumption, safety, engine life, fueling facilities, weight and space requirements of fuel tanks, and the range of a fully-fueled vehicle. Currently, the major competing alternative fuels include ethanol, methanol, propane, and natural gas. Although diesel fuel has been in use for many years, its properties are included in this section. Ethanol and methanol were presented earlier in this chapter.

Propane is a petroleum-based pressurized fuel used as a liquid. It is a constituent of natural gas. Natural gas comes in two forms: compressed natural gas (CNG) and liquefied natural gas (LNG).

Liquid Petroleum Gas

Liquid petroleum (LP) gas is a by-product of crude oil refining, and it is also found in natural gas wells. LP gas is similar chemically to gasoline; however, LP gas is mostly propane with a small percentage of butane (up to 8 percent). LP gas is a vapor above –40°F (–40°C). It is called liquid petroleum because it is stored as a liquid in a pressurized bottle. The pressure increases the boiling point of the gas and prevents it from vaporizing.

Because LP gas burns very clean and creates lower emissions than gasoline, it is often used in fleets such as utility companies, taxis, buses, and for indoor machinery such as forklifts. LP gas has been used for a number of years on farm tractors as an alternative to regular gasoline. Even though LP gas has lower heat energy per volume than gasoline (about 84,000 Btu), it does have a higher octane rating (104). This allows LP gas to be used in engines with higher compression ratios.

LP gas burns clean in the engine because it vaporizes at atmospheric temperatures and pressures. Since it vaporizes so readily, it does not puddle in the intake manifold. This means it emits fewer hydrocarbons and carbon monoxide missions without added emission control devices.

Alternative fuels are those other than gasoline and diesel fuel. LP gas and alcohol are currently in use in automobiles and trucks. Hydrogen is also another example of an alternative fuel.

Liquid petroleum (LP) gas is the term widely used to describe a family of light hydrocarbons called "gas liquids." The most prominent members of this family are propane (C_3H_8) and butane (C_4H_{10}).

Because of its high propane content, many people simply refer to LP gas as propane.

Natural Gas

One of the cleanest burning fuels available is **compressed natural gas (CNG)**. It has 99 percent less carbon monoxide and 85 percent less reactive hydrocarbon emissions than gasoline, with no particulates, and almost no sulfur dioxide. As a result of its low emissions and clean burn, oil change intervals can be increased and spark plugs last longer. A big benefit of natural gas is that it is available in vast quantities in North America.

Vehicles have been designed with gasoline/CNG, diesel/CNG, and dedicated CNG engine applications. CNG vehicles offer several advantages over gasoline:

- The fuel costs less.
- It is the cleanest alternative hydrocarbon fuel.
- It is lighter than air, so it dissipates quickly.
- It has a higher ignition temperature.
- The fuel tanks on natural gas vehicles are safer than those used on gasoline-fueled vehicles. The fuel tanks used for CNG are aluminum or steel cylinders with walls that are 1/2 to 3/4 inch (13 to 19 mm) thick. They can withstand severe crash tests, direct gunfire, dynamite explosions, and burning beyond any standard sheet metal gasoline tank.
- It generally reduces vehicle maintenance since it burns cleanly.
- Natural gas is abundant and readily available in the United States.

At this time, the largest disadvantage of CNG is the lack of an infrastructure for refueling the vehicle. Fuel facilities are needed in greater numbers than are currently in existence due to the relatively shorter range of CNG vehicles. Lack of range is partly due to CNG having about 33,000 to 38,000 Btu at 3,000 psi. However, CNG has an octane rating of 120+. Many utility companies provide CNG filling stations, and some companies sell kits to allow filling from the natural gas line into your house; however, this is a low-pressure tap and it takes a long time to fill the tanks. The space taken by the CNG cylinders and their weight, about 300 pounds (136 kg), would also be considered disadvantages in most applications.

Hydrogen

Hydrogen is one of the most abundant elements in the universe. It is highly flammable and is a promising fuel for the future. It is an ideal fuel, producing no emissions other than water and carbon dioxide (a nonpolluting and nonpoisonous gas). The sun is an example of burning hydrogen. Hydrogen has an engery content of 113,000 to 134,000 Btu per gallon and an octane rating of over 130. Transportation of hydrogen from the plant to the consumer is the biggest obstacle to its use. Hydrogen is currently transported by pipeline or by road via cylinders, tube trailers, and cryogenic tankers, with a small amount shipped by rail or barge. Pipelines are limited to a few areas in the United States where large hydrogen refineries and chemical plants are concentrated. Hydrogen distribution by means of high-pressure cylinders and tube trailers has a range of about 200 miles from the production facility. For longer distances of up to 1,000 miles, hydrogen is usually transported as a liquid in super-insulated, cryogenic, over-the-road tankers, railcars, or barges and is then vaporized for use at the customer site.

E85

Ethanol, or grain alcohol, is produced by fermenting biomass. Ethanol is a renewable resource and contributes nothing in itself to the greenhouse gas loading of the atmosphere. As a motor fuel, E85 is blended in a mixture of 85 percent ethanol and 15 percent unleaded gasoline. However, it can be blended with any amount of gasoline in the tank of a flex-fuel vehicle. E85 has a per-gallon Btu rating of about 80,000 and an octane rating of 100.

Diesel Fuel

Diesel engines are designed to use diesel oil as fuel. **Diesel fuel** is also made from petroleum. At the refinery, the petroleum is separated into three major components: gasoline, middle distillates, and all remaining substances. Diesel fuel comes from the middle distillate group, which has properties and characteristics different from gasoline.

A diesel engine is called a **compression ignition engine**. A four-stroke diesel engine operates in a similar manner to a gasoline engine. The difference is that ignition in a diesel is controlled by the injection of fuel into the cylinder. A diesel engine has a very high compression ratio, usually between 16:1 and 20:1 (Figure 2-13). The pressure of compression (400 to 700 psi or 2800 to 4800 kilopascals) generates temperatures of 1200°F to 1600°F (700°C to 900°C). If the diesel fuel was first drawn into the cylinder and then compressed (as it is in a gasoline engine), the fuel would self-ignite before the piston could reach top dead center (TDC). In a diesel, the fuel is injected by a high-pressure injector at the instant that ignition is desired (Figure 2-14). The high pressures and temperature in the cylinder speeds the pre-flame reaction to start the ignition of the fuel injected into the cylinder. The compressed, hot air in the cylinder causes the fuel to vaporize and burn.

The shape of the fuel spray, turbulence in the combustion chamber, beginning and duration of injection, and the chemical properties of the diesel fuel all affect the power output of the diesel engine. Diesel fuel must meet an entirely different set of standards from gasoline. All diesel fuel must be clean, able to flow at low temperatures, and be of the proper cetane rating. The significant chemical properties of diesel fuel are described in the following:

Shop Manual
Chapter 2,
page 76

Diesel fuel is light oil that is refined as part of the same process that makes gasoline. It has several properties that make it useful as a fuel.

A **compression ignition engine** uses the heat generated by compressing air to ignite the fuel charge. These engines do not use a spark plug to start the combustion process.

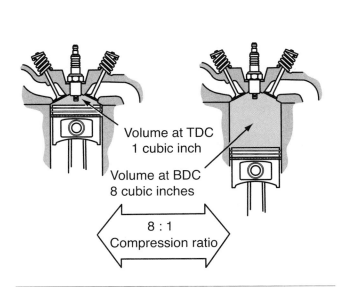

Figure 2-13 The compression ratio is an expression of how much the air-fuel charge will be compressed. This is an example of a compression ratio of 8:1.

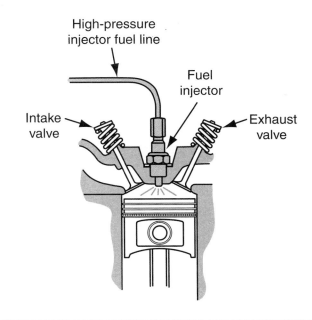

Figure 2-14 The diesel engine uses an injector to spray the fuel into the combustion chamber at the instant combustion is required.

Shop Manual
Chapter 2,
page 76

Density

Diesel engines have a predictable variation in power output depending on the **density** of the fuel used. Relative density of fuels is usually expressed in degrees American Petroleum Institute (API) gravity. A fuel with a high density will have a low API gravity, and vice versa. The formula for determining API gravity is:

$$\text{API Gravity (deg.)} = (141.5/\text{Specific Gravity @ } 60°F) - 131.5$$

Specific gravity is the ratio of the weight of a certain volume of fuel at 60°F to the weight of the same volume of water at 60oF.

Engines using fuels with a high density (low API gravity) will produce more power than those using fuels with lower densities (higher API gravities) because the thermal energy content of the fuel is higher. That is, low API gravity fuels have higher energy content than high API gravity fuels.

A general rule of thumb is that there is approximately a 3 percent to 5 percent decrease in the thermal energy (Btu) content of fuel for every 10-degree increase in API gravity. In addition, there is a 0.7 degree API gravity increase with an increase in temperature. The loss in the energy content of the fuel results in approximately the same percentage of loss in engine power output.

AUTHOR'S NOTE: The use of fuels with higher API gravity will cause an increase in fuel consumption. This is due to the lower thermal energy content per gallon of fuel.

Viscosity

In addition to power variations due to fuel density, variations in power output can be due to differences in fuel **viscosity**. Diesel fuel must be able to flow easily through the fuel system and be sprayed by the injectors. This requires the diesel fuel to have a low viscosity. The viscosity of diesel fuel directly affects the spray pattern of the fuel into the combustion chamber (Figure 2-15). Viscosity must be just high enough to provide lubrication to the fuel injectors and fuel pump, but it must not be so high as to produce large droplets that are hard to burn. On the other hand, fuel with too low viscosity can create the following problems:

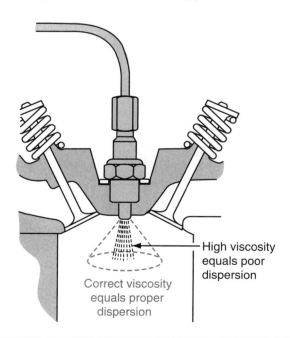

Figure 2-15 The viscosity of the fuel is a major factor contributing to the spray pattern of the fuel.

- Power variations due to changes in density with temperature changes. Viscosity deteriorates as temperatures increase.
- Hard restart with a hot engine.
- Rough idle.
- White exhaust smoke.
- Loss of torque due to the fine mist igniting too rapidly.
- Inadequate lubrication and cooling of the injection pump and nozzles.

Cloud Point and Pour Point

Since diesel fuels contain a wax base called paraffin, ambient temperatures affect diesel fuel more than it does gasoline. As ambient temperatures drop to the wax appearance point (WAP) or **cloud point** of the fuel, the wax crystals grow larger. If cold enough, the crystals will enlarge enough to restrict the flow of fuel through the filters and lines. Eventually, the fuel will stop flowing as a result of the wax crystals plugging the fuel filter or lines. The better the quality of the fuel, the lower the cloud point. As ambient temperatures continue to drop, the fuel will reach its **pour point** where it solidifies and no longer flows.

Volatility

Diesel fuel is not as volatile as gasoline. The amount of carbon residue left by diesel fuel depends on the quality and the volatility of that fuel. Fuel that has a low volatility is much more prone to leaving carbon residue. The small, high-speed diesels found in automobiles require a high quality and highly volatile fuel because they cannot tolerate excessive carbon deposits. Large, low-speed industrial diesels are relatively unaffected by carbon deposits and can run on lower quality fuel.

Cetane Rating

Diesel fuel's ignition quality is measured by the **cetane rating**. Much like the octane number, the cetane number is measured in a single-cylinder test engine with a variable compression ratio. The diesel fuel to be tested is compared to **cetane**. The higher the cetane number assigned to the diesel fuel, the shorter the ignition lag time (delay time) from the point the fuel enters the combustion chamber until it ignites. Cetane rating for diesel fuel and the octane rating for gasoline are not the same. The higher the cetane rating, the more easily the fuel is ignited, whereas the higher the octane rating, the more slowly the fuel is ignited (Figure 2-16).

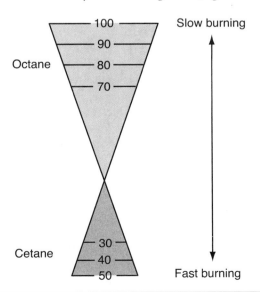

Figure 2-16 Difference between the octane rating of gasoline and the cetane rating of diesel fuel.

Cloud point is the low-temperature point at which the waxes present in most diesel fuel tends to form wax crystals that clog the fuel filter. This is called the cloud point because the fuel will appear cloudy when the wax separates out.

Pour point is the lowest temperature at which the fuel is observed to flow.

The **cetane rating** determines, to a great extent, its ability to start the engine at low temperatures and to provide smooth warm up and even combustion. The cetane number of a diesel fuel describes how easily the fuel will ignite.

Cetane is a colorless, liquid hydrocarbon that has excellent ignition qualities. Cetane is rated at 100.

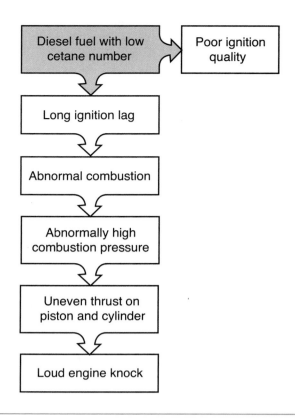

Figure 2-17 Possible problems resulting from use of a fuel with too low of a cetane rating.

In fuels that are readily available, the cetane number ranges from 40 to 55, with values of 40 to 50 being the most common. These cetane values are satisfactory for medium-speed engines whose rated speeds are from 500 to 1,200 rpm and for high-speed engines rated over 1,200 rpm. Low-speed engines rated below 500 rpm can use fuels in the above 30 cetane number range. The cetane number improves with the addition of certain compounds such as ethyl nitrate, acetone peroxide, and amyl nitrate. Amyl nitrate is commercially available for this purpose.

Fuels with a high cetane number will burn as soon as they are injected, so no knock occurs. The lower the cetane number, the higher the temperature required to ignite the fuel. If the cetane number is too low for the conditions, the fuel will take too long to light and will build up in the cylinder. Engine knocking can occur under this condition when the fuel is burned in too much quantity and out of control. Figure 2-17 lists some of the problems caused by fuel with a too low cetane rating.

Flash Point

The **flash point** is the temperature at which the vapors on the surface of the fuel will ignite if exposed to an open flame.

Although **flash point** does not affect diesel engine operation, a lower-than-normal flash point could indicate contamination of the diesel fuel with gasoline or a similar substance.

Sulfur Content

On October 1, 1993, the U.S. Clean Air Act required that all diesel fuels sold for use in highway vehicles contain no more than 0.05 percent sulfur. The sulfur content of diesel fuel is very important to the life of the engine. Sulfur in the fuel creates sulfuric acid during the combustion process that can damage engine components and cause piston ring wear.

Diesel Fuel Grades

Minimum quality standards for diesel fuel grades have been set by ASTM International. The commonly used fuel grades are described.

Grade 1.   Grade 1 (number 1) diesel fuel is thinner, and more volatile, and it is used at very low temperatures. It has the lowest boiling point and the lowest cloud and pour points. As a result, grade 1 is suitable for use during low-temperature operation. Grade 1 may be specified for use in diesel engines involved in frequent changes in load and speed such as those found in city buses and delivery trucks. The API gravity of number 1 diesel is between 48 and 34 degrees, thus it has a Btu rating per gallon of 134,637.

Grade 2.   Grade 2 (number 2) diesel fuel has a lower volatility than number 1 fuel. Number 2 diesel is the most commonly specified fuel to be used for most driving conditions. Number 2 fuel has a higher boiling range, cloud point, and pour point as compared with number 1. The API gravity of number 2 diesel is between 40 and 24 degrees, thus it has a Btu rating per gallon of 139,457.

Grade 4.   Grade 4 has a very high boiling range and is designed for use in low-speed engines running under a constant load. Grade 4 is used mainly in stationary power plants and ships and not as an automotive fuel.

In colder climates, number 1 diesel fuel is blended with number 2 to improve starting. In moderately cold climates, the blend may be 90 percent number 2 to 10 percent number 1. In very cold climates, the blend may be as high as 50/50. Diesel fuel economy can be expected to drop off during the winter months due to the use of number 1 diesel in the fuel blend.

Cleanliness

It is imperative that the fuel used in a diesel engine be clean and free from water. Unlike the case with gasoline engines, the fuel is the lubricant and coolant for the diesel injection pump and injectors. Good quality diesel fuel contains additives such as oxidation inhibitors, detergents, dispersants, rust preventives, and metal deactivators.

Summary

❏ The refining process separates the hydrocarbons of crude oil so they can be used as fuels and solvents.

❏ The combustion process involves the chemical combination of oxygen (O_2) from the air with the hydrogen and carbon from the fuel.

❏ If the combustion process is complete, all the gasoline or hydrocarbons (HC) will be completely combined with all the available oxygen (O_2), this total combination of all components of the fuel is called stoichiometry.

❏ One British thermal unit (Btu) is the amount of heat required to raise one pound of water one Fahrenheit degree. The metric unit of heat is the calorie (cal). One calorie is the amount of heat required to raise the temperature of one gram (g) of water one Celsius degree.

❏ The major factors affecting fuel performance are volatility, sulfur content, deposit control, and octane rating.

❏ Two commonly used methods for determining the octane number of motor gasoline are the motor octane number (MON) method and the research octane number (RON) method. The octane rating posted on pumps in the United States is the average of the two methods and is referred to as (R + M)/2.

Terms to Know

Alternative fuels

(ASTM) American Society for Testing and Materials

British thermal units (Btus)

Calories

Cetane

Cetane rating

Cloud point

Compressed natural gas (CNG)

Compression ignition engine

❑ Oxygenated fuels usually contain alcohol that includes oxygen in its content to lower CO exhaust emissions.

❑ Examples of oxygenated fuels include methanol, ethanol, methyl tertiary butyl ether (MTBE), and ethyl tertiary butyl ether (ETBE).

❑ The significant chemical properties of diesel fuel include density, viscosity, cloud point, pour point, volatility, cetane rating, flash point, and sulfur content.

❑ The higher the cetane number assigned to the diesel fuel, the shorter the ignition lag time (delay time) from the point the fuel enters the combustion chamber until it ignites.

Review Questions

Short Answer Essays

1. Explain the refining process of petroleum-based fuels.
2. Explain how Reid vapor pressure is measured.
3. List and briefly describe the four performance characteristics of gasoline.
4. Describe the common methods of determining the octane number.
5. Describe why lower-octane rated fuels are used in an area of higher elevations.
6. Explain the purpose of oxygenated fuels.
7. List and describe the common oxygenated fuel blends.
8. Describe the different alternative fuels discussed in this chapter.
9. Explain the differences between diesel fuel grades.
10. Explain the cetane rating of diesel fuel.

Fill-in-the-Blanks

1. The ratio of air to fuel, where complete combustion is accomplished, is referred to as _____ .

2. Since cold temperatures reduce the normal vaporization of gasoline, refiners will _____ the RVP of their gasoline to allow the engine to start easier in cold weather.

3. The_____ _____ is a measure of how easily the gasoline can be ignited.

4. The combustion process involves the chemical combination of _____ from the air with the _____ and _____ from the fuel.

5. The most common alcohol-gasoline mixture is _____ .

6. One _____ _____ _____ is the amount of heat required to raise one pound of water one Fahrenheit degree.

7. A diesel fuel with a high _____ will have a low API gravity,

8. The viscosity of diesel fuel directly affects the _____ _____ of the fuel into the combustion chamber.

9. The higher the cetane rating, the _____ the fuel is ignited.

10. The best known standard for measuring vapor pressure is called _____ _____ _____ .

Multiple Choice

1. *Technician A* says that the air-fuel ratio in which all combustion is complete is referred to as stoichiometric.
   Technician B says the stoichiometric quantities for all hydrocarbon fuels are the same.
   Who is correct?
   - **A.** A only
   - **B.** B only
   - **C.** Both A and B
   - **D.** Neither A nor B

2. Seasonal blends of gasoline are being discussed.
   Technician A says for fuel used in the winter, the RVP must be lowered.
   Technician B says if winter blend gasoline is used in an engine during warm weather, vapor lock may occur.
   Who is correct?
   - **A.** A only
   - **B.** B only
   - **C.** Both A and B
   - **D.** Neither A nor B

3. *Technician A* says sulfur is added to the fuel blend to increase the life of the seals used in the vehicle's fuel system.
   Technician B says deposit control additives have been required by law since 1995.
   Who is correct?
   - **A.** A only
   - **B.** B only
   - **C.** Both A and B
   - **D.** Neither A nor B

4. Detonation is described as:
   - **A.** A smooth flame front moving across the combustion chamber, burning all of the air-fuel mixture.
   - **B.** A controlled burn during the combustion process.
   - **C.** Uncontrolled burn of the air-fuel mixture as a result of a second flame front.
   - **D.** None of the above.

5. The AKI index posted on retail fuel pumps is an average of what two methods of determining octane numbers of motor gasoline?
   - **A.** The SG and Boil Point methods
   - **B.** The API and the ATMI methods
   - **C.** The AKI and CFR methods
   - **D.** The MON and RON methods

6. The intent of using oxygenated fuels is to:
   - **A.** Increase the fuel economy of the engine.
   - **B.** Increase the power output of the engine.
   - **C.** Reduce carbon monoxide (CO) emissions.
   - **D.** Reduce the possibility of vapor lock.

7. Oxygenated fuels are being discussed. *Technician A* says in order to use methyl alcohol, the fuel system needs to be adjusted to provide an air-fuel ratio of about 6.5:1.
   Technician B says ethanol has an oxygen content of approximately 35 percent.
   Who is correct?
   - **A.** A only
   - **B.** B only
   - **C.** Both A and B
   - **D.** Neither A nor B

8. *Technician A* says the higher the octane number gasoline, the faster it burns.
   Technician B says the higher the cetane number of diesel fuel, the slower it burns.
   Who is correct?
   - **A.** A only
   - **B.** B only
   - **C.** Both A and B
   - **D.** Neither A nor B

9. *Technician A* says LP gas burns clean in the engine because it vaporizes at atmospheric temperatures and pressures.
   Technician B says the disadvantage of compressed natural gas (CNG) is its high output of carbon monoxide.
   Who is correct?
   - **A.** A only
   - **B.** B only
   - **C.** Both A and B
   - **D.** Neither A nor B

10. Diesel fuel is being discussed. *Technician A* says engines using fuels with a high density will produce less power than those using fuels with lower density.
    Technician B says the viscosity of diesel fuel directly affects the spray pattern of the fuel into the combustion chamber.
    Who is correct?
    - **A.** A only
    - **B.** B only
    - **C.** Both A and B
    - **D.** Neither A nor B

Introduction to the Computer

Upon completion and review of this chapter, you should be able to:

❑ Describe the principle of analog and digital voltage signals.

❑ Explain the principle of computer communications.

❑ Describe the basics of logic gate operation.

❑ Describe the basic function of the central processing unit (CPU).

❑ Explain the basic method by which the CPU is able to make determinations.

❑ List and describe the differences in memory types.

❑ Describe the function of adaptive strategy.

❑ List and describe the function of various semiconductor devices.

❑ List and describe the functions of the various inputs by the computer.

❑ List and describe the operation of output actuators.

❑ Explain the principle of multiplexing.

❑ Describe the operation of common multiplexing systems.

Introduction

This chapter introduces the basic theory and operation of the digital **computer** used to control many of the vehicle's systems. The use of computers on automobiles has expanded to include control and operation of several functions including fuel delivery, ignition systems, emission systems, climate control, lighting circuits, cruise control, anti-lock braking, electronic suspension systems, and electronic shift transmissions. Some of these functions are the responsibility of the power train control module (PCM), while others are functions of what is known as a body computer module (BCM).

A **computer** is an electronic device that stores and processes data. It is capable of controlling other devices.

> **AUTHOR'S NOTE:** When computer controls were first installed on the automobile, the aura of mystery surrounding these computers was so great that some technicians were afraid to work on them. Most technicians coming into the field today have grown up around computers and the fear is not as great. Regardless of your comfort level with computers, knowledge is key to understanding their function. Although it is not necessary to understand all of the concepts of computer operation in order to service the systems they control, knowledge of the digital computer will help you feel more comfortable when working on these systems.

A computer processes the physical conditions that represent information (data). The operation of the computer is divided into four basic functions:

1. *Input*: A voltage signal sent from an input device. This device can be a sensor or a switch activated by the driver or technician.

2. *Processing*: The computer uses the input information and compares it to programmed instructions. The logic circuits process the input signals into output demands.

3. *Storage*: The program instructions are stored in an electronic memory. Some of the input signals are also stored for later processing.

4. *Output*: After the computer has processed the sensor input and checked its programmed instructions, it will put out control commands to various output devices. These output devices may be the instrument panel display or a system actuator. The output of one computer can also be used as an input to another computer.

Understanding these four functions will help today's technician organize the troubleshooting process. When a system is tested, the technician will be attempting to isolate the problem to one of these functions.

Analog and Digital Principles

Remembering the basics of electricity, voltage does not flow through a conductor; current flows and voltage is the pressure that "pushes" the current. However, voltage can be used as a signal, for example, difference in voltage levels, frequency of change, or switching from positive to negative values can be used as a signal.

A **program** is a set of instructions the computer must follow to achieve desired results.

The computer is capable of reading only voltage signals. The **program** used by the computer is "burned" into ignition control (IC) chips using a series of numbers. These numbers represent various combinations of voltages that the computer can understand. The voltage signals to the computer can be either analog or digital. Many of the inputs from the sensors are analog variables. An analog voltage signal is continuously variable within a defined range. For example, engine coolant temperature sensors do not change abruptly. The temperature varies in infinite steps from low to high. The same is true for several other inputs such as engine speed, vehicle speed, fuel flow, and so on.

Compared to an analog voltage representation, digital voltage patterns are square shaped because the transition from one voltage level to another is very abrupt (Figure 3-1). A digital signal is produced by an on/off or high/low voltage. The simplest generator of a digital signal is a switch (Figure 3-2). If 5 volts is applied to the circuit, the voltage sensor will read close to 5 volts (a high voltage value) when the switch is open. Closing the switch will result in the voltage sensor reading close to 0 volts. This measuring of voltage drops sends a digital signal to the computer. The voltage values are represented by a series of digits, which create a **binary code**.

Binary code is represented by the number 1 and 0. Any number and word can be translated into a combination of 1s and 0s.

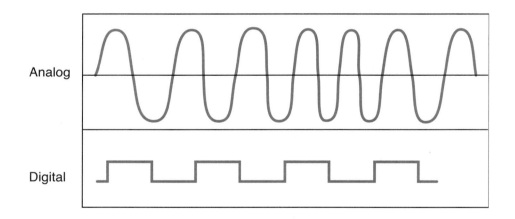

Figure 3-1 Analog voltage signals are constantly variable. Digital voltage patterns are either on or off. Digital signals are referred to as a square sine wave.

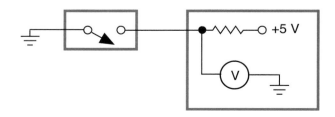

Figure 3-2 Simplified voltage sensing circuit that indicates if the switch is opened or closed.

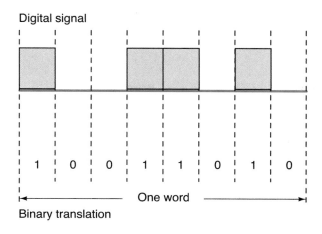

Digital signal

1 0 0 1 1 0 1 0

One word

Binary translation

Figure 3-3 Each binary 1 and 0 is one bit of information. Eight bits equals one byte.

Binary Numbers

A transistor that operates as a relay is the basis of the digital computer. As the input signal switches from off to on, the transistor output switches from cutoff to saturation. The on and off output signals represent the binary digits 1 and 0.

The computer converts the digital signal into binary code by translating voltages above a given value to 1 and voltages below a given value to 0. As shown in Figure 3-3, when the switch is open and 5 volts is sensed, the voltage value is translated into a 1 (high voltage). When the switch is closed, lower voltage is sensed and the voltage value is translated into a 0. Each 1 or 0 represents one **bit** of information. Note, high voltage being translated to a 1 and low voltage to a 0 is given for explanation purposes. In some systems, a low voltage is represented by a binary 1 while a high voltage is equal to 0.

In the binary system, whole numbers are grouped from right to left. Because the system uses only two digits, the first portion must equal a 1 or a 0. To write the value of 2, the second position must be used. In binary, the value of 2 would be represented by 10 (one two and zero ones). To continue, a 3 would be represented by 11 (one two and one one). Figure 3-4 provides a conversion of binary numbers to digital base ten numbers. If a thermistor is sensing 150°, the binary code would be 10010110. If the temperature increases to 151°, the binary code changes to 10010111.

A **bit** is one 0 or one 1 of the binary code. Eight bits is called a byte.

Decimal number	Binary number code 8 4 2 1	Binary to decimal conversion
0	0000	= 0 + 0 = 0
1	0001	= 0 + 1 = 1
2	0010	= 2 + 0 = 2
3	0011	= 2 + 1 = 3
4	0100	= 4 + 0 = 4
5	0101	= 4 + 1 = 5
6	0110	= 4 + 2 = 6
7	0111	= 4 + 2 + 1 = 7
8	1000	= 8 + 0 = 8

Figure 3-4 Binary number code conversion to base ten numbers.

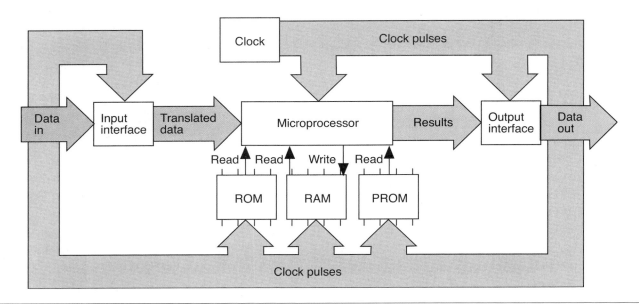

Figure 3-5 Interaction of the main components of the computer. All of the computers monitor clock pulses.

The computer contains a crystal oscillator or **clock** that delivers a constant time pulse. The clock maintains an orderly flow of information through the computer circuits by transmitting one bit of binary code for each pulse (Figure 3-5). In this manner, the computer is capable of distinguishing between the binary codes such as 101 and 1001.

Signal Conditioning and Conversion

The input and/or output signals may require conditioning in order to be used. This conditioning may include amplification and/or signal conversion.

Some input sensors produce a very low voltage signal of less than 1 volt. These signals also have an extremely low current flow, therefore this type of signal must be amplified, or increased, before it is sent to the microprocessor. This amplification is accomplished by the amplification circuit in the input conditioning chip inside the computer (Figure 3-6).

For the computer to receive information from the sensor and to give commands to actuators, it requires an **interface**. The computer will have two interface circuits: input and output. The digital computer cannot accept analog signals from the sensors, and it requires an input interface to convert the analog signal to digital. The analog to digital (A/D) converter continually scans the analog input signals at regular intervals. For example, if the A/D converter scans the throttle position sensor (TPS) signal and finds this signal at 5 volts, the A/D converter assigns a numeric value to this specific voltage. The A/D converter then changes this numeric value to a binary code (Figure 3-7).

Also, some of the controlled actuators may require an analog signal. In this instance, an output digital to analog (D/A) converter is used.

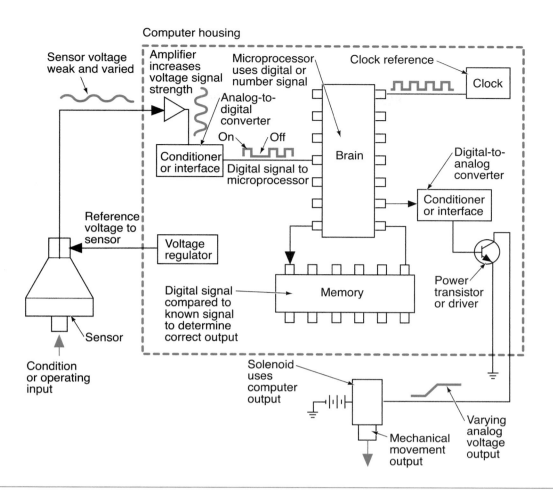

Figure 3-6 Amplification and interface circuits in the computer. The amplification circuit boosts the voltage and conditions it. The interface converts analog inputs into digital signals. The digital-to-analog converter changes the output from digital to analog.

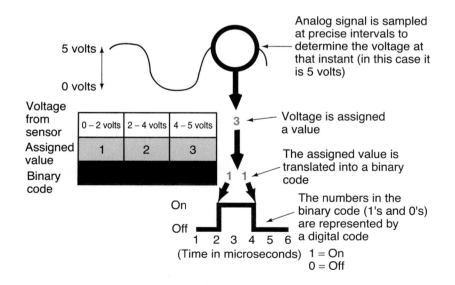

Figure 3-7 The A/D converter assigns a numeric value to input voltages and changes this numeric value to a binary code.

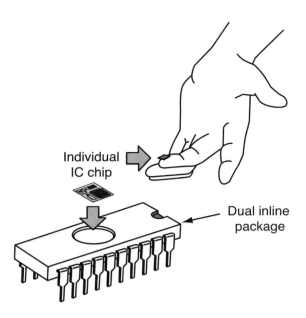

Individual
IC chip

Dual inline
package

Figure 3-8 Microprocessor components are etched on an integrated circuit chip that is small enough to fit on a finger tip.

Central Processing Unit

The **central processing unit (CPU)** is constructed of thousands of transistors that are placed on a small chip (Figure 3-8). The CPU brings information into and out of the computer's memory. The input information is processed in the CPU and checked against the program in memory. The CPU also checks memory for any other information regarding programmed parameters. The information obtained by the CPU can be altered according to the program instructions. The program may have the CPU apply logic decisions to the information. Once all calculations are made, the CPU will deliver commands to make the required corrections or adjustments to the operation of the controlled system.

The program guides the microprocessor in decisionmaking. For example, the program may inform the microprocessor when sensor information should be retrieved and then tell the microprocessor how to process this information. Finally, the program guides the microprocessor regarding the activation of output control devices such as relays and solenoids. The various memories contain the programs and other vehicle data that the microprocessor refers to as it performs calculations. As the microprocessor performs calculations and makes decisions, it works with the memories by either reading information from the memories or writing new information into the memories.

The CPU has several main components (Figure 3-9). The registers used include the accumulator, the data counter, the program counter, and the instruction register. The control unit implements the instructions located in the instruction register. The arithmetic logic unit (ALU) performs the arithmetic and logic functions.

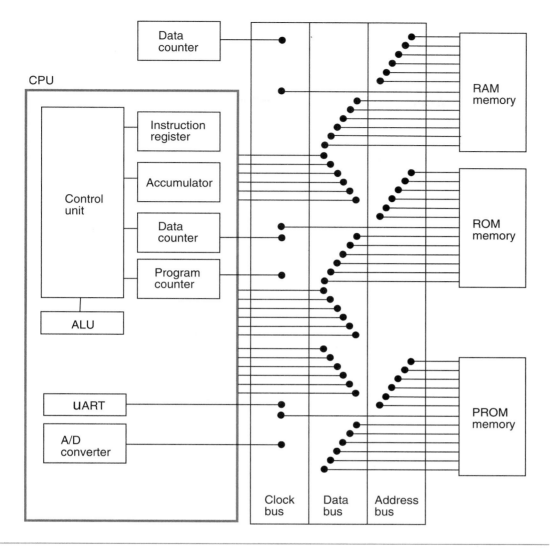

Figure 3-9 Main components of the computer and the CPU.

Computer Memory

The computer requires a means of storing both permanent and temporary memory. The memories contain many different locations. These locations may be compared to file folders in a filing cabinet, with each location containing one piece of information. An address is assigned to each memory location. This address may be compared to the lettering or numbering arrangement on file folders. Each address is written in a binary code, and these codes are numbered sequentially beginning with 0.

While the engine is running, the engine computer receives a large quantity of information from a number of sensors. The computer may not be able to process all this information immediately. In some instances, the computer may receive sensor inputs that the computer requires to make a number of decisions. In these cases, the microprocessor writes information into memory by specifying a memory address and sending information to this address.

When stored information is required, the microprocessor specifies the stored information address and requests the information. When stored information is requested from a specific address, the memory sends a copy of this information to the microprocessor. However, the original stored information is still retained in the memory address.

The memories store information regarding the ideal air-fuel ratios for various operating conditions. The sensors inform the computer about the engine and vehicle operating conditions. The microprocessor reads the ideal air-fuel ratio information from memory and compares this information with the sensor inputs. After this comparison, the microprocessor makes the necessary decision and operates the injectors to provide the exact air-fuel ratio required by the engine.

Several types of memory chips may be used in the computer:

1. *Read only memory (ROM):* Contains a fixed pattern of 1s and 0s that represent permanent stored information. This information is used to instruct the computer on what to do in response to input data. The CPU reads the information contained in ROM, but it cannot write to it or change it. ROM is permanent memory that is programmed in. This memory is not lost when power to the computer is lost. ROM contains formulas, calibrations, ignition-timing tables, and so on. ROM contains the basic operating parameters for the vehicle.

2. *Random access memory (RAM):* Constructed from flip-flop circuits formed into the chip. The RAM will store temporary information that can be read from or written to by the CPU. RAM stores information that is waiting to be acted upon, and it stores output signals that are waiting to be sent to an output device. RAM can be designed as volatile or nonvolatile. In volatile RAM, the data will be retained as long as current flows through the memory. RAM that is connected to the battery through the ignition switch will lose its data when the switch is turned off (see item number 7, nonvolatile RAM).

3. *Keep alive memory (KAM):* A version of RAM, KAM is connected directly to the battery through circuit protection devices. For example, the microprocessor can read and write information to and from the KAM and erase KAM information. However, the KAM retains information when the ignition switch is turned off. KAM will be lost when the battery is disconnected, if the battery drains too low, or if the circuit opens. Nonvolatile RAM will retain its information if current is removed.

4. *Programmable read only memory (PROM):* Contains specific data that pertains to the exact vehicle in which the computer is installed. This information may be used to inform the CPU of the accessories that are equipped on the vehicle. The information stored in the PROM is the basis for all computer logic. The information in PROM is used to define or adjust the *operating perimeters* held in ROM. In many instances, the computer is interchangeable between models of the same manufacturers; however, the PROM is not. Consequently, the PROM may be replaceable and plug into the computer (Figure 3-10).

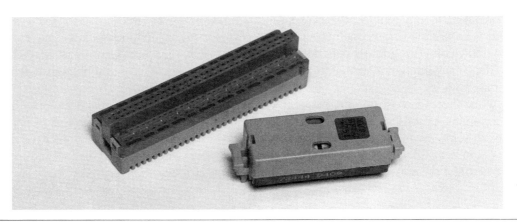

Figure 3-10 Assortment of PROM chips. Many manufacturers design PROMs that are removable by use of a special tool.

The terms *ROM, RAM,* and *PROM* are used fairly consistently in the computer industry, however, the names vary between automobile manufacturers.

Shop Manual
Chapter 3,
page 108

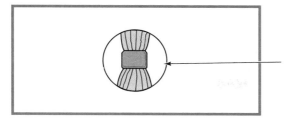

The Erasable PROM has a window such as this that the microcircuitry can be viewed through. This is normally covered by a piece of Mylar™-type material so that the information in it will not be erased by exposing it to ultraviolet light rays.

Figure 3-11 EPROM memory is erased when the ultraviolet ray contacts the microcircuitry.

5. *Erasable PROM (EPROM):* Similar to PROM except that its contents can be erased to allow new data to be installed. A piece of Mylar tape covers a window. If the tape is removed, the microcircuit is exposed to ultraviolet light that erases its memory (Figure 3-11).

6. *Electrically erasable PROM (EEPROM):* Allows changing the information electrically one bit at a time. The flash EEPROM is an IC chip inside the computer. This IC contains the program used by the computer to provide control. It is possible to erase and reprogram the EEPROM without removing this chip from the computer. For example, when a modification to the PCM operating strategy is required, it is no longer necessary to replace the PCM. The flash EEPROM may be reprogrammed through the data link connector (DLC) using the manufacturer's specified diagnostic equipment.

7. *Nonvolatile RAM (NVRAM):* A combination of RAM and EEPROM in the same chip. During normal operation, data is written to and read from the RAM portion of the chip. If the power is removed from the chip, or at programmed timed intervals, the data is transferred from RAM to the EEPROM portion of the chip. When the power is restored to the chip, the EEPROM will write the data back to the RAM.

This process of reprogramming the computer through the EEPROM is called flashing.

Adaptive Strategy

If a computer has adaptive strategy capabilities, the computer can actually learn from past experience. For example, the normal voltage input range from a TPS may be 0.6 volt to 4.5 volts. If a worn TPS sends a 0.4 volt signal to the computer, the microprocessor interprets this signal as an indication of component wear, and the microprocessor stores this altered calibration in the KAM. The microprocessor now refers to this new calibration during calculations, and thus normal engine performance is maintained. If a sensor output is erratic or considerably out of range, the computer may ignore this input. When a computer has adaptive strategy, a short learning period is necessary under the following conditions:

1. After the battery has been disconnected.

2. When a computer system component has been replaced or disconnected.

3. On a new vehicle.

Information Processing

The air charge temperature (ACT) sensor input will be used as an example of how the computer processes information. If the air temperature is low, the air is denser and contains more oxygen per cubic foot. Warmer air is less dense and therefore contains less oxygen per cubic foot. The cold, dense air requires more fuel compared to the warmer air that is less dense. The microprocessor must supply the correct amount of fuel in relation to air temperature and density.

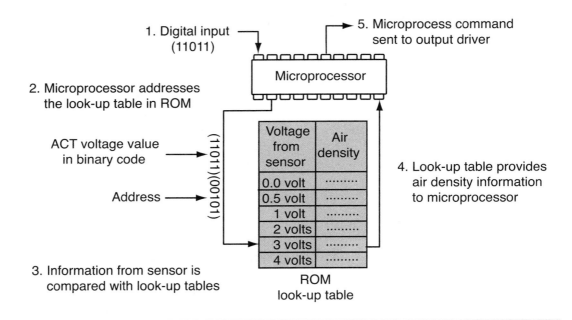

1. Digital input
(11011)

5. Microprocess command
sent to output driver

Microprocessor

2. Microprocessor addresses
the look-up table in ROM

ACT voltage value
in binary code

(11011)(00101)

Address

Voltage from sensor	Air density
0.0 volt	········
0.5 volt	········
1 volt	········
2 volts	········
3 volts	········
4 volts	········

4. Look-up table provides
air density information
to microprocessor

3. Information from sensor is
compared with look-up tables

ROM
look-up table

Figure 3-12 The microprocessor addresses the lookup tables in the ROM, retrieves air density information, and issues commands to the output devices.

An ACT sensor is positioned in the intake manifold where it senses air temperature. This sensor contains a resistive element that has an increased resistance when the sensor is cold. Conversely, the ACT sensor resistance decreases as the sensor temperature increases. When the ACT sensor is cold, it sends a high analog voltage signal to the computer, and the A/D converter changes this signal to a digital signal.

When the microprocessor receives this ACT signal, it addresses the tables in the ROM. The lookup tables list air density for every air temperature. When the ACT sensor voltage signal is very high, the lookup table indicates very dense air. This dense air information is relayed to the microprocessor, and the microprocessor operates the output drivers and injectors to supply the exact amount of fuel required by the engine (Figure 3-12).

Logic Gates

Logic gates are the thousands of field effect transistors (FET) incorporated into the computer circuitry. The FETs use the incoming voltage patterns to determine the pattern of pulses leaving the gate. The following are some of the most common logic gates and their operation. The symbols represent functions and not electronic construction:

1. *NOT gate:* A NOT gate simply reverses binary 1s to 0s and vice versa (Figure 3-13). A high input results in a low output, and a low input results in a high output.

These circuits are called **logic gates** because they act as gates to output voltage signals depending on different combinations of input signals.

The NOT gate is also called an inverter.

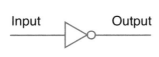

Input   Output

Truth table	
Input	Output
0	1
1	0

Figure 3-13 The NOT gate symbol and truth table. The NOT gate inverts the input signal.

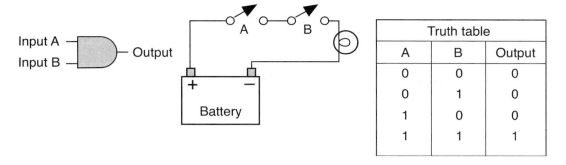

Truth table		
A	B	Output
0	0	0
0	1	0
1	0	0
1	1	1

Figure 3-14 The AND gate symbol and truth table. The AND gate operates similar to switches in series.

2. *AND gate:* The AND gate will have at least two inputs and one output. The operation of the AND gate is similar to two switches in series to a load (Figure 3-14). The only way the light will turn on is if switches A *and* B are closed. The output of the gate will be high only if both inputs are high. Before current can be present at the output of the gate, current must be present at the base of both transistors (Figure 3-15).

3. *OR gate:* The OR gate operates similarly to two switches that are wired in parallel to a light (Figure 3-16). If switch A *or* B is closed, the light will turn on. A high signal to either input will result in a high output.

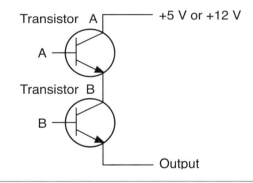

Figure 3-15 The AND gate circuit.

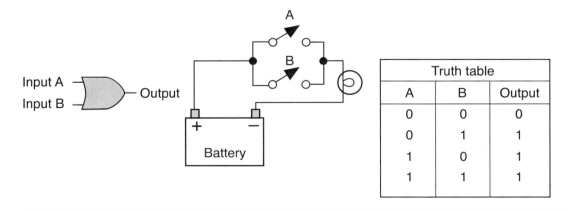

Truth table		
A	B	Output
0	0	0
0	1	1
1	0	1
1	1	1

Figure 3-16 OR gate symbol and truth table. The OR gate is similar to parallel switches.

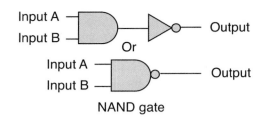

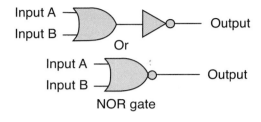

Truth table		
A	B	Output
0	0	1
0	1	1
1	0	1
1	1	0

Truth table		
A	B	Output
0	0	1
0	1	0
1	0	0
1	1	0

Figure 3-17 Symbols and truth tables for NAND and NOR gates. The small circle represents an inverted output on any logic gate symbol.

4. *NAND and NOR gates:* A NOT gate placed behind an AND or OR gate inverts the output signal (Figure 3-17).

5. *Exclusive-OR (XOR) gate:* A combination of gates that will produce a high output signal only if the inputs are different (Figure 3-18).

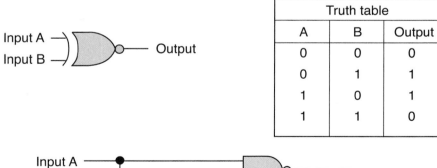

Truth table		
A	B	Output
0	0	0
0	1	1
1	0	1
1	1	0

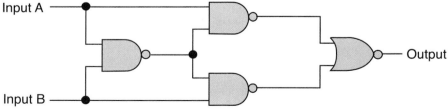

Figure 3-18 XOR gate symbol and truth table. An XOR gate is a combination of NAND and NOR gates.

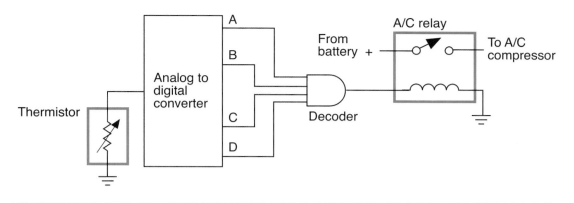

Figure 3-19 Simplified temperature-sensing circuit that will turn on the air-conditioning compressor when inside temperatures reach a predetermined value.

These different gates are combined to perform the processing function. The following are some of the most common combinations:

1. *Decoder circuit:* A combination of AND gates used to provide a certain output based on a given combination of inputs (Figure 3-19). When the correct bit pattern is received by the decoder, it will produce the high voltage signal to activate the relay coil.

2. *Multiplexer (MUX):* The basic computer is not capable of looking at all of the inputs at the same time. A multiplexer is used to examine one of many inputs depending on a programmed priority rating (Figure 3-20). This process is called **sequential sampling**.

3. *Demultiplexer (DEMUX):* DEMUX operates similar to the MUX except that it controls the order of the outputs (Figure 3-21).

The process that the MUX and DEMUX operates on is called **sequential sampling**. This means the computer will deal with all of the sensors and actuators one at a time.

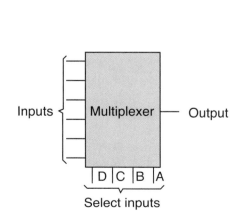

Figure 3-20 Selection at inputs D, C, B, A will determine which data input will be processed.

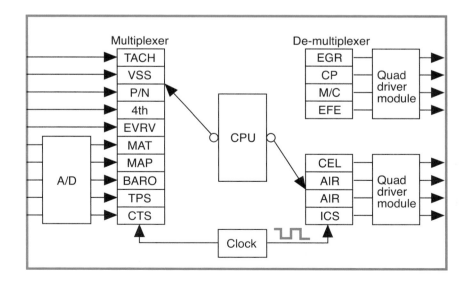

Figure 3-21 Block diagram representation of the MUX and DEMUX circuit.

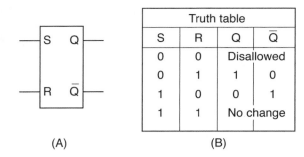

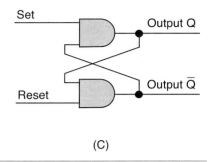

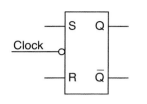

		Truth table	
S	R	Q	$\overline{Q}$
0	0	Disallowed	
0	1	1	0
1	0	0	1
1	1	No change	

(A) (B) (C)

Figure 3-22 (A) RS flip-flop symbol, (B) Truth table, (C) Logic diagram. Variations of the circuit may include NOT gates at the inputs, if used; the truth table outputs would be reversed.

Figure 3-23 Clocked RS flip-flop symbol.

4. *RS and clocked RS flip-flop circuits:* Logic circuits that remember previous inputs and do not change their outputs until they receive new input signals. The illustration (Figure 3-22) shows a basic RS flip-flop circuit. The clocked flip-flop circuit has an inverted clock signal as an input so that circuit operations occur in the proper order (Figure 3-23). Flip-flop circuits are called **sequential logic circuits**.

Flip-flop circuits are called **sequential logic circuits** because the output is determined by the sequence of inputs. A given input affects the output produced by the next input.

5. *Driver circuits:* A driver is a term used to describe a transistor device that controls the current in the output circuit. Drivers are controlled by a computer to operate such controls as fuel injectors, ignition coils, and many other high-current circuits. The currents handled by a driver are not really that high; they are just more than what is typically handled by a transistor. Several types of driver circuits are used on automobiles, such as Quad, Discrete, Peak and Hold, and Saturated Switch driver circuits.

6. *Registers:* A register is a combination of flip-flops that transfer bits from one to another every time a clock pulse occurs (Figure 3-24). It is used in the computer to temporarily store information.

7. *Accumulators:* Registers designed to store the results of logic operations that can become inputs to other modules.

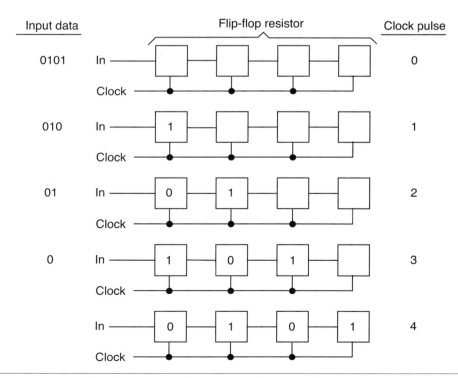

Figure 3-24 It takes four clock pulses to load 4 bits into the register.

EMI Suppression

As manufacturers began to increase the number of electronic components and systems in their vehicles, the problem of **electromagnetic interference (EMI)** had to be controlled. The low-power integrated circuits used on modem vehicles are sensitive to the signals produced as a result of EMI. EMI is produced as current in a conductor is turned on and off. EMI is also caused by static electricity that is created by friction. The friction is a result of the tires and their contact with the road or the result of fan belts contacting the pulleys.

EMI can disrupt the vehicle's computer systems by inducing false messages to the computer. The computer requires messages to be sent over circuits in order to communicate with other computers, sensors, and actuators. If any of these signals are disrupted, the engine and/or accessories may turn off.

EMI can be suppressed by any one of the following methods:

1. Adding a resistance to the conductors. This is usually done to high-voltage systems such as the secondary circuit of the ignition system.

2. Connecting a capacitor in parallel and a choke coil in series with the circuit.

3. Shielding the conductor or load components with a metal or metal-impregnated plastic.

4. Increasing the number of paths to ground by using designated ground circuits. This provides a clear path to ground that is very low in resistance.

5. Adding a clamping diode in parallel to the component.

6. Adding an isolation diode in series to the component.

Electromagnetic interference (EMI) is an undesirable creation of electromagnetism whenever current is switched on and off.

Shop Manual
Chapter 3,
page 93

Semiconductors

Semiconductors include diodes, transistors, and silicon-controlled rectifiers. These semiconductors are often called solid state devices since they are constructed of a solid material. The most common materials used in the construction of semiconductors are silicon or germanium. Both of these materials are classified as **crystals**.

Silicon and germanium have four electrons in their outer orbits. Because of their crystal type of structure, each atom shares an electron with four other atoms (Figure 3-25). The result of this **covalent bonding** is that each atom will have eight electrons in its outer orbit. Since all the orbits are filled, there are no free electrons. Thus, the material (as a category of matter) falls somewhere between conductor and insulator.

Semiconductors are materials that conduct electric current under certain conditions, yet will not conduct under other conditions.

A **crystal** is the term used to describe a material that has a definite atom structure.

When atoms share electrons with other atoms, it is called **covalent bonding**.

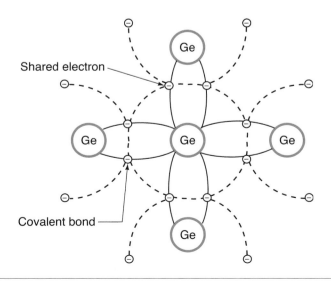

Figure 3-25 Crystal structure of germanium.

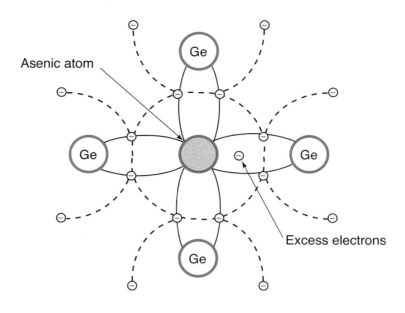

Figure 3-26 Germanium crystal doped with an arsenic atom to produce an N-type material.

When there are free electrons, the material is called an **N-type material**. The N means negative and that it is the negative side of the circuit that pushes electrons through the semiconductor and that the positive side attracts the free electrons.

When there are excessive protons, the material is called a **P-type material**. The P means positive and that it is the positive side of the circuit that attracts the free electrons through the semiconductor.

The absence of an electron is called a **hole.** These holes are said to be positively charged since they have a tendency to attract free electrons into the hole.

Perfect crystals are not used for manufacturing semiconductors; they are doped with impurity atoms. This doping adds a small percentage of another element to the crystal. The doping element can be arsenic, antimony, phosphorous, boron, aluminum, or gallium.

If the crystal is doped by using arsenic, antimony, or phosphorous, the result is a material with free electrons (Figure 3-26). Since materials such as arsenic have five electrons, there is one electron left over. This doped material becomes negatively charged and is referred to as an **N-type material**. Under the influence of an EMF, it will support current flow.

If boron, aluminum, or gallium is added to the crystal, a **P-type material** is produced. Materials like boron have three electrons in their outermost orbit. Since there is one fewer electron, the absence produces a **hole** (Figure 3-27), and the atom becomes positively charged.

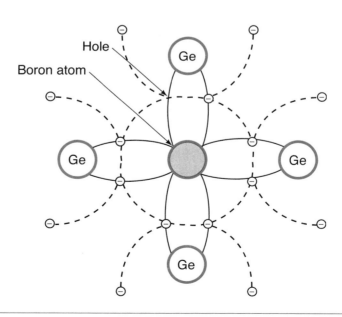

Figure 3-27 Germanium crystal doped with a boron atom to produce a P-type material.

By putting N-type and P-type materials together in a certain order, solid state components are built that can be used for switching devices, voltage regulators, and electrical controls. Because a semiconductor material can operate as both a conductor and an insulator, it is very useful as a switching device. How a semiconductor material works depends on the way current flows, or tries to flow, through it.

Diodes

A **diode** is the simplest semiconductor device. It is formed by joining P-type semiconductor material with N-type material. The N (negative) side of a diode is called the **cathode** and the P (positive) side, the **anode** (Figure 3-28). The point where the cathode and anode join together is called the PN junction. The outer shell of the diode will have a stripe painted around it. This stripe designates which end of the diode is the cathode.

When a diode is made, the positive holes from the P region and the negative charges from the N region are drawn toward the junction. Some charges cross over and combine with opposite charges from the other side. When the charges cross over, the two halves are no longer balanced, and the diode builds up a network of internal charges opposite to the charges at the PN junction. The internal EMF between the opposite charges limits the further diffusion of charges across the junction.

When the diode is incorporated within a circuit and a voltage is applied, the internal characteristics change. If the diode is **forward-biased**, there will be current flow (Figure 3-29). In this state, the negative region will push electrons across the barrier as the positive region pushes holes across. When forward-biased, the diode acts as a conductor.

If the diode is **reverse-biased**, there will be no current flow (Figure 3-30). The negative region will attract the positive holes away from the junction, and the positive region will attract electrons away. This makes the diode act like an insulator.

Shop Manual
Chapter 3,
page 107

A **diode** is an electrical one-way check valve that will allow current to flow in one direction only.

The **cathode** is the negative side of the diode.

The **anode** is the positive side of the diode.

Forward-bias means that a positive voltage is applied to the P-type material and negative voltage to the N-type material.

Reversed-bias means that positive voltage is applied to the N-type material and negative voltage is applied to the P-type material.

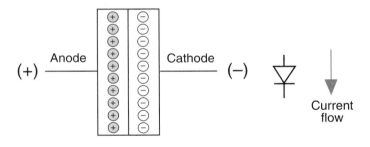

Figure 3-28 A diode and its symbol.

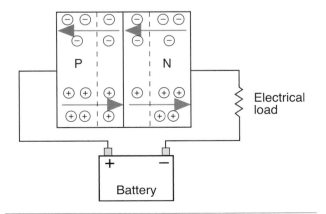

Figure 3-29 Forward-biased voltage causes current flow.

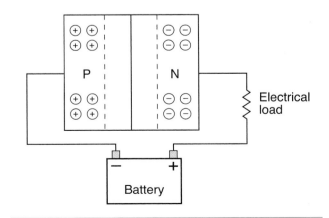

Figure 3-30 Reverse-biased voltage prevents current flow.

When the diode is forward-biased, it will have a small voltage drop across it. On standard silicon diodes, this voltage is usually about 0.6 volt. This is referred to as the **turn-on voltage**.

Zener Diodes

As stated, if a diode is reverse-biased, it will not conduct current. However, if the reverse voltage is increased, a voltage level will be reached at which the diode will conduct in the reverse direction. This voltage level is referred to as **zener voltage**. Reverse current can destroy a simple PN-type diode, but the diode can be doped with materials that will withstand reverse current.

A **zener diode** is designed to operate in reverse bias at the breakdown region. At the point that breakdown voltage is reached, a large current flows in reverse bias. This prevents the voltage from climbing any higher. This makes the zener diode an excellent component for regulating voltage. If the zener diode is rated at 15 volts, it will not conduct in the reverse direction when voltage is below 15 volts. At 15 volts it will conduct, and the voltage will not increase over 15 volts.

Figure 3-31 illustrates a simplified circuit that has a zener diode in it to provide a constant voltage level to the instrument gauge. In this example, the zener diode is connected in series with the resistor and in parallel to the gauge. If the voltage to the gauge must be limited to 7 volts, the zener diode used would be rated at 7 volts. Since the zener diode maintains a constant voltage drop and the total voltage drop in a series circuit must equal the amount of source voltage, voltage that is greater than the zener voltage must be dropped over the resistor. Even though source voltage may vary (as a normal result of the charging system), causing different currents to flow through the resistor and zener diode, the voltage dropped by the zener diode remains the same.

The zener breaks down when system voltage reaches 7 volts. At this point, the zener diode conducts reverse current, causing an additional voltage drop across the resistor. The amount of voltage to the instrument gauge will remain at 7 volts since the zener diode "makes" the resistor drop the additional voltage to maintain this limit.

Here we see the difference between the standard diode and the zener diode. When the zener diode is reverse-biased, the zener holds the available voltage to a specific value.

Light-Emitting Diodes

A **light-emitting diode (LED)** has a small lens built into it so light can be seen when current flows through the diode (Figure 3-32). When the LED is forward-biased, the holes and electrons combine and current is allowed to flow through it. The energy generated is released in the form of light. It is the material used to make the LED that will determine the color of the light emitted.

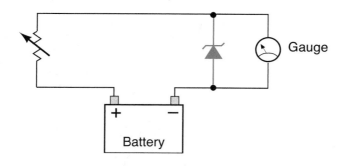

Figure 3-31 Simplified instrument gauge circuit that uses a zener diode to maintain a constant voltage to the gauge. Note the symbol used for a zener diode.

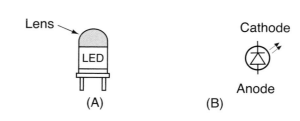

Figure 3-32 (A) A light-emitting diode uses a lens to emit the generated light, (B) Symbol for LED.

Normally the LED requires 1.5 to 2.2 volts to light. The light from the LED is not heat energy as is the case with other lights; it is electrical energy. Because of this, LEDs last longer than lightbulbs.

Photodiodes

A **photodiode** also allows current to flow in one direction only, however, the direction of current flow is opposite that of a standard diode. Reverse current flow only occurs when the diode receives a specific amount of light. These types of diodes can be used in automatic headlight systems.

A **photodiode** allows current to flow in the opposite direction of a standard diode when it receives a specific amount of light.

Clamping Diodes

Whenever the current flow through a coil (such as used in a relay or solenoid) is discontinued, a voltage surge or spike is produced. This surge results from the collapsing of the magnetic field around the coil. The movement of the field across the windings induces a very high voltage spike, which can damage electronic components as it flows through the system. In some circuits, a capacitor can be used as a shock absorber to prevent component damage from this surge. In today's complex electronic systems, a **clamping diode** is commonly used to prevent the voltage spike. By installing a clamping diode in parallel across the coil, a bypass is provided for the electrons during the time the circuit is open (Figure 3-33).

A **clamping diode** is nothing more than a standard diode. The term *clamping* refers to its function.

An example of the use of clamping diodes is on some air-conditioning compressor clutches. Since the clutch operates by electromagnetism, opening of the clutch coil produces a voltage spike. If this voltage spike were left unchecked, it could damage the vehicle's onboard computers. The installation of the clamping diode prevents the voltage spike from reaching the computers.

Relays may also be equipped with a clamping diode, however, some use a resistor to dissipate the voltage spike. The two types of relays are not interchangeable.

Transistors

The word **transistor** is a combination of two words: transfer and resist. The transistor is used to control current flow in the circuit (Figure 3-34). It can be used to allow a predetermined amount of current flow or to resist this flow.

A **transistor** is a three-layer semiconductor (Figure 3-30). It is used as a very fast switching device.

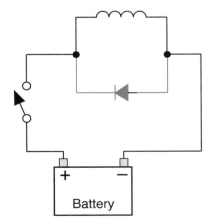

Figure 3-33 A clamping diode in parallel to a coil prevents voltage spikes when the switch is opened.

Figure 3-34 Transistors that are used in automotive applications.

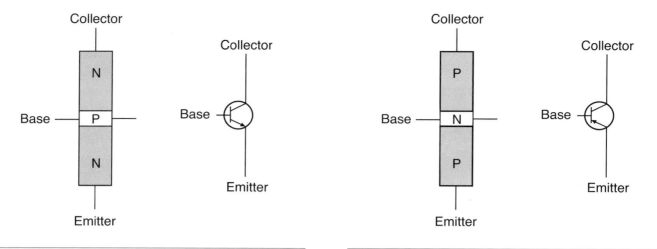

Figure 3-35 An NPN transistor and its symbol. **Figure 3-36** A PNP transistor and its symbol.

Transistors are made by combining P-type and N-type materials in groups of three, therefore, they are actually two diodes that share a common layer. The two possible combinations of transistors are NPN (Figure 3-35) and PNP (Figure 3-36). The three layers of the transistor are designated as **emitter**, **collector**, and **base**. The emitter is the outside layer of the forward-biased diode and has the same polarity as the circuit side to which it is applied. The arrow on the transistor symbol refers to the emitter lead. The arrow points in the direction of positive current flow and to the N-type material. The collector is the outside layer of the reverse-biased diode. The base is the shared middle layer. Each of these different layers has its own lead for connecting to different parts of the circuit. When a transistor is connected to the circuit, the emitter-base junction will be forward-biased and the collector-base junction will be reversed-biased.

In the NPN transistor, the emitter conducts current flow to the collector when the base is forward-biased. The transistor cannot conduct unless the voltage applied to the base leg exceeds the emitter voltage by approximately 0.7 volt. This means both the base and collector must be positive with respect to the emitter. With less than 0.7 volt applied to the base leg (compared to the voltage at the emitter), the transistor acts as an opened switch. When there is about 0.7 volt more voltage at the base then the amount of voltage on the emitter, the transistor acts as a closed switch (Figure 3-37).

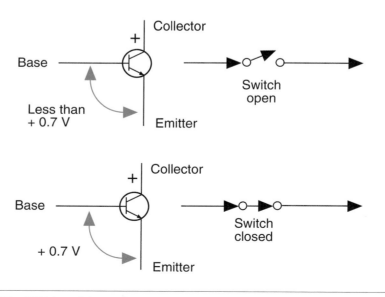

Figure 3-37 NPN transistor action.

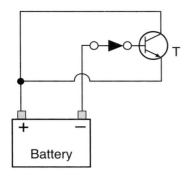

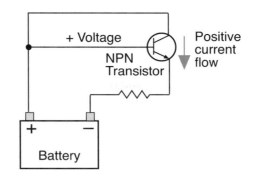

Figure 3-38   NPN transistor with reverse-biased voltage applied to the base. No current flow.

Figure 3-39   NPN transistor with forward-biased voltage applied to the base. Current flows.

When an NPN transistor is used in a circuit, it normally has a reverse bias applied to the base-collector junction. If the emitter-base junction is also reverse-biased, no current will flow through the transistor (Figure 3-38). If the emitter-base junction is forward-biased, current flows from the emitter to the base (Figure 3-39). Since the base is a thin layer and a positive voltage is applied to the collector, electrons flow from the emitter to the collector.

In the PNP transistor, current will flow from the emitter to the collector when the base leg is forward-biased with a voltage that is more negative than that at the emitter (Figure 3-40). For current to flow through the emitter to the collector, both the base and the collector must be negative in respect to the emitter.

Since current can be controlled through a transistor, this component can be used as a very fast electrical switch. It is also possible to control the amount of current flow through the collector. This is because the output current is proportional to the amount of current through the base leg.

Current flow in a PNP transistor is considered as movement of holes, while in the NPN transistor, current flow is the movement of electrons.

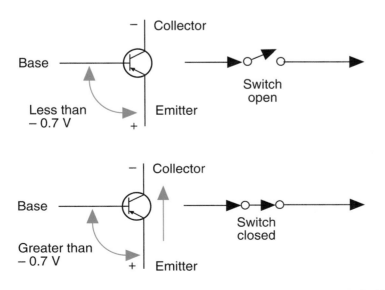

Figure 3-40   PNP transistor action.

A transistor has three operating conditions:

1. **Cutoff**. When reverse-bias voltage is applied to the base leg of the transistor. In this condition, the transistor is not conducting and no current will flow.

2. **Conduction**. Bias voltage difference between the base and the emitter has increased to the point that the transistor is switched on. In this condition, the transistor is conducting. Output current is proportional to that of the current through the base.

3. **Saturation**. This is the point where forward-bias voltage to the base leg is at a maximum. With bias voltage at the high limits, output current is also at its maximum.

These types of transistors are called **bipolar** because they have three layers of silicon, with two of them being the same. Another type of transistor is the **field-effect transistor (FET)**. The FET's leads are listed as source, drain, and gate. The source supplies the electrons and is similar to the emitter in the bipolar transistor. The drain collects the current and is similar to the collector. The gate creates the electrostatic field that allows electron flow from the source to the drain. It is similar to the base.

The FET transistor does not require bias voltage, only a voltage needs to be applied to the gate terminal to receive electron flow from the source to the drain. The source and drain are constructed of the same type of doped material. They can be either N-type or P-type materials. The source and drain are separated by a thin layer of either N-type or P-type material.

If the source is held at 0 voltage and 6 volts are applied to the drain, no current will flow between the two (Figure 3-41). However, if a lower positive voltage to the gate is applied, the gate forms a capacitive field between the channel and itself. The voltage of the capacitive field attracts electrons from the source, and current will flow through the channel to the higher positive voltage of the drain.

This type of FET is called an **enhancement-type FET** because the field effect improves current flow from the source to the drain. This operation is similar to that of a normally open switch. A **depletion-type FET** is like a normally closed switch, whereas the field effect cuts off current flow from the source to the drain.

BIT OF HISTORY

The transistor was developed by a team of three American physicists: Walter Houser Brattain, John Bardeen, and William Bradford Shockley. They announced their achievement in 1948. These physicists won the Nobel Prize in physics for this development in 1956.

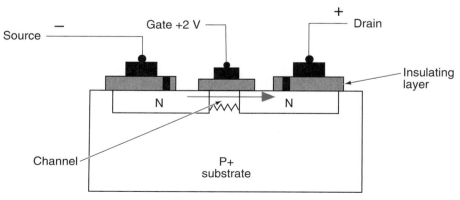

Figure 3-41   A FET uses a positive voltage to the gate terminal to create a capacitive field to allow electron flow.

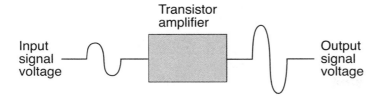

Figure 3-42 A simplified amplifier circuit.

Transistor Amplifiers

A transistor can be used in an amplifier circuit to amplify the voltage. This is useful when using a very small voltage for sensing computer inputs but needing to boost that voltage to operate an accessory (Figure 3-42). The small signal voltage that is applied to the base leg of a transistor may look like that of Figure 3-43a. However, the corresponding signal through the collector may be like that shown in Figure 3-43b. In an amplified circuit:

1. The amplified voltage at the collector is greater than that of the base voltage.

2. The input current increases.

3. The pattern has been inverted.

The first transistor in a **Darlington pair** is used as a preamplifier to produce a large current to operate the second transistor (Figure 3-44). The second transistor is isolated from the control circuit and is the final amplifier. The second transistor boosts the current to the amount required to operate the load component. The Darlington pair is utilized by most control modules used in electronic ignition systems.

Some amplifier circuits use a **Darlington pair**, which is two transistors that are connected together.

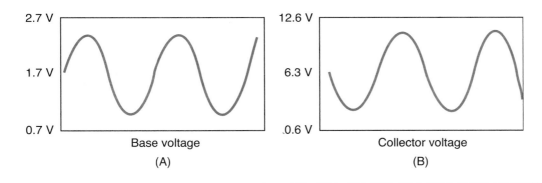

Figure 3-43 The voltage applied to the base (A) is amplified and inverted through the collector (B).

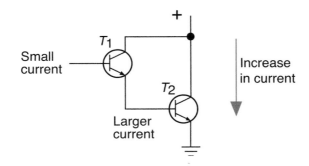

Figure 3-44 A Darlington pair used to amplify current. T_1 acts as a preamplifier that creates a larger base current for T_2, which is the final amplifier that creates a larger current.

Figure 3-45   Phototransistor.

Phototransistors

In a **phototransistor**, a small lens is used to focus incoming light onto the sensitive portion of the transistor (Figure 3-45). When light strikes the transistor, holes and free electrons are formed. These increase current flow through the transistor according to the amount of light. The stronger the intensity of the light, the more current will flow. This type of phototransistor is often used in automatic headlight dimming circuits.

Thyristors

The most common type of **thyristor** used in automotive applications is the silicon-controlled rectifier (SCR). Like the transistor, the SCR has three legs. However, it consists of four regions arranged PNPN (Figure 3-46). The three legs of the SCR are called the anode (or P-terminal), the cathode (or N-terminal), and the gate (one of the center regions).

The SCR requires only a trigger pulse (not a continuous current) applied to the gate to become conductive. Current will continue to flow through the anode and cathode as long as the voltage remains high enough or until the gate voltage is reversed.

The SCR can be connected into a circuit either in the forward or reverse direction. Using Figure 3-46 of a forward-direction connection, the P-type anode is connected to the positive side of the circuit and the N-type cathode is connected to the negative side. The center PN junction blocks current flow through the anode and cathode.

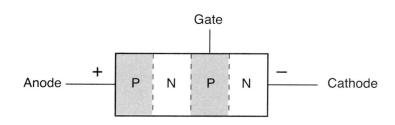

Figure 3-46   A forward direction SCR.

Once a positive voltage pulse is applied to the gate, the SCR turns on. Even if the positive voltage pulse is removed, the SCR will continue to conduct. If a negative voltage pulse is applied to the gate, the SCR will no longer conduct.

The SCR will also block any reverse current from flowing from the cathode to the anode. Since current can flow only in one direction through the SCR, it can rectify AC current to DC current.

Integrated Circuits

An **integrated circuit (IC)** is a complex circuit of thousands of transistors, diodes, resistors, capacitors, and other electronic devices that are formed on a tiny silicon chip (Figure 3-47). As many as 30,000 transistors can be placed on a chip that is $1/4$ inch (6.35 mm) square.

Integrated circuits are constructed by photographically reproducing circuit patterns on a silicon wafer. The process begins with a large scale drawing of the circuit. This drawing can be room size. Photographs of the circuit drawing are reduced until they are the actual size of the circuit. The reduced photographs are used as a mask. Conductive P-type and N-type materials, along with insulating materials, are deposited onto the silicon wafer. The mask is placed over the wafer, and it selectively exposes the portion of material to be etched away or the portions requiring selective deposition. The entire process of creating an integrated circuit chip takes over 100 separate steps. Out of a single wafer 4 inches in diameter, thousands of integrated circuits can be produced.

The small size of the integrated chip has made it possible for the vehicle manufacturers to add several computer-controlled systems to the vehicle without taking up much space. Also, a single computer is capable of performing several functions.

An **integrated circuit (IC)** is a complex circuit of thousands of transistors, diodes, resistors, capacitors, and other electronic devices that are formed on a tiny silicon chip. As many as 30,000 transistors can be placed on a chip that is $1/4$-inch (6.35 mm) square.

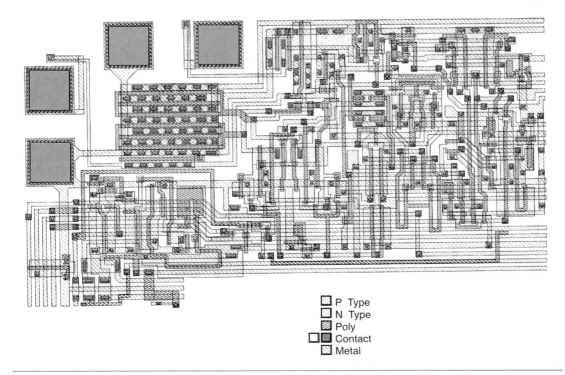

P Type
N Type
Poly
Contact
Metal

Figure 3-47 An enlarged illustration of an integrated circuit with thousands of transistors, diodes, resistors, and capacitors. Actual size can be less than 1/4 inch square.

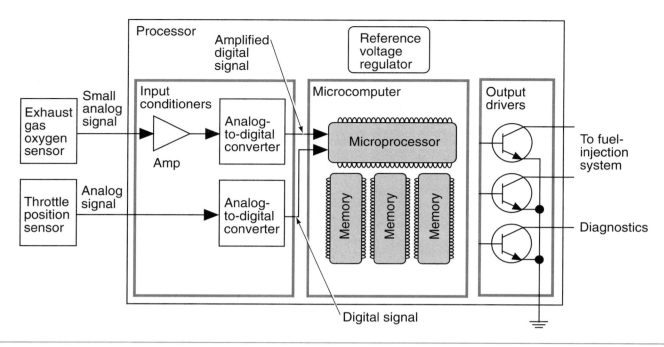

Figure 3-48 The input signals are processed in the microprocessor. The microprocessor directs the output drivers to activate actuators as instructed by the program.

Shop Manual
Chapter 3,
page 101

Sensors convert some measurement of vehicle operation into an electrical signal.

Linearity refers to the sensor signal's being as constantly proportional to the measured value as possible. It is an expression of the sensor's accuracy.

A **thermistor** is a solid state variable resistor made from a semiconductor material that changes resistance in relation to temperature changes.

Inputs

As discussed earlier, the CPU receives inputs that it checks with programmed values. Depending on the input, the computer will control the actuator(s) until the programmed results are obtained (Figure 3-48). The inputs can come from other computers, the driver, the technician, or through a variety of sensors.

Switches can be used as an input for any operation that only requires a yes-no, or on-off, condition. Other inputs include those supplied by means of a sensor, and those signals returned to the computer in the form of feedback.

Sensors

There are many different designs of **sensors**. Some are nothing more than a switch that completes the circuit. Others are complex chemical-reaction devices that generate their own voltage under different conditions. Repeatability, accuracy, operating range, and **linearity** are all requirements of a sensor.

Thermistors. A **thermistor** is used to sense engine coolant or ambient temperatures. By monitoring the thermistor's resistance value, the computer is capable of observing very small changes in temperature. The computer sends a reference voltage to the thermistor (usually 5 volts) through a fixed resistor. As the current flows through the thermistor resistance to ground, a voltage-sensing circuit measures the voltage after the fixed resistor (Figure 3-49). The voltage dropped over the fixed resistor will change as the resistance of the thermistor changes. Using its programmed values, the computer is able to translate the voltage drop into a temperature value.

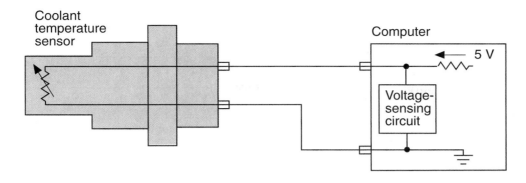

Figure 3-49 A thermistor is used to measure temperature. The sensing unit measures the resitance change and translates the data into temeprature values.

There are two types of thermistors: **negative temperature coefficient (NTC)** thermistors and **positive temperature coefficient (PTC)** thermistors. The NTC is the most commonly used.

Wheatstone Bridges. The illustration (Figure 3-50) shows the construction of the **Wheatstone bridge**. A Wheatstone bridge is nothing more than two simple series circuits connected in parallel across a power supply. Construction design varies between manufacturers, but usually three of the resistors are kept at exactly the same value and the fourth is the sensing resistor. When all four resistors have the same value, the bridge is balanced and the voltage sensor will indicate a value of 0 volts. The output from the amplifier acts as a voltmeter. Remember, since a voltmeter measures electrical pressure between two points, it will display this value. For example, if the reference voltage is 5 volts and the resistors have the same value, then the voltage drop over each resistor is 2.5 volts. Since the voltmeter is measuring the potential on the line between R_S and R_1 and between R_2 and R_3, it will read 0 volts because both of these lines have 2.5 volts on them. If there is a change in the resistance value of the sense resistor, a change will occur in the circuit's balance. The sensing circuit will receive a voltage reading that is proportional to the amount of resistance change.

Negative temperature coefficient (NTC) thermistors reduce their resistance as the temperature increases.

Positive temperature coefficient (PTC) thermistors increase their resistance as the temperature increases.

The **Wheatstone bridge** is a series-parallel arrangement of resistors between an input terminal and ground.

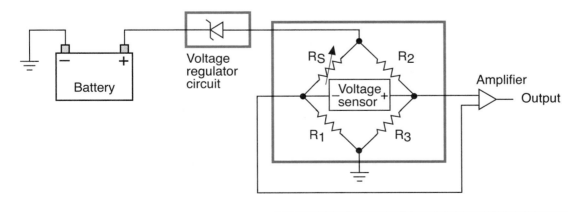

Figure 3-50 Wheatstone bridge.

A common use of a Wheatstone bridge is the hot-wire sensor in a mass air flow (MAF) sensor. The sensor consists of a hot-wire circuit, a cold-wire circuit, and an electronic signal processing area. The hot- and cold-wire circuits form the Wheatstone bridge. The cold-wire circuit is made of a fixed resistor and a thermistor. The amount of voltage dropped across the two resistors is determined by the temperature of the thermistor. The hot-wire circuit is made up of a fixed resistor and a variable resistance heat element (hot wire). The heat element generates heat in proportion to the amount of current flowing through it. This heat, in turn, changes its resistance. As air flows past the hot wire, it moves a small amount of heat from the element. This cooling of the element causes the voltage drop across it to change. The voltage drop across the hot wire is compared to the voltage drop across the fixed resistor in the cold-wire circuit and airflow is determined.

Piezoelectric Devices. Piezoelectric devices are used to measure fluid and air pressures by generating their own voltages. The most commonly found piezoelectric device is the engine knock sensor. The knock sensor measures engine knock, or vibration, and converts the vibration into a voltage signal.

The sensor is a voltage generator and has a resistor connected in series with it. The resistor protects the sensor from excessive current flow in case the circuit becomes shorted. The voltage generator is a thin ceramic disc attached to a metal diaphragm. When engine knock occurs, the vibration of the noise puts pressure on the diaphragm. This puts pressure on the piezoelectric crystals in the ceramic disc. The disc generates a voltage that is proportional to the amount of pressure. The voltage generated ranges from zero to one or more volts. Each time the engine knocks, a voltage spike is generated by the sensor.

Piezoresistive Devices. In construction, a piezoresistive device is similar to a piezoelectric one. However, it operates differently. The sensor acts like a variable resistor. Its resistance value changes as the pressure applied to the crystal changes. A voltage regulator supplies a constant voltage to the sensor. Since the amount of voltage dropped by the sensor will change with the change in resistance, the control module can determine the amount of pressure on the crystals by measuring the voltage drop across the sensor. Piezoresistive sensors are commonly used as gauge-sending units.

Potentiometers. The **potentiometer** usually consists of a wire-wound resistor with a moveable center wiper (Figure 3-51). A constant voltage value (usually 5 volts) is applied to terminal A. If the wiper (which is connected to the shaft or moveable component of the unit that is being monitored) is located close to this terminal, there will be low-voltage-drop represented by a high-voltage signal back to the computer through terminal B. As the wiper is moved toward the C terminal, the sensor signal voltage to terminal B decreases. The computer interprets the different

Shop Manual
Chapter 3,
page 101

A **potentiometer** is a voltage divider that provides a variable DC voltage reading to the computer.

The word Piezoelectric comes from the Greek word "piezo," which means pressure.

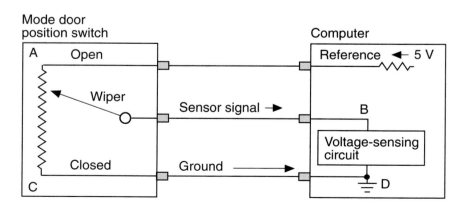

Figure 3-51 A potentiometer sensor circuit measures the amount of voltage drop to determine position.

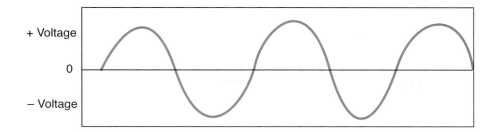

Figure 3-52   Pulse signal sine wave.

Shop Manual
Chapter 3,
page 102

voltage values into different shaft positions. The potentiometer can measure linear or rotary movement. As the wiper is moved across the resistor, the position of the unit can be tracked by the computer.

Since applied voltage must flow through the entire resistance, temperature and other factors do not create false or inaccurate sensor signals to the computer. A rheostat is not as accurate, and its use is limited in computer systems.

Magnetic Pulse Generators.   **Magnetic pulse generators** are commonly used to send data to the computer about the speed of the monitored component. The magnetic pulse generator is also used to inform the computer of the position of a monitored component. This is common in engine controls where the computer needs to know the position of the crankshaft in relation to rotational degrees.

The components of the pulse generator are:

1. A **timing disc** (reluctor) that is attached to the rotating shaft or cable. The number of teeth on the timing disc is determined by the manufacturer and depends on application. The teeth will cause a voltage generation that is constant per revolution of the shaft. For example, a vehicle speed sensor may be designed to deliver 4,000 pulses per mile. The number of pulses per mile remains constant regardless of speed. The computer calculates how fast the vehicle is going based on the frequency of the signal.

2. A **pickup coil** consists of a permanent magnet that has fine wire wound around it.

3. A magnet.

4. A pole piece.

An air gap is maintained between the timing disc and the pickup coil. As the timing disc rotates in front of the pickup coil, the generator sends an A/C signal (Figure 3-52). As a tooth on the timing disc aligns with the core of the pickup coil, it repels the magnetic field. The magnetic field is forced to flow through the coil and pickup core (Figure 3-53). Since the magnetic field is

> **Magnetic pulse generators** use the principle of magnetic induction to produce a voltage signal.
>
> Magnetic pulse generators are also called permanent magnet generators.
>
> The **timing disc** is known as an armature, reluctor, trigger wheel, pulse wheel, or timing core. It is used to conduct lines of magnetic force.
>
> The **pickup coil** is also known as a stator, sensor, or pole piece. It remains stationary while the timing disc rotates in front of it. The changes of magnetic lines of force generate a small voltage signal in the coil.

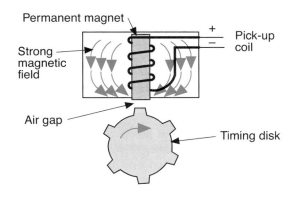

Figure 3-53   A strong magnetic field is produced in the pickup coil as the teeth align with the core.

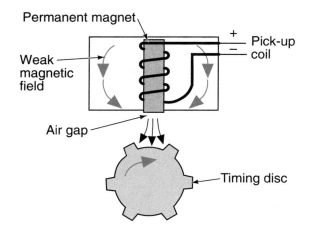

Figure 3-54 The magnetic field expands as the teeth pass the core.

not expanding, a voltage of zero is induced in the pickup coil. As the tooth passes the core, the magnetic field is able to expand (Figure 3-54). The expanding magnetic field cuts across the windings of the pickup coil. This movement of the magnetic field induces a voltage in the windings. This action is repeated every time a tooth passes the core. The moving lines of magnetic force cut across the coil windings and induce a voltage signal.

When a tooth approaches the core, a positive current is produced as the magnetic field begins to concentrate around the coil (Figure 3-55). The voltage will continue to increase as long as the magnetic field is expanding. As the tooth approaches the magnet, the magnetic field becomes smaller, causing the induced voltage to decrease and drop toward zero. When the tooth and core align, there is no more expansion or contraction of the magnetic field (thus no movement) and the voltage drops to zero (Figure 3-56). When the tooth passes the core, the magnetic

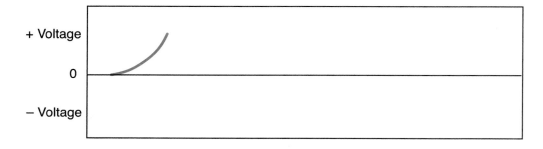

Figure 3-55 A positive voltage swing is produced as the tooth approaches the core.

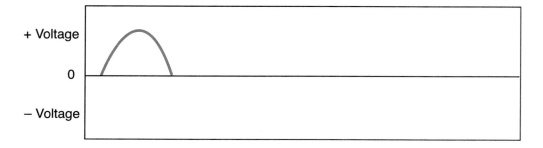

Figure 3-56 When the tooth aligns with the core, there is no magnetic movement and no voltage.

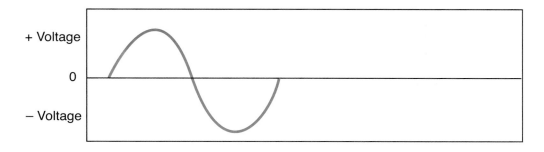

Figure 3-57   A negative waveform is created as the tooth passes the core.

field expands and a negative current is produced (Figure 3-57). The resulting pulse signal is amplified, digitized, and sent to the microprocessor.

AUTHOR'S NOTE: The magnetic pulse (PM) generator operates on basic magnetic principles. Remember that a voltage can only be induced when a magnetic field is moved across a conductor. The magnetic field is provided by the pickup unit, and the rotating timing disc provides the movement of the magnetic field needed to induce voltage.

The toothed disc is mounted on the crankshaft, vibration damper, or distributor shaft in a very precise manner. When the disc teeth align with the pickup coil corresponds to the exact time certain pistons are nearing TDC. This means the zero voltage signal needed to trigger the secondary circuit occurs at precisely the correct time.

Metal Detection Sensors.   Metal detection sensors are found on many early electronic ignition systems. They work much like a magnetic pulse generator with one major difference; the pickup coil of a metal detection sensor does not have a permanent magnet. Instead, the pickup coil is an electromagnet. A low level of current is supplied to the coil by an electronic control unit, inducing a weak magnetic field around the coil. As the reluctor on the distributor shaft rotates, the trigger teeth pass very close to the coil (Figure 3-58). As the teeth pass in and out of the coil's magnetic field, the magnetic field builds and collapses, producing a corresponding change in the coil's voltage. The voltage changes are monitored by the control unit to determine crankshaft position.

Shop Manual
Chapter 3,
page 102

Metal detection sensors operate much like a magnetic pulse generator but use a pickup coil that is an electromagnet.

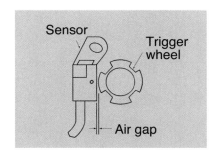

Figure 3-58   A metal deflector sensor.

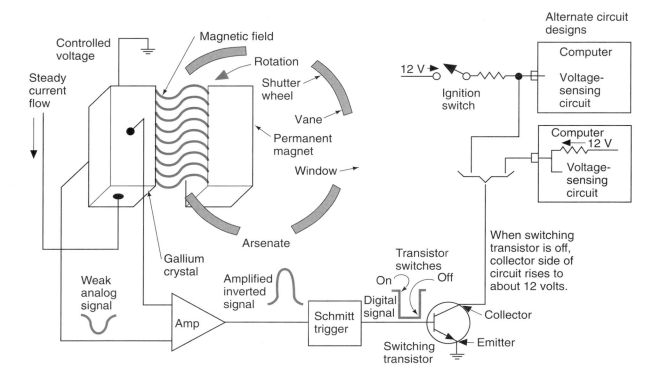

Figure 3-59 Typical circuit of a Hall-effect switch.

Shop Manual
Chapter 3,
page 104

The **Hall-effect switch** operates on the principle that if a current is allowed to flow through thin conducting material that is exposed to a magnetic field, another voltage is produce (Figure 12-34).

Hall-Effect Switches. The **Hall-effect switch** performs the same functions as the magnetic pulse generator, however, its operation is different. It contains a permanent magnet, a thin semiconductor layer made of gallium arsenate crystal (Hall layer), and a shutter wheel (Figure 3-59). The Hall layer has a negative and a positive terminal connected to it. Two additional terminals located on either side of the Hall layer are used for the output circuit.

The permanent magnet is located directly across from the Hall layer so that its lines of flux will bisect at right angles to the current flow. The permanent magnet is mounted so that a small air gap is between it and the Hall layer. A steady current is applied to the crystal of the Hall layer. This produces a signal voltage that is perpendicular to the direction of current flow and magnetic flux. The signal voltage produced is a result of the effect the magnetic field has on the electrons. When the magnetic field bisects the supply current flow, the electrons are deflected toward the Hall layer negative terminal (Figure 3-60). This results in a weak voltage potential being produced in the Hall switch.

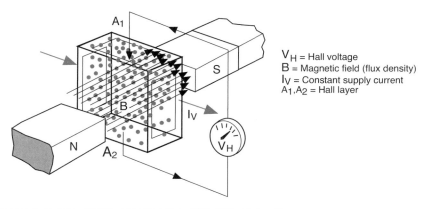

V_H = Hall voltage
B = Magnetic field (flux density)
I_V = Constant supply current
A_1, A_2 = Hall layer

Figure 3-60 The magnetic field causes the electrons from the supply current to gather at the Hall layer negative terminal. This creates a voltage potential.

A **shutter wheel** is attached to a rotational component. As the wheel rotates, the shutters (vanes) will pass in this air gap. When a shutter vane enters the gap, it intercepts the magnetic field and shields the Hall layer from its lines of force. The electrons in the supply current are no longer disrupted, so they return to a normal state. This results in low-voltage potential in the signal circuit of the Hall switch.

The signal voltage leaves the Hall layer as a weak analog signal. To be used by the computer, the signal must be conditioned. It is first amplified because it is too weak to produce a desirable result. The signal is also inverted so that a low-input signal is converted into a high-output signal. It is then sent through a **Schmitt trigger** where it is digitized and conditioned into a clean square wave signal. The signal is finally sent to a switching transistor. The computer senses the turning on and off of the switching transistor to determine the frequency of the signals and calculate speed.

The American physicist Edwin Herbert Hall (1855–1938) discoved the principle of the Hall-effect switch in 1879.

Feedback Signals

If the computer sends a command signal to open a blend door in an automatic climate-control system, a **feedback signal** may be sent back from the actuator to inform the computer the task was performed. The feedback signal will confirm both the door position and actuator operation (Figure 3-61). Another form of feedback is for the computer to monitor voltage as a switch, relay, or other actuator is activated. Changing states of the actuator will result in a predictable change in the computer's voltage-sensing circuit. The computer may set a diagnostic code if it does not receive the correct feedback signal.

Oxygen (O$_2$) Sensors.   One of the most commonly used feedback sensors is the O$_2$ sensor. The O$_2$ sensor is mounted in the exhaust gas stream and provides the PCM with a measurement

The **shutter wheel** consists of a series of alternating windows and vanes. It creates a magnetic shunt that changes the strength of the magnetic field from the permanent magnet.

A **Schmitt trigger** is an A/D converter.

Feedback signal means that data concerning the effects of the computer's commands are fed back to the computer as an input signal.

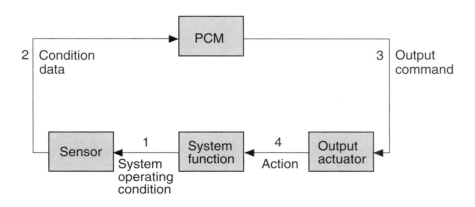

Figure 3-61   Principle of feedback signals.

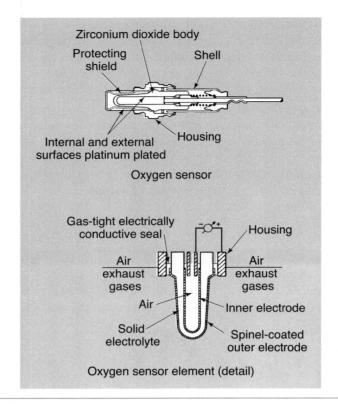

Figure 3-62 Oxygen (O₂) sensor design.

of the oxygen in the engine's exhaust. The sensor is constructed of a zirconium dioxide ceramic thimble covered with a thin layer of platinum (Figure 3-62).

When the thimble is filled with oxygen-rich outside air and the outer surface of the thimble is exposed to oxygen-depleted exhaust gases, a chemical reaction in the sensor produces a voltage. The generation of voltage is similar to the same activity that takes place in a battery, except at much lower voltages. The voltage output varies with the level of oxygen present in the exhaust. As oxygen in the exhaust decreases, the voltage output increases. Likewise, as the oxygen level in the exhaust increases, the output voltage decreases (Figure 3-63).

Some O_2 sensors have a single wire that is connected from the oxygen-sensing element to the computer, and this wire acts as a signal wire. If an O_2 sensor has two wires, the second wire is a ground wire connected back to the computer.

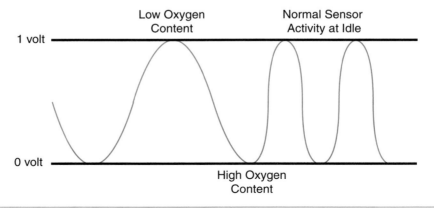

Figure 3-63 A high voltage value from the O₂ sensor indicates low oxygen content in the exhaust. A low voltage indicates high oxygen content.

Most O_2 sensors today use an internal heater element. Since the O_2 sensor does not produce a satisfactory signal until it reaches about 600°F (315°C), the internal heater provides faster sensor warm-up time and helps to keep the sensor hot during prolonged idle operation. The internal O_2 sensor heater maintains higher sensor temperatures, which helps to burn deposits off the sensor. When the O_2 sensor has an internal heater, the sensor can be placed farther away from the engine in the exhaust stream, giving engineers more flexibility in sensor location. These O_2 sensors are identified by having three or four wires. The third wire is connected to an electric heating element in the sensor. Voltage is supplied from the ignition switch (or a relay that is energized when the engine is running) to this heater. In a four-wire sensor, there is an individual circuit for the signal wire, the heater wire, and the two ground wires. In these four-wire sensors, the heater and the sensing element have individual ground wires. A replacement O_2 sensor must have the same number of wires as the original sensor.

A lean air-fuel ratio provides excess quantities of oxygen in the exhaust stream because the mixture entering the cylinders has an excessive amount of air in relation to the amount of fuel. Therefore, air containing oxygen is left over after the combustion process. When the exhaust stream has high oxygen content, oxygen from the atmosphere is also present inside the O_2 sensor element. When oxygen is present on both sides of the sensor element, the sensor produces a low voltage.

A rich air-fuel ratio contains excessive fuel in relation to the amount of air entering the cylinders. A rich air-fuel mixture produces very little oxygen in the exhaust stream because the oxygen in the air is all mixed with fuel, and excess fuel is left over after combustion is completed. When the exhaust stream with very low oxygen content strikes the O_2 sensor, there is high oxygen content from the atmosphere inside the sensor element. With different oxygen levels on the inside and outside of the sensor element, the sensor produces up to 1 volt. As the air-fuel ratio cycles from lean to rich, the O_2 sensor voltage changes in a few milliseconds.

In a gasoline fuel system, the stoichiometric (ideal) air-fuel mixture is 14.7:1. At the stoichiometric air-fuel ratio, combustion is most efficient, and nearly all the oxygen in the air is mixed with fuel and burned in the combustion chambers. Computer-controlled fuel-injection systems maintain the air-fuel ratio at stoichiometric under most operating conditions. As the air-fuel ratio cycles slightly rich and lean from the stoichiometric ratio, the O_2 sensor voltage cycles from high to low. Once the engine is at normal operating temperature, the O_2 sensor voltage should vary between 0.3 volt and 0.8 volt if the air-fuel ratio is at, or near, stoichiometric.

Some vehicles are equipped with a titania-type O_2 sensor. This sensor design is similar to the zirconia-type sensor, but the sensing element is made from titania rather than zirconia.

The titania-type sensor modifies voltage, whereas the zirconia-type sensor generates voltage. The computer supplies battery voltage to the titania-type sensor, and this voltage is lowered by a resistor in the circuit. The resistance of the titania varies as the air-fuel ratio cycles from rich to lean. When the air-fuel ratio is rich, the titania resistance is low, and this provides a higher voltage signal to the computer. If the air-fuel ratio is lean, the titania resistance is high, and a lower voltage signal is sent to the computer (Figure 3-64).

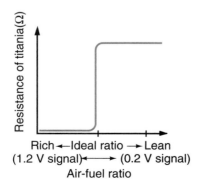

Figure 3-64 Titania-type O_2 sensor resistance and voltage signal.

The titania-type O_2 sensor provides a satisfactory signal almost immediately after a cold engine is started. This action provides improved air-fuel ratio control during engine warm up.

Outputs

Once the computer's programming instructs that a correction or adjustment must be made in the controlled system, an output signal is sent to an actuator. This involves translating the electronic signals into mechanical motion.

An output driver is used within the computer to control the actuators. The driver usually applies the ground circuit of the actuator. The ground can be applied steadily if the actuator must be activated for a selected amount of time. For example, if the PCM inputs indicate that the A/C clutch relay needs to be energized, the driver keeps the relay energized steadily until inputs inform the PCM that A/C clutch operation is not needed. Then, the ground for the relay coil is removed.

Other systems require the actuator to either be turned on and off very rapidly or for a set amount of cycles per second. It is duty cycled if it is turned on and off a set amount of cycles per second. The duty cycle is the percentage of on time to total cycle time. For example, if a duty-cycled actuator cycles ten times per second, one actuator cycle is completed in one tenth of a second. If the actuator is turned on for 30 percent of each tenth of a second and off for 70 percent, it is referred to as a 30 percent duty cycle (Figure 3-65).

If the actuator is cycled on and off very rapidly (but not at a set duty cycle), the pulse width can be varied to provide the programmed results. Pulse width is the length of time in milliseconds that an actuator is energized. For example, the computer program will select a fuel delivery requirement for a cylinder based on inputs from several sensors. The delivery of the fuel through the injector is achieved through pulse-width modulation of the injector driver. If more fuel is required, the pulse width is increased, which increases the length of on time. As the fuel quantity needs to be reduced, the pulse width is decreased (Figure 3-66).

Shop Manual
Chapter 3,
page 99

Actuators perform the actual work commanded by the computer. They can be in the form of a motor, relay, switch, or solenoid.

Actuators

Most computer-controlled **actuators** are electromechanical devices that convert the output commands from the computer into mechanical action. These actuators are used to open and close switches, control vacuum flow to other components, and operate doors or valves depending on the requirements of the system.

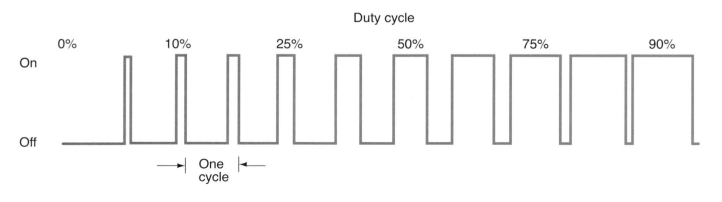

Figure 3-65 Duty cycle is the percentage of on time per cycle. Duty cycle can be changed, however total cycle time remains constant.

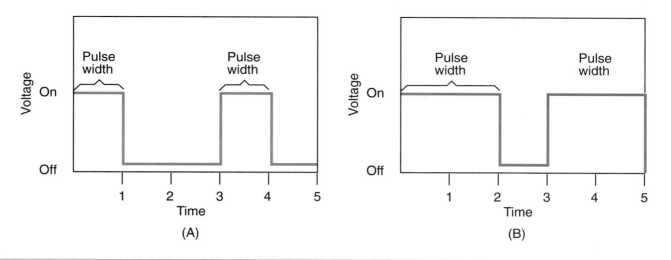

Figure 3-66 Pulse width is the duration of on time. (A) Pulse width modulation logic used to achieve fuel flow, (B) Pulse width modulation logic used to achieve increased fuel flow.

Relays. A relay allows control of a high-current draw circuit by a very low-current draw circuit. The computer usually controls the relay by providing the ground for the relay coil (Figure 3-67). The use of relays protects the computer by keeping the high current from passing through it. For example, the electromagnetic clutch used on an A/C compressor requires a high-current draw to operate. Instead of having the computer operate the clutch directly, it will energize a relay. With the relay energized, a direct circuit from the battery to the clutch is completed.

Solenoids. Computer control of the solenoid is usually provided by applying the ground through the output driver. A solenoid is commonly used as an actuator because it operates well under duty cycling conditions. Many newer PCMs will use high-side drivers to control some solenoids. One of the most common uses of the solenoid is the fuel injector. Another common use is to control vacuum to other components.

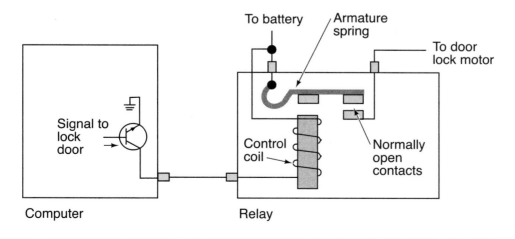

Figure 3-67 The computer's output driver applies the ground for the relay coil.

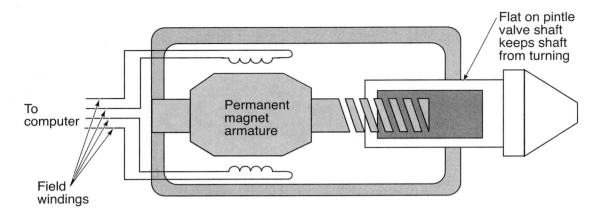

Figure 3-68 Typical stepper motor.

A **stepper motor** contains a permanent magnet armature with two, four, or more field coils.

Motors. Many computer-controlled systems use a **stepper motor** to move the controlled device to whatever location is desired (Figure 3-68). By applying voltage pulses to selected coils of the motor, the armature will turn a specific number of degrees. When the same voltage pulses are applied to the opposite coils, the armature will rotate the same number of degrees in the opposite direction.

Some applications require the use of a permanent magnet field servo motor (Figure 3-69). The polarity of the voltage applied to the armature windings determines the direction the motor rotates. The computer can apply a continuous voltage to the armature until the desired result is obtained.

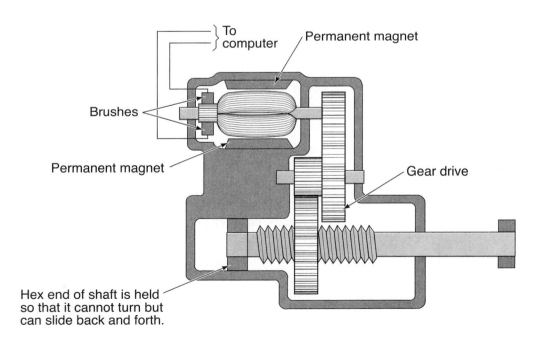

Figure 3-69 Reversible permanent magnet motor.

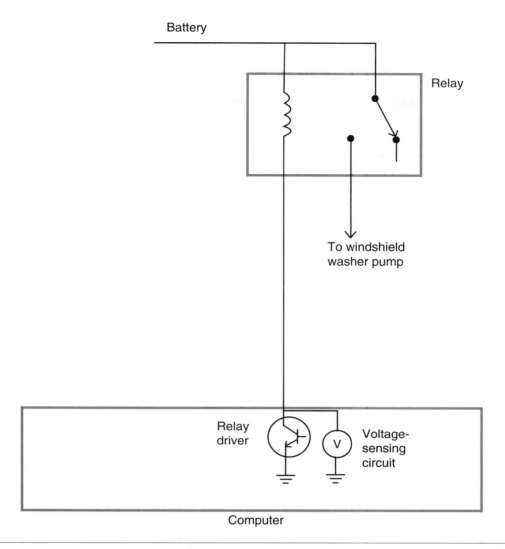

Battery

Relay

To windshield
washer pump

Relay
driver

Voltage-
sensing
circuit

Computer

Figure 3-70 Computers using low-side drivers may be able to monitor the circuit for proper operation. When the relay coil is not energized, the sense circuit should see a high voltage (12 volts). When the relay is turned on, the voltage should go low (0 volts).

High-Side and Low-Side Drivers

Usually the computer will control an actuator by the use of **low-side drivers**. These drivers will complete the path to ground through a FET transistor to control the output device. The computer may monitor the voltage on this circuit to determine if the actuator operates when commanded (Figure 3-70). Monitoring of the system can be done by either measuring voltage on the circuit or by measuring the current draw of the circuit.

Many newer vehicles are now using **high-side drivers**. High-side drivers consist of a Metal Oxide Field Effect Transistor (MOSFET) that is controlled by a bipolar transistor. The bipolar transistor is controlled by the microprocessor. The advantage of the high-side driver is that it can provide quick response self-diagnostics for shorts, opens, and thermal conditions. It also reduces vehicle wiring.

Low-side drivers are used to complete the path to ground to turn on an actuator.

High-side drivers control the output device by varying the positive (12-volt) side of the circuit.

High-side driver diagnostic capabilities include the ability to determine a short circuit or open circuit condition. The high-side driver will take the place of a fuse in the event of a short circuit condition. When it senses a high current condition, it will turn off the power flow and then store a diagnostic trouble code (DTC) in memory. The driver will automatically reset once the short circuit condition is removed. In addition, the high-side driver monitors its temperature. The driver reports the junction temperature to the microprocessor. If a slow acting resistive short occurs in the circuit, the temperature will begin to climb. Once the temperature reaches 300°F (150°C) the driver will turn off and set a DTC.

The high-side driver is also capable of detecting an open circuit, even if the system is turned off. This is done by reading a feedback voltage to the microprocessor when the driver is off. A 5-volt, 50 μA current is fed through the circuit, which also has a resistor wired in parallel. Low voltage (less than 2.25 volts) will indicate a normal circuit. If the voltage is high (above 2.25 volts), a high resistance or open circuit is detected. If the open circuit is detected, a DTC is set.

Multiplexing

Multiplexing
provides the ability to use a single circuit to distribute and share data between several control modules throughout the vehicle. Because the data is transmitted through a single circuit, bulky wiring harnesses are eliminated.

The common acronym for multiplexing is **MUX**.

The term **bus** refers to the data transportation from one module to another.

Vehicle manufacturers will use **multiplexing** systems to enable different control modules to share information. A **MUX** wiring system uses **bus** data links that connect each module. Each module can transmit and receive digital codes over the bus data links (Figure 3-71). The signal sent from a sensor can go to any one of the modules and can be shared by the other modules. Before multiplexing, if information from the same sensing device is needed by several controllers, a wire from each controller needed to be connected in parallel to that sensor. If the sensor signal is analog, the controllers need an analog-to-digital (A/D) converter to be able to "read" the sensor information. By using multiplexing, the need for separate conductors from the sensor to each module is eliminated, and the number of drivers in the controllers is reduced.

Digital signals are used by controllers to communicate messages, both internally and with other controllers. A chip is used to prevent the digital codes from overlapping by allowing only one code to be transmitted at a time. Each digital message is preceded by an identification code that establishes its priority. If two modules attempt to send a message at the same time, the message with the higher priority code is transmitted first.

The major difference between a multiplexed system and a non-multiplexed system is the way data is gathered and processed. In non-multiplexed systems, the signal from a sensor is sent as an analog signal through a dedicated wire to the computer or computers. At the computer, the

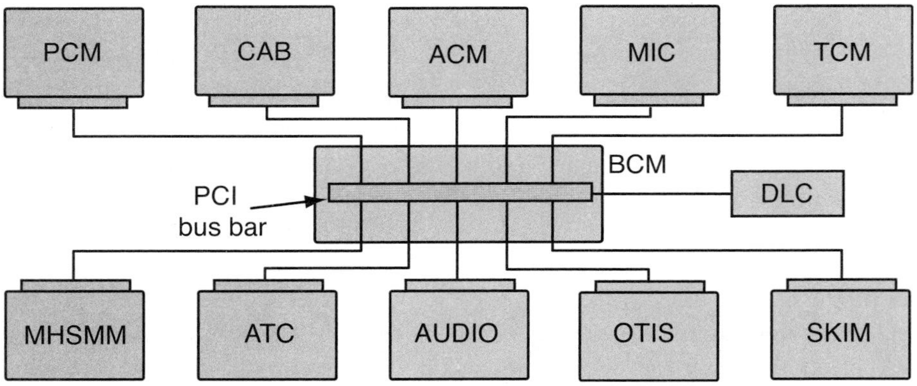

Figure 3-71 Computers use multiplexing to reduce the number of conductors that would be required.

signal is changed from an analog to a digital signal. Because each sensor requires its own dedicated signal wire, the number of wires required to feed data from all of the sensors and transmit control signals to all of the output devices is great.

In a MUX system, the signal is sent to a computer where it is converted from analog to digital if needed. Since the computer or control module of any system can only process one input at a time, it calls for input signals as it needs them. By timing the transmission of data from the sensors to the control module, a single data wire can be used. Between each transmission of data to the control module, the sensor is electronically disconnected from the control module.

The following are some examples of how data bus messages are transmitted. One of the earliest multiplexing systems was developed by Chrysler in 1988 and used through the 2003 model year. This system is called Chrysler Collision Detection (CCD). The term "collision" refers to the collision of data occurring simultaneously. This bus circuit uses two wires. The advantages of the CCD system include:

1. Reduction of wires.
2. Reduction of drivers required in the computers.
3. Reduced load across the sensors.
4. Enhanced diagnostics.

The CCD system uses a twisted pair of wires to transmit the data in digital form. One of the wires is called the bus (+) and the other is bus (−). Negative voltages are not used. The (+) and (−) means one is more positive than the other when the bus is in the dominate state. All modules that are connected to the CCD bus system have a special CCD chip installed (Figure 3-72). In most vehicles (but not all), the body control module (BCM) will provide the bias voltage to

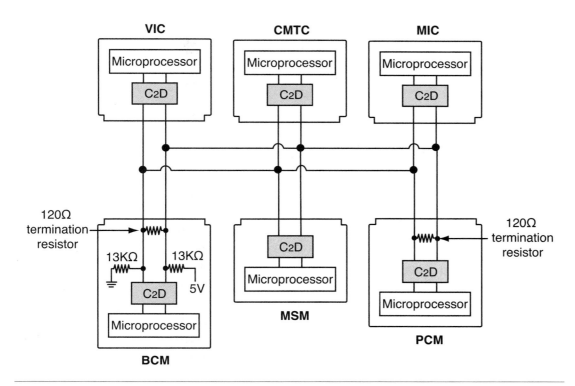

Figure 3-72   Each module on the CCD bus system has a CCD (C₂D) chip.

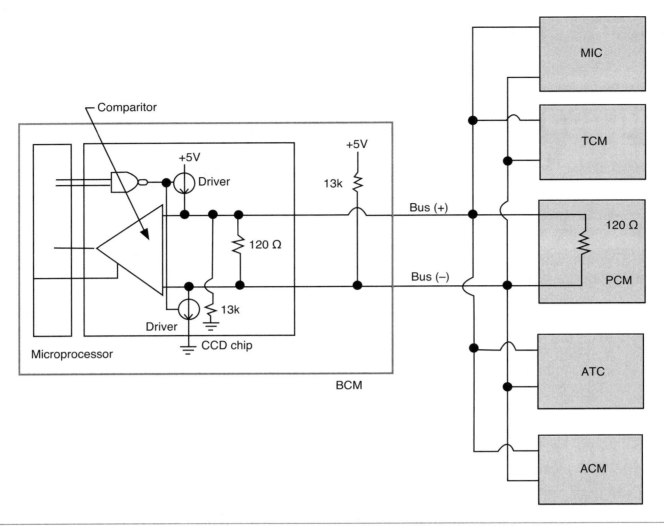

Figure 3-73   CCD bus circuit.

power the bus circuits. Since the BCM powers the system, its internal components are illustrated (Figure 3-73). The other modules will operate the same as the BCM to send messages.

The bias voltage on the bus (+) and bus (–) circuits is approximately 2.50 volts when the system is idle (no data transmission occurring). This is accomplished through a regulated 5-volt circuit and a series of resistors. The regulated 5 volts sends current through a 13K-ohm resistor to the bus (–) circuit (Figure 3-74). The current is then sent through the two 120-ohm resistors that are wired in parallel and to the bus (+) circuit. Finally, the current is sent to ground through a second 13K-ohm resistor. A simplified schematic of this biasing circuit is shown along with the normal voltage drops that occur as a result of the resistors (Figure 3-75). The two 120-ohm resistors are referred to as **termination resistors**. One is internal to the BCM while the other is located in the PCM.

With the bus circuits at the proper voltage levels, communication can occur. The comparator in the CCD chip acts as a voltmeter. If the positive lead of the voltmeter is connected to the bus (+) circuit and the negative lead is connected to the bus (–) circuit, the voltmeter will read the voltage difference between the circuits. At idle, the difference is 0.02 volts. When a module needs to send a message, the microprocessor will use the NAND gate to turn on and off the two

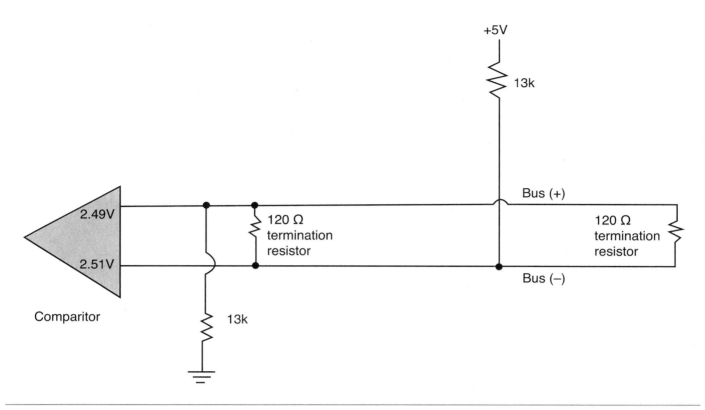

Figure 3-74 The bus is supplied 2.5 volts through the use of pull-up and pull-down resistors.

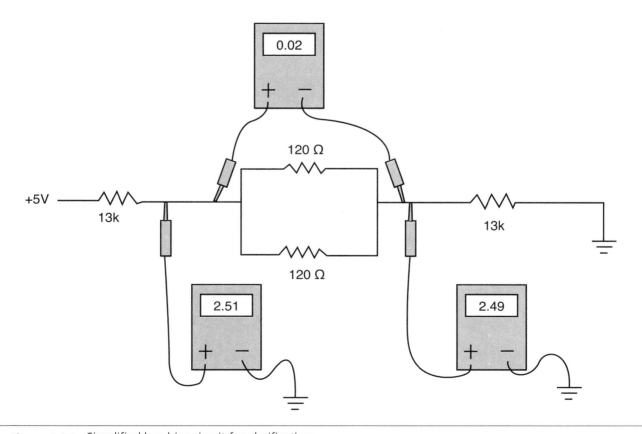

Figure 3-75 Simplified bus bias circuit for clarification.

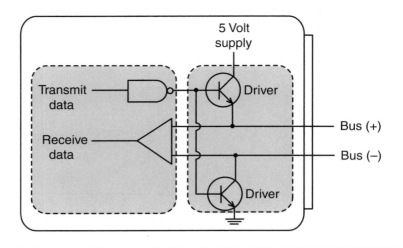

Figure 3-76 For a message to be transmitted, the drivers are activated which pulls up bias on bus (+) and pulls down bias on bus (−).

drivers at the same time (Figure 3-76). The driver to the bus (+) circuit provides an alternate 5 volts to the bus (+) circuit. The comparitor will measure this voltage. At the same time, the driver to the bus (−) circuit provides an alternate ground path for the original 5 volts. This alternate ground bypasses the termination resistors and the second 13K-ohm resistor. Since the first 13K-ohm resistor is now the only one in the circuit, all of the voltage is dropped over it and the comparitor will see low voltage on the bus (−) circuit.

Since the drivers are turned on and off at a rate at a rate of 7,812.5 times per second, the voltage will not go to a full 5 volts on bus (+) nor to 0 volts on bus (−). However, bus (+) voltage is pulled higher *toward* 5 volts and bus (−) is pulled lower *toward* 0 volts (Figure 3-77). Once the voltage difference measured by the comparator is greater than 0.060 volts, the computers will recognize a bit value change. When the bus circuit is idle (0.02 voltage difference), the bit value is a 1. Once the voltage difference increases, the bit value is changed to a 0.

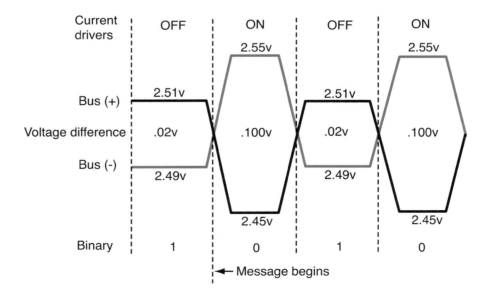

Figure 3-77 As the drivers are activated, the voltage difference between bus (+) and bus (−) increase over their voltage values at idle. The difference in voltage determines if a binary 1 or 0 is being transmitted.

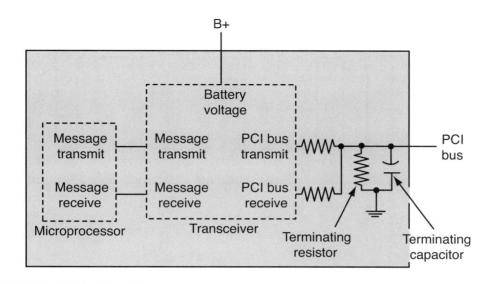

B+

Figure 3-78 Bias and termination is supplied by each module on the PCI bus system.

Beginning in the 1998 model year, DaimlerChrysler began to phase out the CCD bus system and replace it with a new Programmable Communications Interface (PCI) bus system. Since this system is similar to those of other manufacturers, it will be used for discussion purposes.

The PCI system is a single wire, bidirectional communication bus. Each module on the bus system supplies its own bias voltage and has its own termination resistors (Figure 3-78). Like the CCD system, the modules of the PCI system are connected in parallel. As a message is sent, a variable pulse width modulation (VPWM) voltage between 0 and 7.75 volts is used to represent the 1 and 0 bits (Figure 3-79). The voltage signal is not a clean digital signal. Rather, the voltage traces appear to be trapezoidal in shape because the voltage is slowly ramped up and down to prevent magnetic induction.

AUTHOR'S NOTE: The reference to "slowly ramped up and down" is relative. In the PCI bus, an average of 10,400 bits is transmitted per second. So, relatively speaking, the voltage is slow to go to 7.75 volts and slow to return to 0 volts.

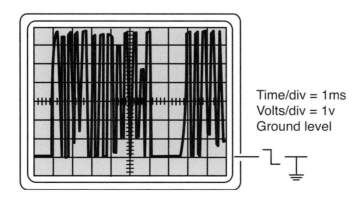

Time/div = 1ms
Volts/div = 1v
Ground level

Figure 3-79 Lab scope trace of PCI bus voltages.

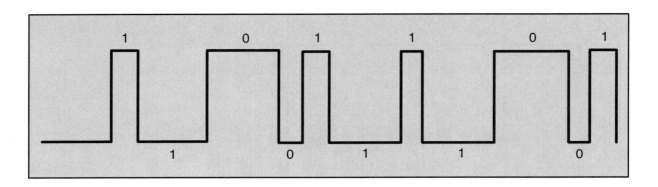

Figure 3-80 The VPWM determines the bit value.

The length of time the voltage is high or low determines if the bit value is a 1 or a 0 (Figure 3-80). The typical PCI bus message will have the following elements (Figure 3-81):
* *Header*—One to three bytes of information concerning the type, length, priority, target module, and sending module.
* *Data byte(s)*—The message that is being sent. This can be up to eight bytes in length.
* *Cyclic Redundancy Check (CRC) byte*—Detects if the message has been corrupted or if there are any other errors.
* *In-Frame Response (IFR) byte(s)*—If the sending module requires an acknowledgment or an immediate response from the target module, this request will be received with the message. The IFR is the target module sending the requested information to the original sending module.

Figure 3-82 illustrates the type of information that is sent over the PCI bus system.

Another example of a bus system is the Simplified Wiring System (SWS) used by Mitsubishi (Figure 3-83, page 90). The top trace is the messages being sent by different modules. The bottom trace is the timer circuit that notifies the modules when transmission may begin. In this system, the modules will transmit their messages in a set order.

Finally, a very popular bus system that has been used in Europe is the Controller Area Network (CAN) bus. Most CAN bus-equipped vehicles use the CAN bus for communications between modules only and not for diagnostics with a scan tool. Scan tool diagnostics is usually performed over the ISO K-line. These CAN bus systems are ISO 9141.

New U.S. regulations are requiring the use of the CAN bus systems under industry standard J2284. This will be used as the new protocol for communications with a scan tool on U.S.-sold vehicles. Although the CAN bus system has been used since the 1980s, J2284 makes it unique because for the first time it will be used for diagnostics.

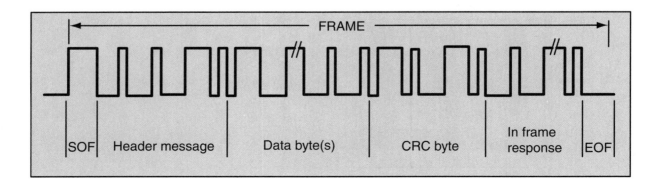

Figure 3-81 Components of a typical PCI bus message.

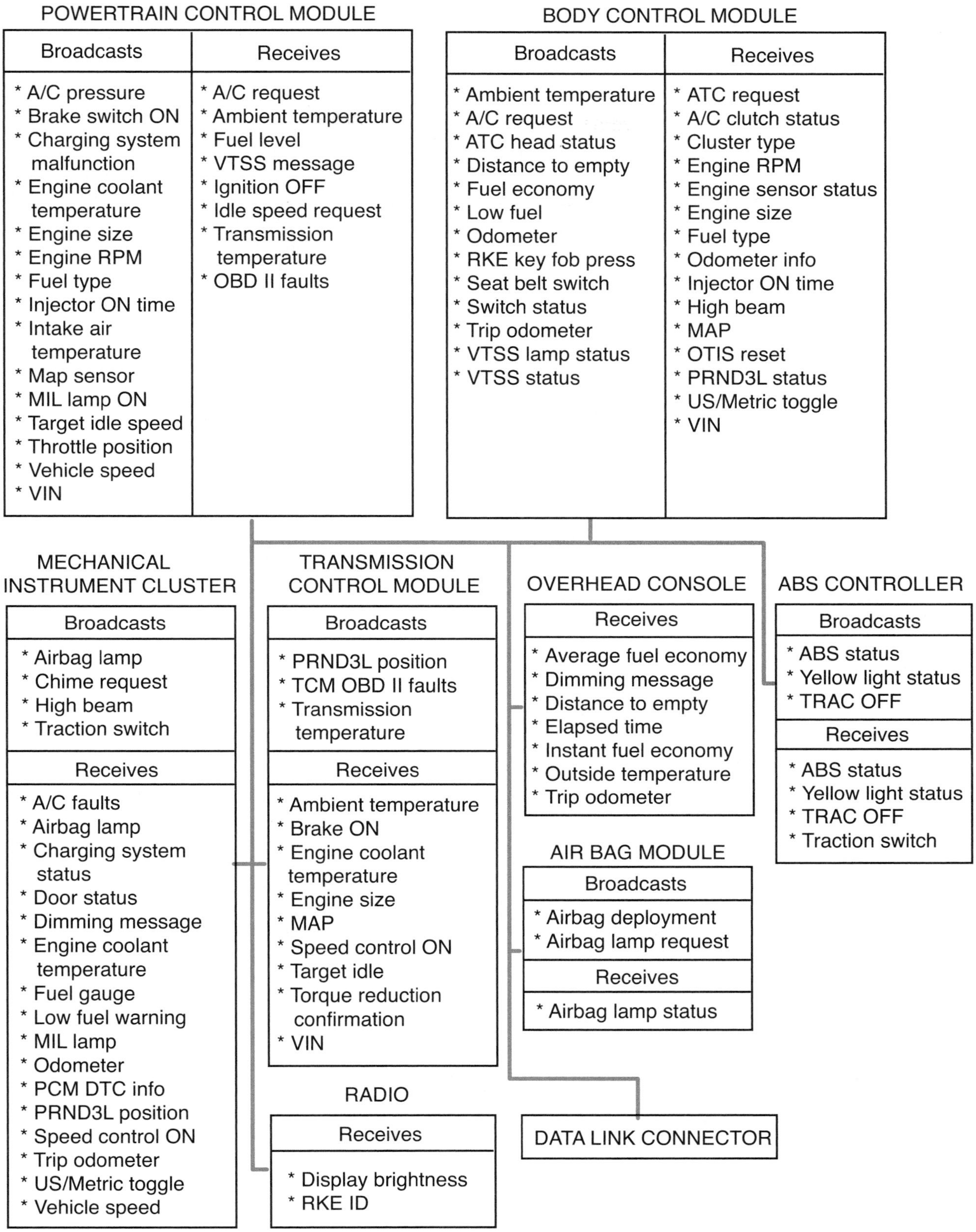

POWERTRAIN CONTROL MODULE

Broadcasts	Receives
* A/C pressure	* A/C request
* Brake switch ON	* Ambient temperature
* Charging system	* Fuel level
malfunction	* VTSS message
* Engine coolant	* Ignition OFF
temperature	* Idle speed request
* Engine size	* Transmission
* Engine RPM	temperature
* Fuel type	* OBD II faults
* Injector ON time	
* Intake air	
temperature	
* Map sensor	
* MIL lamp ON	
* Target idle speed	
* Throttle position	
* Vehicle speed	
* VIN	

BODY CONTROL MODULE

Broadcasts	Receives
* Ambient temperature	* ATC request
* A/C request	* A/C clutch status
* ATC head status	* Cluster type
* Distance to empty	* Engine RPM
* Fuel economy	* Engine sensor status
* Low fuel	* Engine size
* Odometer	* Fuel type
* RKE key fob press	* Odometer info
* Seat belt switch	* Injector ON time
* Switch status	* High beam
* Trip odometer	* MAP
* VTSS lamp status	* OTIS reset
* VTSS status	* PRND3L status
	* US/Metric toggle
	* VIN

MECHANICAL INSTRUMENT CLUSTER

Broadcasts
* Airbag lamp
* Chime request
* High beam
* Traction switch

Receives
* A/C faults
* Airbag lamp
* Charging system
status
* Door status
* Dimming message
* Engine coolant
temperature
* Fuel gauge
* Low fuel warning
* MIL lamp
* Odometer
* PCM DTC info
* PRND3L position
* Speed control ON
* Trip odometer
* US/Metric toggle
* Vehicle speed

TRANSMISSION CONTROL MODULE

Broadcasts
* PRND3L position
* TCM OBD II faults
* Transmission
temperature

Receives
* Ambient temperature
* Brake ON
* Engine coolant
temperature
* Engine size
* MAP
* Speed control ON
* Target idle
* Torque reduction
confirmation
* VIN

OVERHEAD CONSOLE

Receives
* Average fuel economy
* Dimming message
* Distance to empty
* Elapsed time
* Instant fuel economy
* Outside temperature
* Trip odometer

AIR BAG MODULE

Broadcasts
* Airbag deployment
* Airbag lamp request

Receives
* Airbag lamp status

ABS CONTROLLER

Broadcasts
* ABS status
* Yellow light status
* TRAC OFF

Receives
* ABS status
* Yellow light status
* TRAC OFF
* Traction switch

RADIO

Receives
* Display brightness
* RKE ID

DATA LINK CONNECTOR

Figure 3-82 Chart of messages received and broadcast by each module on the PCI base.

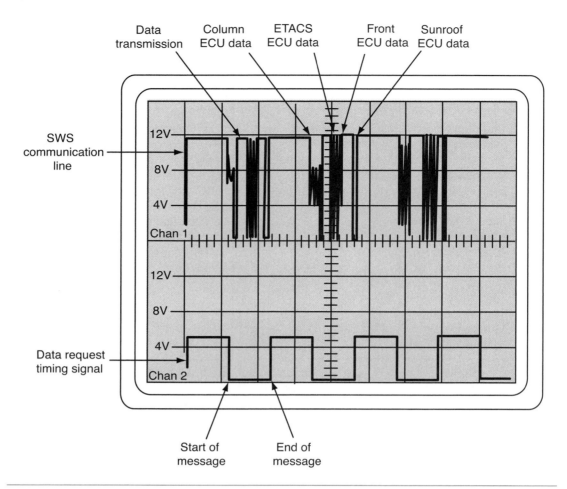

Figure 3-83 Lab scope traces of the SWS bus system show data transmission and the timer signal.

The CAN bus system uses terminology such as CAN B and CAN C. The letters B and C distinguish the speed of the bus. CAN B is a medium-speed bus with a speed of 83,300 bits per second. The CAN C bus has a speed of 500,000 bits per second. A vehicle can be equipped with both of these bus networks. In addition, a new Diagnostics CAN C bus is used to connect the scan tool. Diagnostic CAN C (which can be called by many different names) has a speed of 500,000 bits per second.

The circuitry of the CAN bus usually consists of a pair of twisted wires. For digital data to transfer, voltage is simultaneously pulled high on one circuit and pulled low on the other. The wires for the CAN bus system are twisted to reduce electromagnetic interference (EMI). This requires 33 to 50 twists per meter. To maintain the twist, the bus wire pair are in adjacent cavities at connectors. Wires are routed to avoid parallel paths with high-current sources, such as ignition coil drivers, motors, and high-current PWM circuits. Wherever possible, the bus wires use a different connector from those carrying high current.

On a CAN bus system, each module provides its own bias. Because each module provides its own bias, communication between groups of modules is still possible if an open occurs in the bus circuit. The CAN bus transceiver has drivers internal to the transceiver chip to supply the voltage and ground to the bus circuit.

Each CAN bus system has distinct advantages and limitations. The high-speed CAN C bus is only functional when the ignition is on. This bus can transfer data at a "real time" rate. The medium-speed CAN B bus cannot transfer data as quickly but it can remain active when the ignition is turned off if individual modules require it to be active. The individual requirements of each module determine to which bus system it is connected. The simultaneous use of two separate bus systems on one vehicle allows the optimum characteristics of each system to be used.

Vehicle systems that exchange data in real time use CAN C. Typically, these modules would be the anti-lock brake module and the power train control module. The manufacturer may also include the transmission control module and other modules that require real time information. Other modules that may need to transfer data with the ignition turned off use the CAN B bus.

The CAN B bus system is very fault tolerant and can operate with one of its conductors shorted to ground or both of its conductors shorted together. As long as there is a voltage potential between one of the CAN B circuits, chassis ground communication may still be possible. Due to its high speed, CAN C is not fault tolerant.

The CAN C bus becomes active when the ignition is turned on. When the bus is biased, the voltage is approximately 2.5 volts. When both CAN C(+) and CAN C(−) are equal, the bus is recessive. In this state, the logic is "1." When bus C(+) is pulled high and bus C(−) is pulled low, the bus is considered dominant and the logic is "0" (Figure 3-84). To be dominant, the voltage difference between CAN C(+) and CAN C(−) must be at least 1.5 volts and not more than 3.0 volts. To be recessive, the voltage difference between the two circuits must not be more than 50 millivolts.

The optimum CAN C bus termination is 60 ohms. Two CAN C modules each provide 120 ohms of termination. Since the modules are wired in parallel, total resistance is 60 ohms. The two modules that provide termination are located the farthest apart for all vehicle combinations. To provide termination, the modules that provide termination resistance have two 60-ohm resistors that are connected in series. The end of one resistor is connected to CAN + and the end of the other resistor is connected to CAN −. A center tap between the two resistors is connected through

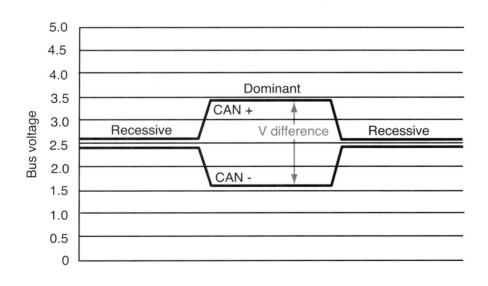

Figure 3-84 Voltages on the CAN C bus.

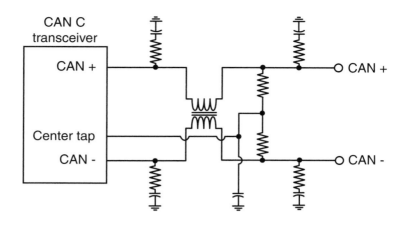

Figure 3-85   Termination resistance of a CAN C module.

a capacitor to ground. This center tap may also be connected to the transceiver micro if necessary (Figure 3-85).

The CAN B bus can be active whether the ignition is on or off. When CAN B (+) is approximately 0 to 0.2 volt and CAN B (−) is 4.8 to 5 volts, the bus is idle or recessive. In this state, the logic is "1." When CAN B(+) is pulled between 3.6 and 5 volts and CAN B(−) is pulled low to 1.4 to 0 volts, the bus is considered dominant and the logic is "0" (Figure 3-86). When CAN B(+) is approximately 0 volts and CAN B(−) is near battery voltage, the bus is asleep.

The termination resistance of CAN B bus is determined by the number of modules installed on the vehicle. Each module will supply its own termination. Internal to CAN B modules are two termination resistors. The resistors connect CAN B(+) and CAN B(−) to their respective transceiver termination pins (Figure 3-87). To provide termination and bias, the transceiver internally connects the CAN B(+) resistor to ground and the CAN B(−) resistor to a 5-volt source. When the

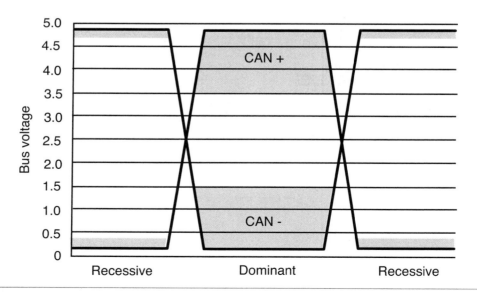

Figure 3-86   Typical CAN B bus voltages.

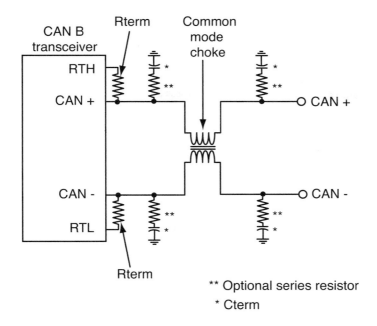

Figure 3-87 CAN B bus module termination resistance.

CAN B bus goes into sleep mode, the termination pin connected to CAN B(–) switches from 5 volts to battery voltage by the transceiver.

Some of the network messages are defined by the following:

- *Cyclic:* A message launched on a periodic schedule. An example is the ignition on status broadcasts on the CAN B bus every 100 ms.

- *Spontaneous:* An application-driven message.

- *Cyclic and Change:* A message launched on a periodic schedule as long as the signal is not changing. The message is re-launched whenever the signal changes.

- *By Active Function (BAF):* A message that is only transmitted at a specific rate when the message does not equal a default value.

A **central gateway (CGW)** module is used where all three CAN bus networks (CAN B, CAN C, and Diagnostic CAN C) connect together. Similar to a router in a computer network, this module allows data exchange between the different buses. The CGW can take a message on one bus and transfer that message to the other bus. It can hand one message to the other without changing the message. If several messages are being sent simultaneously, some messages are captured in a buffer and the message with the highest priority is sent first. The other messages are transmitted in order of priority. The CGW also monitors the CAN network for failures and can log a Network DTC ("U" Code) if it detects a malfunction.

The CGW is also the gateway to the CAN network for the scan tool. The scan tool is connected to the gateway using its own CAN bus circuit known as Diagnostic CAN C. Because CAN C is used for diagnostics, data can be exchanged with the scan tool at a real-time rate (500 kbps). As a result, a scan tool that is compatible with the CAN bus system is required for vehicle diagnosis.

Since a variety of generic scan tools may be connected to the vehicle, mandated regulations prohibit scan tools from containing any termination resistance. If termination resistance resided in the tool, the variance in the tool termination could affect CAN C. For this reason, the entire termination for the Diagnostic CAN C resides in the CGW. The configuration of the resistors is similar to dominate CAN C modules except the two termination resistors in series are 30 ohms rather than 60 ohms. The CGW provides the full 60-ohm termination for the Diagnostic CAN C.

The **central gateway (CGW)** module is the manager of the CAN bus system. It is responsible for arbitration of the various bus message.

Summary

❏ A computer is an electronic device that stores and processes data and is capable of operating other devices.

❏ The operation of the computer is divided into four basic functions: input, processing, storage, and output.

❏ Binary numbers are represented by the numbers 1 and 0. A transistor that operates as a relay is the basis of the digital computer. As the input signal switches from off to on, the transistor output switches from cutoff to saturation. The on and off output signals represent the binary digits 1 and 0.

❏ Logic gates are the thousands of field-effect transistors that are incorporated into the computer circuitry. The FETs use the incoming voltage patterns to determine the pattern of pulses that leave the gate. The most common logic gates are NOT, AND, OR, NAND, NOR, and XOR gates.

❏ There are several types of memory chips used in the body computer: ROM, RAM, and PROM are the most common types.

❏ Electromagnetic interference (EMI) is produced as current in a conductor is turned on and off. EMI is also caused by static electricity that is created by friction.

❏ A diode is used as an electrical check valve that allows current to flow in one direction only.

❏ If the diode is forward-biased, there will be current flow. If the diode is reverse-biased, there will be no current flow.

❏ A transistor is a three-layer semiconductor that can be used as a very fast switching device or to control current flow.

❏ The three layers of the transistor are designated as emitter, collector, and base.

❏ A transistor has three operating conditions: cutoff, conduction, and saturation.

❏ An integrated circuit is a complex circuit of thousands of transistors, diodes, resistors, capacitors, and other electronic devices that are formed on a tiny silicon chip.

❏ Inputs provide the computer with system operation information or driver requests.

❏ Switches can be used as an input for any operation that only requires a yes-no or on-off condition.

❏ Sensors convert some measurement of vehicle operation into an electrical signal. There are many different designs of sensors: thermistors, Wheatstone bridge, potentiometers, magnetic pulse generator, and Hall-effect switches.

❏ Actuators are devices that perform the actual work commanded by the computer. They can be in the form of a motor, relay, switch, or solenoid.

❏ High-side drivers consist of a Metal Oxide Field Effect Transistor (MOSFET) that is controlled by a bipolar transistor.

❏ Multiplexing is a system in which electrical signals are transmitted by a peripheral serial bus instead of conventional wires. This allows several devices to share signals on a common conductor.

Review Questions

Short Answer Essays

1. What is binary code?
2. Describe the basics of NOT, AND, and OR logic gate operation.
3. List and describe the four basic functions of the computer.
4. What is the difference between ROM, RAM, and PROM?
5. Explain the principle of multiplexing.
6. How does the Hall-effect switch generate a voltage signal?
7. Describe the basic function of a stepper motor.
8. What is meant by feedback as it relates to computer control?
9. What is the difference between duty cycle and pulse width?
10. What are the purposes of the interface?

Fill-in-the-Blanks

1. In binary code, the number 4 is represented by _____ .
2. The _____ is a crystal that electrically vibrates when subjected to current at certain voltage levels.
3. _____ are registers designed to store the results of logic operations.
4. The _____ _____ _____ is the heart of the computer.
5. _____ contains specific data that pertains to the exact vehicle in which the computer is installed.
6. _____ convert some measurement of vehicle operation into an electrical signal.
7. Negative temperature coefficient (NTC) thermistors _____ their resistance as the temperature increases.
8. _____ switches operate on the principle that if a current is allowed to flow through thin conducting material exposed to a magnetic field, another voltage is produced.
9. Magnetic pulse generators use the principle of _____ _____ to produce a voltage signal.
10. _____ means that data concerning the effects of the computer's commands are fed back to the computer as an input signal.

Multiple Choice

1. *Technician A* says during the processing function, the computer uses input information and compares it to programmed instructions.
   Technician B says during the output function, the computer will put out control commands to various output devices.
   Who is correct?
   A. A only   **C.** Both A and B
   B. B only   **D.** Neither A nor B

2. *Technician A* says analog means the voltage signal is either on-off, yes-no, or high-low.
   Technician B says digital means the voltage signal is infinitely variable within a given range.
   Who is correct?
   A. A only   **C.** Both A and B
   B. B only   **D.** Neither A nor B

3. Logic gates are being discussed: *Technician A* says NOT gate operation is similar to that of two switches in series to a load.
   Technician B says an AND gate simply reverses binary 1s to 0s and vice versa.
   Who is correct?
   A. A only   **C.** Both A and B
   B. B only   **D.** Neither A nor B

4. Computer memory is being discussed: *Technician A* says ROM can be written to by the CPU.
   Technician B says RAM will store temporary information that can be read from or written to by the CPU.
   Who is correct?
   A. A only   **C.** Both A and B
   B. B only   **D.** Neither A nor B

5. *Technician A* says volatile RAM is erased when it is disconnected from its power source.
   Technician B says nonvolatile RAM will retain its memory if removed from its power source.
   Who is correct?
   A. A only   **C.** Both A and B
   B. B only   **D.** Neither A nor B

6. *Technician A* says EPROM memory is erased if the tape is removed and the microcircuit is exposed to ultraviolet light.
   Technician B says electrostatic discharge will destroy the memory chip.
   Who is correct?
   A. A only   **C.** Both A and B
   B. B only   **D.** Neither A nor B

7. *Technician A* says negative temperature coefficient thermistors reduce their resistance as the temperature decreases.
   Technician B says positive temperature coefficient thermistors increase their resistance as the temperature increases.
   Who is correct?
   A. A only   **C.** Both A and B
   B. B only   **D.** Neither A nor B

8. *Technician A* says magnetic pulse generators are commonly used to send data to the computer concerning the speed of the monitored component.
   Technician B says an on-off switch sends a digital signal to the computer.
   Who is correct?
   A. A only   **C.** Both A and B
   B. B only   **D.** Neither A nor B

9. Speed sensors are being discussed. *Technician A* says the timing disc is stationary and the pickup coil rotates in front of it.
   Technician B says the number of pulses produced per mile increases as rotational speed increases.
   Who is correct?
   A. A only   **C.** Both A and B
   B. B only   **D.** Neither A nor B

10. *Technician A* says a Hall-effect switch uses a steady supply current to generate a signal.
    Technician B says a Hall-effect switch consists of a permanent magnet wound with a wire coil.
    Who is correct?
    A. A only   **C.** Both A and B
    B. B only   **D.** Neither A nor B

Engine Operation and the Ignition System

Upon completion and review of this chapter, you should be able to:

❏ Describe the basic laws of physics involved with engine operation.

❏ Define the four-stroke cycle theory.

❏ Define mechanical, volumetric, and thermal efficiencies and describe factors that affect each.

❏ Describe the three major functions of an ignition system.

❏ Detail the operating conditions of an engine that affect ignition timing.

❏ Describe the function of the two major electrical circuits used in ignition systems and identify their common components.

❏ Describe the operation of ignition coils, spark plugs, and ignition cables.

❏ Describe the various types of spark timing systems, including electronic switching systems and their related engine position sensors.

❏ Describe the function of common sensors used in computer-controlled DI and EI systems.

❏ Explain spark control in computerized DI systems.

❏ Describe the difference between DI and EI systems.

❏ Explain spark control in common EI systems.

❏ Describe the advantages of electronic ignition (EI) systems.

❏ Describe the operation of common EI systems.

❏ Describe the coil secondary-to-spark plug wiring connections on an EI system, including an explanation of how the spark plugs fire.

❏ Explain the purpose of the camshaft sensor signal in an EI system.

❏ Describe the operation of coil-on-plug (COP) ignition systems

Introduction

Today's engines are designed to meet the demands of the automobile buying public along with many government-mandated emissions and fuel economy regulations. High performance, fuel economy, reduced emissions, and reliability are the focus of today's automotive engine manufacturers. To meet these demands, manufacturers are producing engines using lightweight blocks and cylinder heads, nontraditional materials such as powdered metals and composites, and computerized engine designs. Today's technician is called upon to properly diagnose and service these advanced engines.

The **internal combustion engines** that are used in automotive applications utilize several laws of physics and chemistry to operate. Although engine sizes, designs, and construction vary greatly, they all operate on the same basic principles. This chapter discusses these basic principles.

A properly operating ignition system is vital for optimum engine performance. This chapter will detail the operation of the ignition system. This system has evolved greatly from the early days of the automobile. Today all ignition systems are computer controlled.

In addition, proper understanding of engine operating principles is mandatory for diagnosing emission faults. The mechanical condition of the engine must be considered whenever an emission or driveability complaint is encountered. Refer to *Today's Technician: Engine Repair and Building* for complete coverage of this topic.

Internal combustion engines burn their fuels within the engine.

Engine Operation

Thermodynamics is the study of the relationship between heat energy and mechanical energy.

One of the many laws of physics utilized within the automotive engine is **thermodynamics**. The driving force of the engine is the expansion of gases. Gasoline (a liquid fuel) will change states to a gas if it is heated or burned. When it changes states, it also expands as the molecules of the gas collide with each other and bounce apart. Increasing the temperature of the gasoline molecules increases their speed of travel, causing more collisions and expansion.

Gasoline must be mixed with oxygen before it can burn. In addition, the air-fuel mixture must be burned in a confined area in order to produce sufficient power. Gasoline that is burned in an open container produces very little power, but if the same amount of fuel is burned in an enclosed container, it will expand with force.

Compressing the gasoline and air mixture within the combustion chamber generates heat. Igniting the compressed mixture causes the heat, pressure, and expansion to multiply. This process releases the energy of the gasoline so it can produce work. The igniting of the mixture is a controlled burn, not an explosion. The controlled combustion releases the fuel energy at a controlled rate in the form of heat energy. The heat, and consequential expansion of molecules, increases the pressure inside the combustion chamber. Typically, the pressure works on top of a piston that is connected to a crankshaft. As the piston is driven, it causes the crankshaft to rotate.

Torque is a rotating force around a pivot point.

The engine produces **torque** that is applied to the drive wheels. As the engine drives the wheels to move the vehicle, a certain amount of work is done. The rate of work being performed is measured in **horsepower**.

Horsepower is the measure of the rate of work.

Energy and Work

In engineering terms, in order to have work, there must be motion. Using this definition, work can be measured by combining distance and weight and can be expressed as foot-pound (ft.-lb.) or Newton-meter (Nm). These terms describe how much weight can be moved how far. A foot-pound is the amount of energy required to lift one pound of weight one foot in distance. The amount of work required to move a 500 pound weight 5 feet is 2,500 foot-pounds (3,390 Nm). In the metric system, the unit used to measure force is called Newton-meters (Nm). One foot pound is equal to 1.355 Nm. Torque is measured as the amount of force in newtons multiplied by the distance that the force acts in meters.

There is a distinction between foot-pounds and pound-foot. Pound-foot is used in measuring torque. However, the amount of force used is the same; only the direction of force is different.

Work is a unit of energy in that energy is the cause and work is the effect. Basically, energy is anything that is capable of resulting in motion. Common forms of energy include electrical, chemical, heat, radiant, and mechanical. Energy can be classified into two types: **potential energy** and **kinetic energy**.

Potential energy indicates energy that is available to be used for a purpose, but is not in use at this point in time.

Energy cannot be created or destroyed. However, it can be stored, controlled, and changed to other forms of energy. For example, the vehicle's battery stores chemical energy that is changed to electrical energy when a load is applied. An automotive engine transforms some form of energy into mechanical work.

Engine Cycles

Kinetic energy is energy that is in motion, it is also called working energy.

Most automotive engines are referred to as **reciprocating engines**. The in-line movement of the piston in the cylinder produces power. This linear motion is then converted to rotary motion by a crankshaft. Most automotive and truck engines are four-**stroke** cycle engines. For example, if the piston is at the top of its travel and then is moved to the bottom of its travel, one stroke has occurred. Another stroke occurs when the piston is moved from the bottom of its travel to the top again. In the four-stroke engine, four strokes are required to complete one **cycle**.

The internal combustion engine must draw in an air-fuel mixture, compress the mixture, ignite the mixture, and then expel the exhaust. This is accomplished in four piston strokes (Figure 4-1). The process of "drawing" in the air-fuel mixture is actually accomplished by atmospheric pressure pushing it into a low-pressure area created by the downward movement of the piston.

The first stroke of the cycle is the intake stroke (Figure 4-1a). As the piston moves down from **top dead center (TDC)**, the intake valve is opened so the vaporized air-fuel mixture can be pushed into the cylinder by atmospheric pressure. During this time the exhaust valve is closed. As the piston moves downward in its stroke, a vacuum is created (low pressure). Since high pressure seeks low pressure, the air-fuel mixture is pushed past the open intake valve and into the cylinder. After the piston reaches **bottom dead center (BDC)**, the intake valve is closed and the stroke is completed. Closing the intake valve after BDC allows for an additional amount of air-fuel mixture to enter the cylinder, increasing the **volumetric efficiency** of the engine. Even though the piston is at the end of its stroke and no more vacuum is created, the additional mixture enters the cylinder since it weighs more than air alone.

The compression stroke begins as the piston starts its travel back to TDC (Figure 4-1b). The intake and exhaust valves are both closed, trapping the air-fuel mixture in the combustion chamber area above the piston. The movement of the piston toward TDC compresses the mixture. As the molecules of the mixture are pressed tightly together, they begin to heat. When the piston reaches TDC, the mixture is fully compressed and a spark is induced in the cylinder by the ignition system. Compressing the mixture provides for improved burning and more intense combustion.

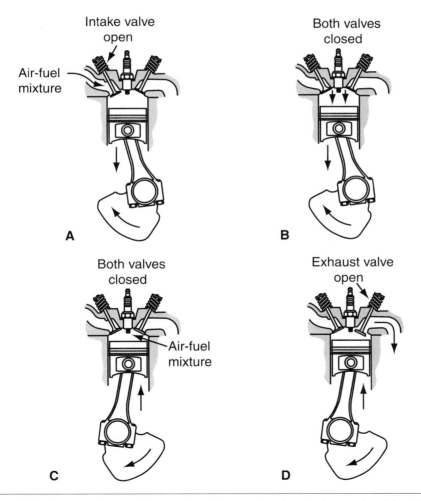

Figure 4-1   The four strokes of an automotive engine. (A) Intake stroke, (B) compression stroke, (C) power stroke, and (D) exhaust stroke.

Reciprocating engines get their name from the up-and-down, or back-and-forth, motion of the piston in the cylinder. This engine type is also referred to as the piston engine.

A **stroke** is the movement of the piston from one end of its travel to the other.

A **cycle** is a sequence that is repeated.

Top dead center (TDC) means the piston is at the very top of its stroke.

Bottom dead center (BDC) means the piston is at the very bottom of its stroke.

Volumetric efficiency is a measurement of the amount of air-fuel mixture that actually enters the combustion chamber compared to the total amount that could enter.

When the spark occurs in the compressed mixture, the rapid burning causes the molecules to expand, causing the beginning of the power stroke (Figure 4-1c). The expanding molecules create a pressure above the piston and push it downward. The downward movement of the piston in this stroke is the only time the engine is productive concerning power output. During the power stroke, the intake and exhaust valve remains closed.

The exhaust stroke of the cycle begins when the piston reaches BDC of the compression stroke (Figure 4-1d). Just prior to the piston reaching BDC, the exhaust valve is opened. The upward movement of the piston back toward TDC pushes out the exhaust gases from the cylinder past the exhaust valve and into the vehicle's exhaust system. A few degrees before the piston reaches TDC, the intake valve is opened. After the piston passes TDC, the exhaust valve is closed. The number of degrees of crankshaft revolution that the intake and exhaust valves are open is referred to as valve overlap. The cycle is then repeated again as the piston begins the intake stroke.

BIT OF HISTORY

Jean Joseph Lenoir created the first workable internal combustion engine in 1860. Dr. Nikolaus Otto designed the first successful four-stroke engine in 1866. All previous internal combustion engines did not compress the air-fuel mixture. They attempted to draw the mixture in during a downward movement of the piston and then ignite it. The expansion of the gases would force the piston down the remainder of its travel. This design was used in an attempt to make the pistons double acting (a power stroke each way). Dr. Otto's engine used the downward stroke to ingest the air-fuel mixture, then used an upward stroke to compress it. Many laughed at his idea since only one stroke was used to produce power. However, when we compare Otto's engine to other engines of that era we see that his was lighter, able to run almost twice as fast, and required only 7 percent of the cylinder displacement to produce the same amount of horsepower with the same amount of fuel. Both Lenoir's and Otto's engines were powered by cooking gas. In 1876, Dr. Otto successfully converted his engine to burn gasoline.

Engine Efficiency

Efficiency is a measure of a device's ability to convert energy into work.

Mechanical efficiency is a comparison of the power actually delivered by the crankshaft to the power developed within the cylinders at the same rpm.

There are several different measurements used to describe the **efficiency** of an engine. Efficiency is mathematically expressed as output divided by input. The terms used to define the input and output must be the same. Some of the most common efficiencies of concern are mechanical, volumetric, and thermal efficiencies.

Mechanical efficiency is a comparison between brake horsepower and indicated horsepower. In other words, it is a ratio of the power actually delivered by the crankshaft to the power developed within the cylinders at the same rpm. The formula used to calculate mechanical efficiency is:

Mechanical efficiency = brake horsepower / indicated horsepower

Ideally, mechanical efficiency would be 100 percent. However, the power delivered will always be less due to the power lost in overcoming friction between moving parts in the engine. The mechanical efficiency of a 4-stroke internal combustion engine is approximately 90 percent. If the engine's brake horsepower is 120 hp and the indicated horsepower is 140 hp, the mechanical efficiency of the engine is 85.7 percent.

Volumetric efficiency is a measurement of the amount of air-fuel mixture that actually enters the combustion chamber compared to the amount that could be ingested at that speed. As the volumetric efficiency increases, the power developed by the engine increases proportionately. Ideally, 100 percent volumetric efficiency is desired, but actual efficiency is reduced due to pumping losses. It must be realized that a given mass of air-fuel mixture occupies different volumes under different conditions. Atmospheric pressures and temperatures will affect the volume of the mixture.

Pumping losses are influenced by engine speed. For example, an engine may have a volumetric efficiency of 75 percent at 1,000 rpm and increase to an efficiency of 85 percent at 2,000 rpm, then drop to 60 percent at 3,000 rpm. As the engine speed increases, volumetric efficiency may drop to 50 percent.

If the airflow is drawn in at a low engine speed, the cylinders can be filled close to capacity. A required amount of time is needed for the airflow to pass through the intake manifold and pass the intake valve to fill the cylinder. As engine speed increases, the amount of time the intake valve is open is not long enough to allow the cylinder to be filled. Because of the effect of engine speed on volumetric efficiency, engine manufacturers use intake manifold design, valve port design, camshaft timing, and exhaust tuning to improve the engine's breathing. The use of turbochargers and superchargers (which increase the pressure in the engine's intake system) will bring volumetric efficiency over 100 percent.

Pumping losses are the result of restrictions to the flow of air-fuel into the engine. These restrictions are the result of intake manifold passages, throttle plate opening, valves, and cylinder head passages. There are some machining techniques used by builders of performance engines that will reduce these restrictions. In addition, the use of superchargers or turbochargers increases pressure in the engine's intake system and increases the volumetric efficiency.

Thermal efficiency is a measurement comparing the amount of energy present in a fuel and the actual energy output of the engine. Thermal efficiency is measured in British thermal units (Btu). The formula used to determine a gasoline engine's thermal efficiency has a couple of constants. First, one horsepower equals 42.4 Btus per minute. Second, gasoline has approximately 110,000 Btus per gallon. Knowing this, the formula is as follows:

Thermal efficiency = (bhp × 42.4 Bpmin) / (110,000 Bpg × gpmin)

Where: bhp = brake horsepower
Bpmin = Btu per minute
Bpg = Btu per gallon
gpmin = gallons used per minute

> **Thermal efficiency** is a measurement comparing the amount of energy present in a fuel and the actual energy output of the engine.

As a whole, gasoline engines waste about two-thirds of the heat energy available in gasoline. The engine's cooling system carries approximately one-third of the heat energy away. This is required to prevent the engine from overheating. Another third is lost in hot exhaust gases. The remaining third of heat energy is reduced by about 5 percent due to friction inside of the engine. Another 10 percent is lost due to friction in drivetrain components. Due to all of the losses of heat energy, only about 19 percent is actually applied to the driving wheels.

Of the 19 percent applied to the driving wheels, an additional amount is lost due to the rolling resistance of the tires against the road. Resistance to the vehicle moving through the air requires some more of this energy. By the time the vehicle is actually moving, the overall vehicle efficiency is about 15 percent.

Other Engine Designs

Throughout the history of the automobile, the gasoline 4-stroke internal combustion engine has been the standard. As technology improved, other engine designs have been used. This section of the chapter will provide an overview of these other engine designs.

Miller-Cycle Engine. Some late-model Mazda Millennias use this modification of the four-stroke engine. The Miller-cycle engine uses a supercharger to supply highly compressed air into the combustion chambers. The air first passes through an intercooler to cool it, then past the intake valves. The intake valves of the Miller-cycle engine remain open longer than that of the conventional four-stroke engine. This is done to prevent the upward movement of the piston from compressing the air-fuel charge in the combustion chamber until the piston has traveled 1/5 of its distance during the compression stroke. Since the intake valve is still open during this initial piston movement, the supercharger is able to force additional air into the combustion chamber. The greater volume of air in the combustion chamber results in a longer combustion time and a longer duration of downward forces on the piston during the power stroke. In addition, the shorter compression stroke results in cooler combustion chamber temperatures. This design allows the engine to produce a high horsepower and torque output even with a smaller engine displacement.

Stratified Charge Engine. The **stratified charge engine** uses a precombustion chamber (on top of the main cylinder) to ignite the main combustion chamber (Figure 4-2). The advantage of this system is increased fuel economy and reduced emission levels.

The air-fuel mixture is stratified to produce a small, rich mixture at the spark plug while providing a lean mixture to the main chamber. During the intake stroke, a very lean air-fuel mixture enters the main combustion chamber. At the same time, a very rich mixture is pushed past the auxiliary intake valve and into the precombustion chamber (Figure 4-3). At the completion of the compression stroke, the spark plug fires to ignite the rich mixture in the precombustion chamber. The burning rich mixture will then ignite the lean mixture in the main combustion chamber. The result is an engine that can run efficiently on an air-fuel ratio of 20:1.

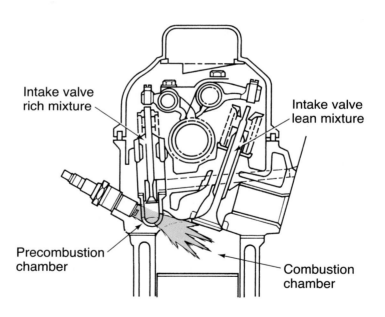

Intake valve rich mixture

Intake valve lean mixture

Precombustion chamber

Combustion chamber

Figure 4-2 Stratified engine design.

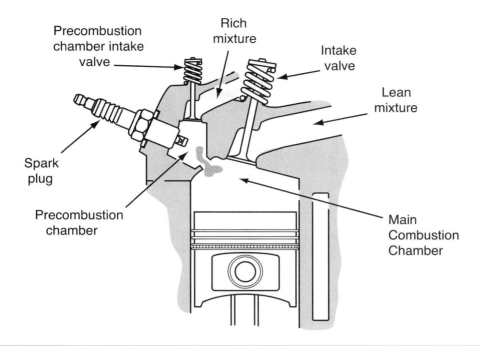

Figure 4-3 The stratified engine uses the rich mixture in the precombustion chamber to ignite the lean mixture in the main chamber.

AUTHOR'S NOTE: Honda's CVCC engine, released in the early 1980s models of Civic, Accord, and City, is a form of stratified charge engine. The CVCC system had conventional inlet and exhaust valves, and a third, supplementary inlet valve, which charged a volume around the spark plug. The spark plug and CVCC inlet was isolated from the main cylinder by a perforated metal plate; upon ignition, flame fronts shot into the very lean main charge through these perforations, ensuring complete ignition.

Another design uses a combustion chamber in the shape of a lazy 8 (Figure 4-4). This makes it possible to create two axial vortices (swirls) of similar intensity but in opposite direc-

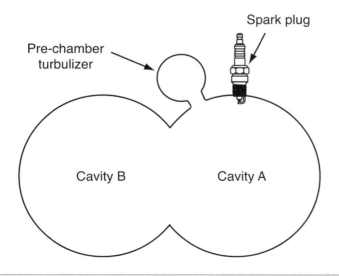

Figure 4-4 The two-stage combustion chamber.

tions of rotation. The swirl in combustion chamber cavity A contains a rich mixture. The spark plug (1) and pre-chamber turbulizer (2) are located close to this area. The swirl in combustion chamber cavity B consists of air. The equality of torque of these swirls doesn't allow mixing of the rich charge and the air until the combustion process starts. This makes it possible to formulate a deeply stratified charge in the combustion chamber. At all loading conditions, the rich mixture is ignited around the spark plug. After ignition, the flame moves to the center of the combustion chamber by swirls in cavity A, ensuring the burning of the rich mixture. As a consequence of the combustion of the rich charge, the pressure increases and the flame is transferred from cavity A to cavity B (Figure 4-5). Running in tangential direction to swirls A and B, initially the flame of burning gases turbulizes and ignites the rich layer of the stratified mixture (first stage of the combustion process), then it transfers unburned products to cavity B by gasodynamical process. After this, the afterburning of partially reacted products of the rich charge take place in cavity B (second stage of the combustion process).

The turbulizer has two main functions. First, it accelerates the burning of the rich charge in cavity A by generating turbulence. The second function is to transport partially burned products to the air zone (cavity B) for completion of the combustion process. The two-stage combustion process of air-fuel mixture is achieved first in the rich part (cavity A), in the main part of the central zone of the combustion chamber (torch action zone), and then in the lean burn zone (cavity B).

The Japanese automakers are continuing to research the stratified charge engine and design a gasoline direct injection (GDI) system. Mitsubishi's development of its GDI engine provides a fuel and induction system that can create a precisely layered air-fuel mixture. Volvo has teamed

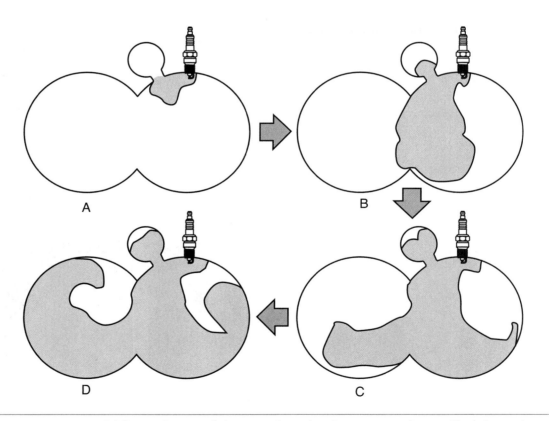

Figure 4-5 Schlieren pictures of the staged combustion process of a stratified charge in the oppositely rotated swirls. It can be clearly seen that the flame first propagates inside the cavity occupied with rich mixture and is then transferred to the other cavity.

with Mitsubishi to use this engine in their vehicles. The benefits of the GDI stratified-charge engines include:

- *Lower exhaust emissions.* Stratified-charge engines tend to burn in a more progressive fashion. End gases are ignitable, resulting in a 10 percent improvement in vehicle emissions.

- *Resistance to pre-ignition.* Even with an extremely lean air-fuel mixture (20:1), pre-ignition can occur. Since a homogeneous mixture is impossible to achieve in spark ignition gasoline engines, engineers select port, valve, and piston crown designs that will reliably structure the air-fuel intake charge to prevent auto ignition of end gases.

- *Improved throttle response.* A direct injection, stratified-charge engine does not use a throttle plate. Engine output is determined by fuel flow (not airflow as is the case in conventional spark ignition engines) which can be delivered in a more progressive and efficient manner.

- *Improved fuel economy.* A stratified-charge engine's ability to safely tolerate lean mixtures can translate into a gain in fuel economy.

Ignition Systems

One of the requirements for an efficient engine is the correct amount of heat delivered into the cylinders at the right time. This is the responsibility of the ignition system. The ignition system supplies properly timed high-voltage surges to the spark plugs. These voltage surges cause an arc across the electrodes of a spark plug, and this heat begins the combustion process inside the cylinder. For each cylinder in an engine, the ignition system has three main jobs. First, it must generate an electrical spark that has enough heat to ignite the air-fuel mixture in the combustion chamber. Second, it must maintain that spark long enough to allow for the combustion of all the air and fuel in the cylinder. Third, it must deliver the spark to each cylinder so that combustion can begin at the right time during the compression stroke.

When the combustion process is completed, a very high pressure is exerted against the top of the piston. This pressure pushes the piston down into its power stroke. This pressure is the force that gives the engine power. In order for an engine to maximize efficiency, the spark to the combustion chamber must be timed to deliver peak combustion pressures when the piston reaches between 10 to 23 degrees after top dead center (ATDC).

> **AUTHOR'S NOTE:** The amount of degrees ATDC that peak combustion pressures are actually obtained depends on combustion chamber design and other factors. Manufacturers will determine when this pressure is to be achieved, and it will differ between designs. In our discussion of ignition systems, we will refer to 23 degrees ATDC since this is the latest point that peak pressures can be obtained for the engine to run efficiently.

Because combustion of the air-fuel mixture within a cylinder takes a short period of time, usually measured in thousandths of a second (milliseconds), the combustion process must begin before the piston is in its power stroke. Therefore, the delivery of the spark must be timed to arrive at some point before the piston reaches top dead center (TDC) on the compression stroke.

Determining how much before TDC the spark should begin gets complicated by the fact that the speed of the piston increases (as it moves from its compression stroke to its power stroke); however, the time needed for combustion remains fairly constant. This means that the

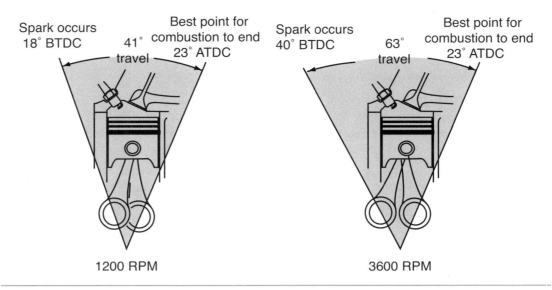

Spark occurs
18° BTDC

Best point for
combustion to end
23° ATDC

41°
travel

1200 RPM

Spark occurs
40° BTDC

Best point for
combustion to end
23° ATDC

63°
travel

3600 RPM

Figure 4-6 As engine speed increases, ignition must begin earlier to end by 23 degrees ATDC.

spark should be delivered earlier as the engine's speed increases (Figure 4-6). However, as the engine has to provide more power to do more work, the load on the crankshaft tends to slow down the acceleration of the piston and, as a result of the slower piston speed, the spark needs to be somewhat delayed. So, based on speed and load, spark timing is balanced to provide optimum engine performance.

Calculating when the spark should begin gets more complicated with the fact that the rate of combustion varies according to certain factors. Higher compression pressures tend to accelerate the combustion process. Higher octane gasoline ignites less easily and requires more burning time. Increased vaporization and turbulence tend to decrease combustion times. Other factors, including intake air temperature, humidity, and barometric pressure, also affect combustion. Because of all of these complications, delivering the spark at the right time is a difficult task.

To fully realize the complexity involved with delivering the spark at the correct time, consider the following. If a V8 engine is rotating at 3,000 revolutions per minute (rpm) and the ignition system must fire 4 spark plugs per revolution, the ignition system must supply 12,000 sparks per minute. These plug firings must also occur at the proper instant, without misfiring. If the ignition system misfires or does not fire the spark plugs at the proper time, fuel economy, engine performance, and emission levels are adversely affected.

Basic Circuitry

All ignition systems consist of two interconnected electrical circuits (Figure 4-7): a **primary circuit** (low voltage) and a **secondary circuit** (high voltage).

Depending on the type of ignition system, components in the primary circuit may include the following:
- battery
- ignition switch
- ballast resistor or resistance wire (some systems)
- starting bypass (some systems)
- ignition coil primary winding
- triggering device
- switching device or control module
- ground

Shop Manual
Chapter 4,
page 127

All the components that regulate the current in the coil primary windings are referred to as the **primary circuit**. This would include the ignition switch, primary coil windings, the switching device, and all conductors that connect these components.

The **secondary circuit** carried high voltage to the combustion chamber.

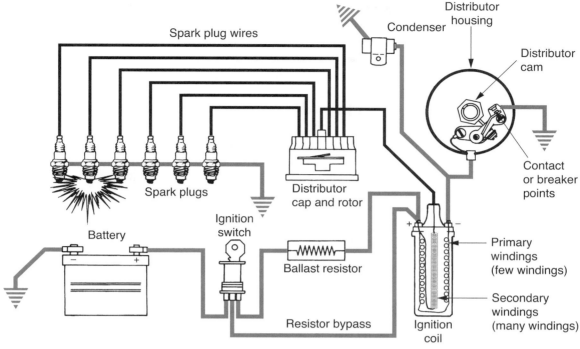

PRIMARY CIRCUIT ━━━━━
SECONDARY CIRCUIT ━━━━━

Spark plug wires

Condenser

Distributor housing

Distributor cam

Contact or breaker points

Spark plugs

Distributor cap and rotor

Ignition switch

Battery

Ballast resistor

Primary windings (few windings)

Secondary windings (many windings)

Resistor bypass

Ignition coil

Figure 4-7   The ignition switch is two circuits. The primary circuit carries the low voltage, while the secondary circuit carries the high voltage.

The secondary circuit may include these components:
- ignition coil secondary winding
- distributor cap and rotor (some systems)
- ignition (spark plug) cables
- spark plugs

Primary Circuit Operation.   When the ignition switch is in the RUN position, current from the battery flows to the primary winding of the ignition coil. This current may be provided from the ignition switch or by the closed contacts of a control relay. From there, the current passes through some type of switching device and to ground. The switching device can be mechanically or electronically controlled by a triggering device. The current flow in the ignition coil's primary winding creates a magnetic field. The switching device interrupts this current flow at predetermined times. When current flow is interrupted, the magnetic field in the primary winding collapses. This collapse induces a high-voltage surge into the secondary winding of the ignition coil. The secondary circuit of the system begins at this point. In order to have electromagnetic induction, three factors are required:

1. Magnetic field
2. Conductor
3. Motion

Secondary Circuit Operation.   The exact manner in which the secondary circuit delivers the high-voltage surges to the combustion chamber depends on the system design. Most ignition systems in use until 1984 used some type of **distributor** to accomplish this job. However, in an effort to reduce emissions, improve fuel economy, and boost component reliability, most auto manufacturers are now using distributorless or **electronic ignition (EI)** systems.

The **distributor** controls the primary circuit and is responsible for distributing the secondary spark to the correct combustion camber.

In the SAE J1930 terminology, the term **electronic ignition (EI)** replaces all previous terms for distributorless ignition systems.

In a **distributor ignition (DI)** system, high voltage from the secondary winding passes through an ignition cable running from the coil to the distributor. The distributor then distributes the high voltage to the individual spark plugs through a set of ignition cables. The cables are arranged in the distributor cap according to the **firing order** of the engine. A **rotor**, which is driven by the distributor shaft, rotates and completes the electrical path from the secondary winding of the coil to the individual spark plugs. The distributor delivers the spark to match the compression stroke of the piston. The distributor assembly may also have the capability of advancing or retarding ignition timing.

The **distributor cap** is mounted on top of the distributor assembly and an alignment notch in the cap fits over a matching lug on the housing. The cap can only be installed in one position, thereby assuring the correct firing sequence.

The rotor is positioned on top of the distributor shaft, and a projection inside the rotor fits into a slot in the shaft. This allows the rotor to be installed in only one position. A metal strip on the top of the rotor makes contact with the center distributor cap terminal, and the outer end of the strip rotates past the cap terminals. This action completes the circuit between the ignition coil and the individual spark plugs according to the firing order.

Ignition Components

All ignition systems share a number of common components. Some, such as the battery and ignition switch, perform simple functions. The battery supplies battery voltage to the ignition primary circuit. Current flows when the ignition switch is in the START or the RUN position.

Ignition Coils.   The ignition coil (Figure 4-8) is the heart of the ignition system. It works in the manner of a **pulse transformer** that builds up the low battery voltage of approximately 12.6 volts to a voltage that is high enough to **ionize** (electrically charge) the spark plug gap and ignite

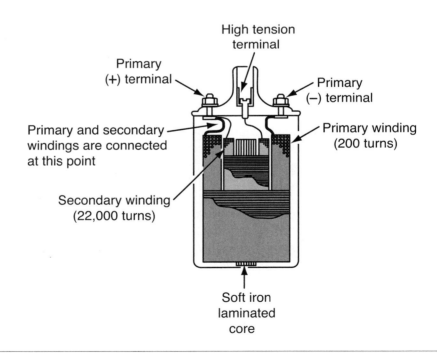

Figure 4-8   Cross-section view of a typical ignition coil.

In the SAE J1930 terminology, the term **distributor ignition (DI)** replaces all previous terms for electronically controlled distributor-type ignition systems.

Firing order is the sequence in which the spark plugs are fired.

The **rotor** is attached to the top of the distributor shaft and directs secondary voltage from the coil to the terminal of the distributor cap.

The **distributor cap** is fastened to the distributor and has a terminal in the center to receive the secondary voltage from the coil and a terminal tower for each cylinder to send the spark to the spark plug.

Shop Manual
Chapter 4,
page 143

A **pulse transformer** steps up low-voltage pulses to a higher voltage value by using induction principles.

Ionize means to electrically charge.

the air-fuel mixture. The coil is capable of producing approximately 30,000 to 60,000 volts. However, how high a voltage is produced is dependent on many factors. The coil will produce only enough voltage required to overcome these factors: plug gap, air-fuel ratio, plug wire resistance, engine speed, and compression ratio. The margin of voltage which can be produced above that which is required to fire the spark plug represents the **reserve voltage** (electrical reserve) built into the ignition system. The typical amount of secondary coil voltage required to jump the spark plug gap is approximately 10,000 volts. Most coils used in basic ignition systems have a maximum secondary voltage of 25,000 volts. This reserve voltage is necessary to compensate for high cylinder pressures and increased secondary resistances as the spark plug gap increases through use. The maximum available voltage must always exceed the required firing voltage or ignition misfire will occur. If there is an insufficient amount of voltage available to ionize the spark plug gap, the spark plug will not fire.

The center of the ignition coil contains a core of laminated soft iron or steel. Adding a core to a coil increases the magnetic strength of the coil. The core is surrounded by approximately 22,000 turns of very fine wire. This winding is called the **secondary coil windings**. Surrounding the secondary coil windings is the **primary coil windings**. The primary windings are made of approximately 200 turns of 20-gauge wire. Because of the build up of heat through the windings, due to resistance to current flow, some coils contain oil or tar to help cool them. If oil or tar is not used then an air-cooled, epoxy-sealed E coil is used. It is called an **E coil** because of the shape of the core is in the figure of an E.

The primary windings extend through the case of the coil and are identified by some form of marking. The markings differ from manufacturer to manufacturer. Some of the most common markings are: BAT and DIST, or positive (battery) and negative (distributor), or + and −. The naming of the terminals as positive and negative refers to the most positive and the least positive terminals. These labels assure that the polarity through the coil is correct. The plug polarity is center pole negative. Negative plug polarity requires less voltage to ionize the plug then does positive polarity. This is because of the relative temperatures between the center electrode (hot) and the ground electrode (cold). Electrical current will more easily jump from hot to cold than from cold to hot.

> **AUTHOR'S NOTE:** If the primary wires are incorrectly connected to the coil, the coil will have reverse polarity and the voltage required to fire the plug will be increased by 40 percent. This may cause poor engine performance and a high speed miss.

The negative terminal is attached to a switching device (either mechanical or electronic) that opens and closes the primary circuit. With the ignition switch in the RUN position (or control relay energized), voltage should be present at both sides of the primary windings. Once the engine is rotating, the switching device will complete and break the primary circuit to ground. When the circuit is grounded, current starts to flow into a coil winding, resulting in an opposing current (**inductive reactance**) being created in the windings of the coil. Inductive reactance is similar to resistance since it resists any increase in current flow in a coil. This results in a slow buildup to maximum magnetic field strength or **saturation**.

Since there are two windings within the ignition coil, whenever there is a change in the magnetic field of one winding, it affects the other winding. If the current to the primary winding is stopped, then the magnetic field collapses and cuts across the secondary winding. This creates a high voltage in the secondary winding. Also there is a buildup of about 250 volts within the pri-

---

The difference between the required voltage and the maximum available voltage is referred to as secondary **reserve voltage**.

The **secondary coil windings** are one of two windings in the ignition coil. This winding has several thousand turns and is the location low voltage will be transformed to a high voltage.

The **primary coil windings** are the second set of winding in the ignition coil. The primary winds will have about 200 turns to create a magnetic field to induce voltage into the secondary winding.

Because of the buildup of heat through the windings due to resistance to current flow, some coils contain oil to help cool them. If oil is not used, then an air-cooled, epoxy-sealed **E coil** is used.

Opposing current as the result of self-induction is called **inductive reactance**. Inductive reactance is similar to resistance since it resists any increase in current flow in a coil.

The point where there can be no more buildup of magnetic strength is referred to as **saturation**.

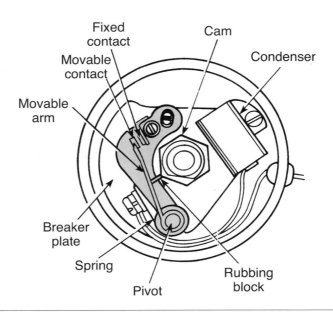

Figure 4-9 Older ignition systems used breaker points as the switching device in the primary circuit.

Mutual induction is an induction of voltage in an adjacent coil by changing current in a primary coil.

mary winding. This process of creating high voltage in the secondary windings by changing the magnetic field in the primary windings is called **mutual induction**.

Figure 4-9 illustrates a basic ignition system that uses mechanical contact points for the switching device to control current flow in the primary windings. When the switch is closed, current flowing through the primary windings of the coil generates a high magnetic field inside the coil. When the switch opens the primary winding's return path, the magnetic field collapses and induces a high voltage current into the secondary windings. This high voltage is then directed to the correct spark plug to ignite the air-fuel mixture in that cylinder. For each spark plug, the coil must be charged then discharged.

Shop Manual
Chapter 4,
page 162

A BIT OF HISTORY

A turning point in the automotive industry came in 1902 when the Humber, an automobile from England, used a magnetoelectric ignition system.

The **spark plug** is designed to transfer the high-voltage current to the combustion chamber, where the electrical spark that is produced is the motivating force behind establishing combustion of the air-fuel mixture.

The number of ignition coils used in an ignition system varies with the type of ignition system found on a vehicle. On most ignition systems with a distributor, only one ignition coil is used. The high voltage of the secondary winding is directed by the distributor to the various spark plugs in the system. Therefore, there is one secondary circuit with a continually changing path. Distributorless ignition systems have several secondary circuits, each with an unchanging path.

Spark Plugs. For ignition to start in a gasoline engine, a spark is delivered to the combustion chamber. In order to produce this spark, a gap (between .020 and .080 inch) is established by the **spark plug** that must be bridged by the high-voltage current (Figure 4-10).

Because of the environment the spark plug is subject to, it is exposed to more stress than most other components in the engine. The spark plug must deliver high voltage, with split second accuracy, thousands of times per second, in temperatures that can exceed 2,500° F (1370° C). Also,

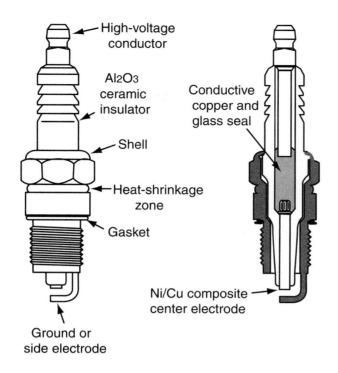

High-voltage conductor

Al$_2$O$_3$ ceramic insulator

Shell

Heat-shrinkage zone

Gasket

Ground or side electrode

Conductive copper and glass seal

Ni/Cu composite center electrode

Figure 4-10 Construction of a typical spark plug.

the spark plug must be designed to seal the cylinder from compression loss. To perform this task, there are four important aspects to the spark plug:

1. *Thread diameter.* The spark plug is manufactured with one of the following thread sizes—10, 12, 14, or 18 mm diameters. The most common sizes for automotive use are 14 mm and 18 mm.

2. *Reach.* It is important that the proper **reach** be used when replacing a plug (Figure 4-11). A spark plug must reach into the combustion chamber far enough so that the spark gap will be properly positioned in the combustion chamber without interfering with the turbulence of the air-fuel mixture or reducing combustion action. If the reach is too short, the engine could ping (pre-ignition) as a result of carbon buildup on the exposed cylinder head threads. Also the shrouded plug gap may cause a misfire. If the reach is too long, the piston may contact the spark plug and cause severe damage. In addition, the air-fuel mixture may pre-ignite due to overheating of the exposed threads.

Spark plug **reach** is the distance between the end of the spark plug threads and the seat or sealing surface of the plug. Plug reach determines how far the plug reaches through the cylinder head.

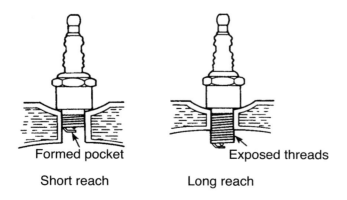

Formed pocket

Short reach

Exposed threads

Long reach

Figure 4-11 Spark plug reach determines the location of the electrode in the combustion chamber.

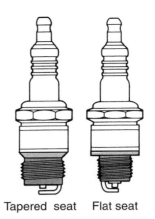

Tapered seat   Flat seat

Figure 4-12   Spark plug seats are either tapered or flat.

3. *Method of sealing.* There are two methods used to seal the spark plug in the cylinder head (Figure 4-12). The first is the use of a compressible gasket against a flat surface. The second uses a tapered seat that provides for a wedge seal between the head and the spark plug. The two methods of sealing are not interchangeable.

4. *Heat range.* The spark plug must be able to retain some of the heat developed during combustion to burn away carbon and oil deposits that form on the electrode. If the **heat range** of the spark plug is too low, these deposits will accumulate and foul the spark plug, causing the engine to misfire. If the heat range is too high, the insulator tip will become overheated to the point of igniting the air-fuel mixture prematurely. The heat range of a spark plug is determined by how far the center insulator goes up into the shell around the center electrode (Figure 4-13). The greater the distance, the longer it will take to dissipate heat and the hotter the plug will get. Whenever replacing a spark plug, it is important that only the heat range designated by the manufacturer be installed. There are many operational factors that affect the temperature at which the spark plug must work (Figure 4-14). Insulator tip temperature is shown on the left axis of each chart while the different factors are listed on the bottom axis.

The **heat range** of the spark plug refers to its heat dissipating properties.

Spark plugs are often referred to as "hot" or "cold" plugs based on their heat range.

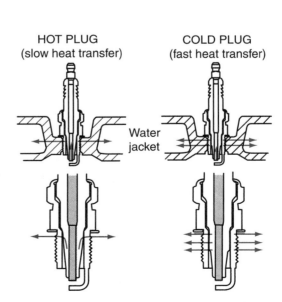

Figure 4-13   The heat range is determined by how far the center insulator goes up into the shell. The longer the distance, the hotter the plug.

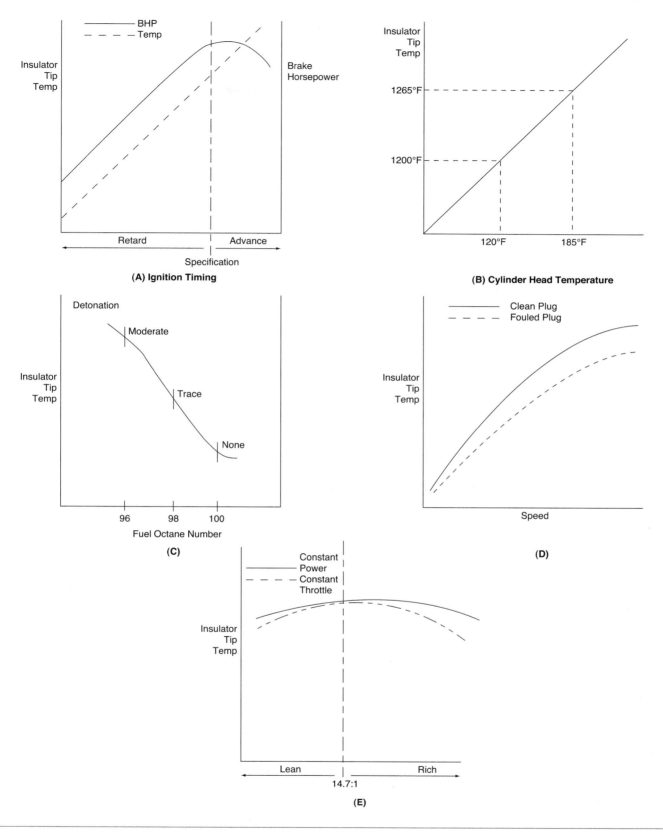

Figure 4-14 Factors that affect the spark plug include (A) timing advance, (B) engine coolant temperature, (C) detonation, (D) spark plug condition, and (E) air-fuel ratio.

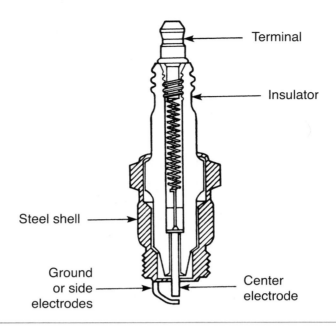

Figure 4-15 The internal resistor reduces the amount of current flow through the secondary circuit. This in turn will reduce radio interference caused by the plug firing.

Resistor plugs are spark plugs with resistors built into the electrode core.

Most spark plugs are classified as **resistor plugs** (Figure 4-15). The purpose of this resistor is to reduce the electromagnetic radiation created within the ignition system. If this radiation is not reduced by use of resistor plugs or resistive plug wires, then radio and television interference would result. In addition, the extra resistance increases spark plug life by reducing current.

BIT OF HISTORY

Louis S. Clark of the Auto Car Company designed porcelain spark plug insulation in 1902.

Shop Manual
Chapter 4,
page 128

The ribs inside of the distributor cap are called "antiflashover ribs."

Distributor Caps and Rotors.   The distributor cap and rotor distribute the high voltage that is created in the secondary windings of the coil to the correct spark plug (Figure 4-16). These units are manufactured from Bakelite, alkyd plastic, phenol resin, or fiberglass-reinforced polyester resin plastic. Inside the cap are ribs between the side tower inserts that prevent sparks from arcing to the wrong tower insert.

The high voltage created in the coil when the switching device opens is sent to the center tower of the distributor cap through the secondary cable. The high voltage is then sent to inside the cap and to the rotor. The rotor is turned by the distributor shaft. The high voltage is then directed to each cylinder according to the firing order.

AUTHOR'S NOTE: An air gap of a few thousandths of an inch exists between the tip of the rotor electrode and the spark plug electrode inside the cap. This gap is necessary in order to prevent the two electrodes from making contact. If they did make contact, both would wear out rapidly. This gap cannot be measured when the distributor is assembled; therefore, the gap is usually described in terms of the voltage needed to create an arc between the electrodes.

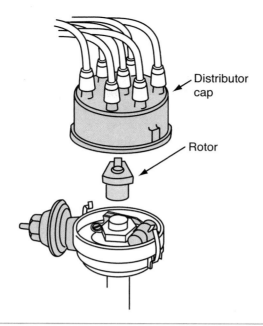

Figure 4-16   Typical distributor cap and rotor.

Shop Manual
Chapter 4,
page 127

Ignition Cables.   Ignition cables make up the secondary wiring. The cables are not solid wire; rather, they contain carbon-impregnated fiber cores that act as resistors in the secondary circuit (Figure 4-17). They eliminate radio and television interference, increase firing voltages, and reduce spark plug wear by decreasing current. Metal terminals on each end of the spark plug

Ignition cables
carry the high
voltage from the
distributor or the
multiple coils (EI
systems) to the spark
plugs.

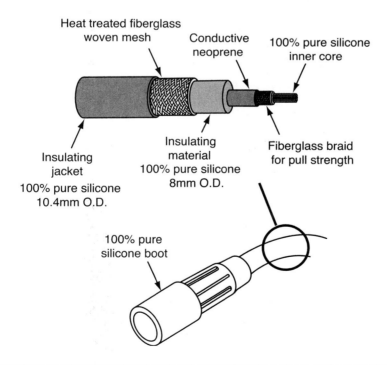

Figure 4-17   Construction of a typical ignition wire.

wires contact the spark plug and the distributor cap terminals. Insulated boots on the ends of the cables strengthen the connections as well as prevent dust and water infiltration and voltage loss.

Shop Manual
Chapter 4,
page 158

Ignition Timing

The secondary ignition spark is fired when the coil primary circuit is opened. As the piston approaches TDC on the compression stroke, the air-fuel mixture in the cylinder is compressed and ready to be ignited. The precise **ignition timing** for the ignition of the compressed air-fuel mixture is critical for maximum power, increased fuel economy, reduced emissions, and reduced engine wear. Ignition timing is specified by referring to the position of the number one piston in relationship to crankshaft rotation. Ignition timing reference marks can be located on a pulley or flywheel to indicate the position of the number one piston (Figure 4-18).

AUTHOR'S NOTE: If the spark is sent at the time that the piston is at TDC, it is called zero degrees advance. If the spark is sent to the cylinder after the piston has passed TDC, it is called firing after top dead center (ATDC). Most engines are designed to receive the spark as the piston approaches TDC. This allows the air-fuel mixture to burn as the piston starts its downward stroke (power stroke). When the spark is sent as the piston approaches TDC it is called before top dead center (BTDC).

Ignition timing refers to the precise time a spark is sent to the cylinder relative to the piston position.

A spark that is delivered late as compared to piston travel is called **retarded timing**.

A spark sent earlier than it is required is called **advanced timing**.

Retarded timing results in the spark arriving too late. For example, if the specification is 12 degrees BTDC but the spark is sent at 8 degrees BTDC, the spark is slow to get to the cylinder in reference to piston position. **Advanced timing** results in the spark arriving too soon, for

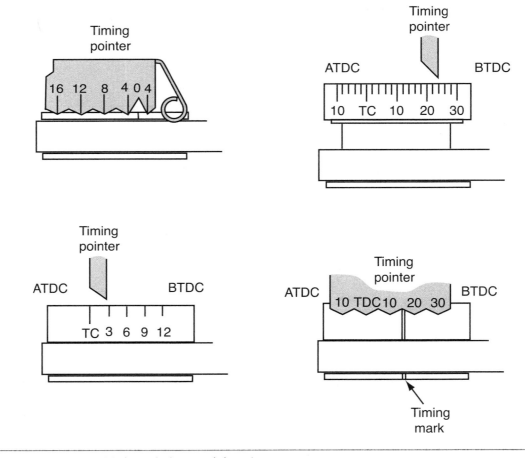

Figure 4-18 Various timing mark locations.

example, if the spark is sent at 10 degrees BTDC when the specification requires 6 degrees. In this case, the spark is too early into the cylinder in reference to piston position.

If the spark timing is overly advanced, detonation may be experienced. Too much spark advance ignites the spark plug too soon, resulting in an increase in pressure. Because the pressure spike from detonation is so severe and of very short duration, it can actually shock the boundary layer of gas that surrounds the piston. This thin layer isolates the flame and causes it to be quenched as the flame approaches the cooler piston. This combination of actions will normally protect the piston and chamber from absorbing all of the heat of combustion. However, under extreme conditions, the shock wave from the detonation spike can cause that boundary layer to break down, which then allows excessive heat transfer into the piston and combustion chamber surfaces.

Detonation can be caused by other influences such as low-octane fuel, carbon buildup, high engine temperatures, or a faulty EGR system. However, check and adjust the ignition initial or **base timing** before condemning any of the other possibilities.

If optimum engine performance is to be maintained, the ignition timing of the engine must change as the operating conditions of the engine change. All of the different operating conditions affect the speed of the engine and the load on the engine. All ignition timing changes are made in response to these primary factors.

Engine Speed.   At higher engine speeds, the crankshaft turns through more degrees in a given period of time. The time that is required to burn the air-fuel mixture in the cylinder is about 2 to 3 milliseconds. This requires that the air-fuel mixture burn to be completed by 23 degrees ATDC on the power stroke of the piston. As engine speed increases, the timing of the spark must be advanced in order to complete the burn by 23 degrees ATDC.

Engine Load.   The load on an engine is related to the work it must do. Driving up hills or pulling extra weight increases engine load. Under load, the pistons accelerate more slowly and the engine runs less efficiently. A good indication of engine load is the amount of vacuum in the intake manifold. Under light loads and with the throttle plate(s) partially opened, a high vacuum exists in the intake manifold. The amount of air-fuel mixture drawn into the manifold and cylinders is small. This means the air and fuel molecules are relatively far apart. On compression, this thin mixture produces less combustion pressure, and combustion time is slow. To complete combustion by 23 degrees ATDC, ignition timing must be advanced to allow for a longer burning time. Advancing the spark timing under light engine loads results in better fuel economy and reduced emissions because the fuel burns more completely.

Under heavy loads, when the throttle is opened fully, a larger mass of the air-fuel mixture can be drawn in, and the vacuum in the manifold is low. Combustion is fast because the air and fuel molecules are close together. The fast burn results in high combustion pressures. In such a case, the ignition timing must be advanced less, or retarded, to prevent complete burning from occurring before 23 degrees ATDC.

AUTHOR'S NOTE: Earlier, it was stated that combustion time is fairly consistent. In our thought of time there is not much difference between 2 ms and 3 ms, however, in relationship to piston travel 1 ms time difference can affect combustion quality.

A BIT OF HISTORY

The first use of automatic spark advance occurred in 1900. In spite of this development, many cars, for many years, had driver-operated timing advance controls.

Normal combustion temperatures exceed 1,800°F. An aluminum piston that is subjected to this temperature would melt. However, in an engine the piston does not melt due to thermal inertia and to the boundary layer of a few molecules thick next to the piston top.

The timing that is set when the engine is at curb idle is called initial or **base timing**.

According to the SAE, **rise time** is the amount of time (measured in microseconds) for the output of the coil to rise from 10 percent to 90 percent of its maximum output.

Rise Time.   The faster the **rise time**, the easier it is to fire a fouled plug. An ignition spark that reaches its maximum voltage too quickly has no time to flow through fouled spark plug deposits to ground. Typical rise times for various ignition systems are:

1. Point type: 200 microseconds (μs)
2. Electronic ignition: 20 to 50 μs
3. Capacitor discharge: 1 to 3 μs.

Firing Order.   Up to this point, the primary focus of discussion has been ignition timing as it relates to any one cylinder. However, the function of the ignition system extends beyond timing the arrival of a spark to a single cylinder. It must perform this task for each cylinder of the engine in a specific sequence.

Each cylinder of an engine produces power once every 720 degrees of crankshaft rotation. Each cylinder must have a power stroke at its own appropriate time during the rotation. To make this possible, the pistons and rods are arranged in a precise fashion. This is called the engine's firing order. The firing order is arranged to reduce rocking and imbalance problems. Because the potential for this rocking is determined by the design and construction of the engine, the firing order varies from engine to engine. Vehicle manufacturers simplify identifying each cylinder by numbering them (Figure 4-19). Regardless of the particular firing order used, the number one cylinder always starts the firing order, with the rest of the cylinders following in a fixed sequence.

COMMON CYLINDER NUMBERING AND FIRING ORDER

IN-LINE

4-Cylinder	6-Cylinder
① ② ③ ④	① ② ③ ④ ⑤ ⑥
Firing Order 1–3–4–2	Firing 1–5–3–6–2–4
1–2–4–3	Order

V CONFIGURATION

V6	V8
⑤ ③ ① Right Bank	① ② ③ ④ Right Bank
⑥ ④ ② Left Bank	⑤ ⑥ ⑦ ⑧ Left Bank
Firing 1–4–5–2–3–6	Firing 1–5–4–8–6–3–7–2
Order	Order
② ④ ⑥ Right Bank	① ② ③ ④ Right Bank
① ③ ⑤ Left Bank	⑤ ⑥ ⑦ ⑧ Left Bank
Firing 1–6–5–4–3–2	Firing 1–5–4–2–6–3–7–8
Order	Order
① ② ③ Right Bank	② ④ ⑥ ⑧ Right Bank
④ ⑤ ⑥ Left Bank	① ③ ⑤ ⑦ Left Bank
Firing 1–2–3–4–5–6	Firing 1–8–4–3–6–5–7–2
Order	Order
① ② ③ Right Bank	② ④ ⑥ ⑧ Right Bank
④ ⑤ ⑥ Left Bank	① ③ ⑤ ⑦ Left Bank
Firing 1–4–2–3–5–6	Firing 1–8–7–2–6–5–4–3
Order	Order

Figure 4-19   Cylinder numbering and firing orders.

The ignition system must be able to monitor the rotation of the crankshaft and the relative position of each piston in order to determine which piston is on its compression stroke. It must also be able to deliver a high-voltage surge to each cylinder at the proper time during its compression stroke. How the ignition system does these things depends on the design of the system and is discussed later in this chapter.

A BIT OF HISTORY

The man credited with the development of the ignition system, as we know it today, was Charles Kettering. The contact point ignition system is also known as Kettering ignition. He is also credited with the development of electric starters for Cadillac in 1911, quick drying paint, and tetraethyl leaded gasoline.

Solid State Ignition

From the fully mechanical breaker point system, ignition technology progressed to solid state ignitions (Figure 4-20). Breaker points were replaced with electronic triggering and switching

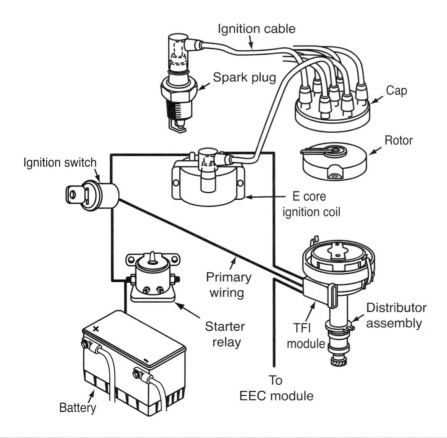

Figure 4-20 One type of electronic distributor-type ignition system.

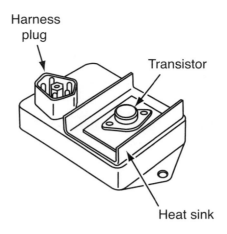

Figure 4-21 Example of an electronic control module.

devices. The electronic switching components are normally inside a separate housing known as an electronic control unit (ECU) or control module (Figure 4-21).

Solid state ignition systems control the primary circuit, using an NPN transistor instead of breaker contact points. The transistor's emitter is connected to ground and takes the place of the fixed contact point. The collector is connected to the negative (–) terminal of the coil, taking the place of the movable contact point. When the triggering device supplies a small amount of current to the base of the switching transistor, the collector and emitter act as if they are closed contact points (a conductor), allowing current to build in the coil primary circuit. When the current to the base is interrupted by the switching device, the collector and emitter act as an open contact (an insulator), interrupting the coil primary current. An example of how this works is shown in Figure 4-22, which is a simplified diagram of a solid state ignition system.

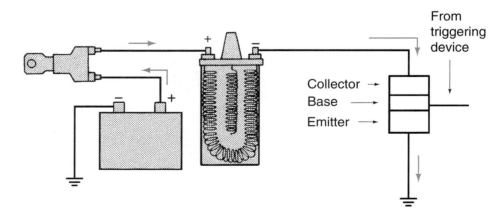

Figure 4-22 NPN transistor used as a switching device in an electronic ignition system.

Computer-Controlled Ignition Systems

Computer-controlled ignition systems offer continuous spark timing control through a network of engine sensors and a central microprocessor. Based on the inputs it receives, the central microprocessor (computer) makes decisions regarding spark timing and sends signals to the ignition module to fire the spark plugs according to those inputs and according to the programs in its memory.

> **AUTHOR'S NOTE:** The ignition module may be a separate component located external to the engine controller, or it may be located on a board within the engine controller.

Some manufacturers with computer-controlled ignition systems use a distributor to distribute secondary voltage to the spark plugs. This allows them to continue using one ignition coil. Both the distributorless and distributor-type computer-controlled ignition systems are discussed in this section. Both systems rely on common sensor inputs.

Common Sensors

Computer-controlled DI and EI ignition systems use a series of sensors to determine the correct timing of the spark. Although the sensors on different systems are designed to provide the same basic information, the type of sensor used can vary between manufacturers and even between different engine applications.

Shop Manual
Chapter 4,
page 130

Engine Crankshaft Position Sensors. Since the time when the ignition primary circuit must be opened is related to the position of the pistons and the crankshaft, the position of the crankshaft is used to control the flow of current to the base of the switching transistors in the ignition coil drivers. A number of different types of sensors are used to monitor the position of the crankshaft. These engine position sensors and generators serve as triggering devices and include magnetic pulse generators, metal detection sensors, Hall-effect sensors, and photoelectric sensors.

Usually, the crankshaft position sensor supplies the PCM with information concerning the position of two pistons as they approach TDC. However, the PCM does not know which piston is approaching TDC compression stroke. The function of the camshaft position sensor (CMP) is to identify the piston approaching TDC compression stroke.

Introduced in early 1982, the Hall-effect switch is now the most commonly used type of engine crankshaft and camshaft position sensor. There are several good reasons for this. Unlike a magnetic pulse generator, the Hall-effect sensor produces an accurate digital voltage signal throughout the entire rpm range of the engine, especially at low engine speeds such as during cranking. Furthermore, a Hall-effect switch produces a square wave signal that is more compatible with the digital signals required by onboard computers. Refer to Chapter 3 on the principles of the Hall-effect switch.

An example of using a Hall-effect switch to determine piston position and speed is to use a slotted flexplate (Figure 4-23). The slots are spaced at manufacturer set intervals. In the illustration of a V6 flexplate there are three groups of slots. Each group is 120 degrees apart from each other. Each group has four slots spaced 20 degrees apart from each other. As the slots pass below a Hall-effect crankshaft position sensor, a switched voltage between high and low is produced (Figure 4-24).

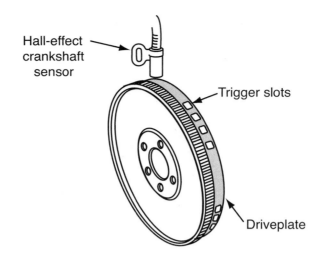

Figure 4-23 Some crankshaft position sensors use slotted driveplates.

Figure 4-24 The switched voltage from the crankshaft position sensor informs the PCM of pistons approaching TDC. Frequency of signals determines the engine speed.

Another type of crankshaft position sensor is the photoelectric sensor. The parts of this sensor include a light-emitting diode, a light sensitive phototransistor, and a slotted disc called a light beam interrupter (Figure 4-25). This type of sensor is used on DI systems. A slotted disc is attached to the distributor shaft. The LED and the photo cell are situated over and under the disc opposite each other. As the slotted disc rotates between the LED and photo cell, the intermittent flashes of light are translated into voltage pulses by the photo cell. When the voltage signal occurs, the control unit turns the primary system on. When the disc interrupts the light and the voltage signal ceases, the control unit turns the primary system off, causing the magnetic field in the coil to collapse and sending a surge of voltage to a spark plug.

Camshaft Position Sensor.   Since EI systems use multiple coils, the computer must know which coil to fire. This is the function of the camshaft position sensor (CMP). Based on the pattern of the sensor signal, the computer identifies which cylinder is the next one to fire (Figure 4-26). Remember, the crankshaft position sensor will inform the computer that two cylinders are approaching TDC. The camshaft sensor identifies which one is one the compression stroke. Hall-effect sensors and magnetic pulse generators can also be used as camshaft reference sensors

Knock Sensor.   The **knock sensor** is threaded into the engine block, intake manifold, or cylinder head. Many V block engines contain two knock sensors for improved detonation control. The knock sensor contains a piezoelectric sensing element, which changes a vibration to a voltage signal. As stated earlier, the resonance combustion detonation occurs is at about 6,400 hertz (depending on engine design). The knock sensor is tuned to 6,400 hertz to pick up spark knock.

When the engine detonates, a vibration is present in the engine block and cylinder head castings. The knock sensor changes this vibration to a voltage signal and sends the signal to the

A light-sensitive phototransistor is often called a photocell.

The **knock sensor** is a piezoelectric device that generates a voltage based on the pressure caused by detonation.

In some applications, the knock sensor is called a detonation sensor.

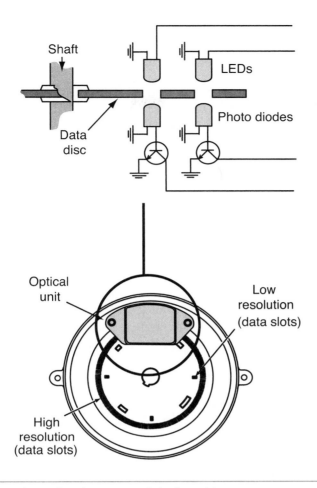

Figure 4-25 A photoelectric-type crankshaft position sensor.

CAM CRANK SYNCHRONIZATION

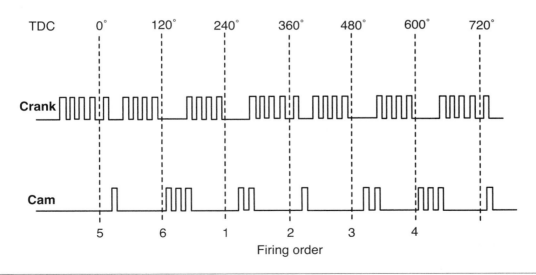

Figure 4-26 The camshaft sensor pulses are used to identify which spark plugs need to be fired next.

computer. When this signal is received, the computer reduces spark advance to eliminate the detonation. A typical knock sensor signal would be 300 millivolts (mV) to 500 mV depending on the severity of detonation.

MAP/MAF Sensor.   The PCM can determine the load on the engine using the manifold absolute pressure (MAP) or mass air flow (MAF) sensors. This information is used to determine the amount of spark timing advance required. These sensors are discussed in Chapter 6.

Idle Air Temperature Sensor.   Although this sensor is primarily used in the pulse width calculation for the injection system, it is also used for the ignition system. If the sensor indicates that air has a high ambient temperature, the PCM will retard the ignition timing to prevent detonation during WOT operation. As the air temperature increases, the density of the air decreases. This means that the fuel and air mixture will not bond very well, and it will result in excessive end gases in the combustion chamber. Since the air is already warm, it is easier for the end gas to ignite.

Throttle Position Sensor.   The throttle poison sensor informs the computer of the position and rate of change of the throttle plate. If the TPS indicates the engine is at idle (along with rpm and MAP/MAF sensors) the computer will use **spark scatter** to control idle quality. Once the TPS indicates the throttle is no longer closed, spark scatter is stopped and ignition timing advance strategies are used. The rate of change is used to determine spark advance or retard for the current condition (acceleration or deceleration).

Computer-Controlled DI System Operation

The main difference between the basic distributor system and the DI system is the elimination of any mechanical or vacuum advance devices from the distributor. In the DI systems, the distributor's purpose is to generate the primary circuit's switching signal and distribute the secondary voltage to the spark plugs. Timing advance is controlled by a microprocessor, or computer. In fact, some of these systems have even removed the primary switching function from the distributor by using a crankshaft position sensor. In this case, the sole function of the distributor is to distribute secondary voltage to the spark plugs. Using the computer to sense engine load, engine speed, engine temperature, ambient temperature, and to control spark delivery allows for precise ignition timing for increased fuel economy and reduced emissions.

Based on sensor inputs, the computer signals an ignition module to collapse the primary circuit, allowing the secondary circuit to fire the spark plugs. Timing control is selected by the computer's program. During engine starting, computer control is bypassed and the mechanical setting of the distributor controls spark timing. Once the engine is started and running, spark timing is controlled by the computer. This scheme or strategy allows the engine to start regardless of whether the electronic control system is functioning properly or not.

The computer continuously monitors existing conditions, adjusting timing to match what its memory tells it is the ideal setting for those conditions. It can do this very quickly, making thousands of decisions in a single second. The control computer typically has the following types of information permanently programmed into it:

- *Speed-related spark advance.* As engine speed increases to a particular point, there is a need for more advanced timing. As the engine slows, the timing should be retarded or have less advance. The computer bases speed-related spark advance decisions on engine speed and signals from the TP sensor.

Spark scatter refers to altering the ignition timing rapidly, and it is used to control idle quality. Since the idle air control motor is not capable of making very fine adjustments, fuel pulse width and spark scatter make fine changes in idle speed.

Shop Manual
Chapter 4,
page 167

- *Load-related spark advance.* This is used to improve power and fuel economy during acceleration and heavy load conditions. The computer defines the load and the ideal spark advance by processing information from the TP sensor, MAP/MAF, and engine speed sensors.

- *Warm-up spark advance.* This is used when the engine is cold, since a greater amount of advance is required while the engine warms up.

- *Special spark advance.* This is used to improve fuel economy during steady driving conditions. During constant speed and load conditions, the engine will be more efficient with more advance timing.

- *Spark advance due to barometric pressure.* This is used when barometric pressure exceeds a preset calibrated value.

All of this information is looked at by the computer to determine the ideal spark timing for all conditions. The calibrated or programmed information in the computer is contained in software look-up tables (Figure 4-27). In this three-dimensional map, arrows point in the directions of increased speed, load, and spark timing advance. The line intersections represent the spark advance for all combinations of load and speed. These values are stored digitally as a look-up table in the controller's memory.

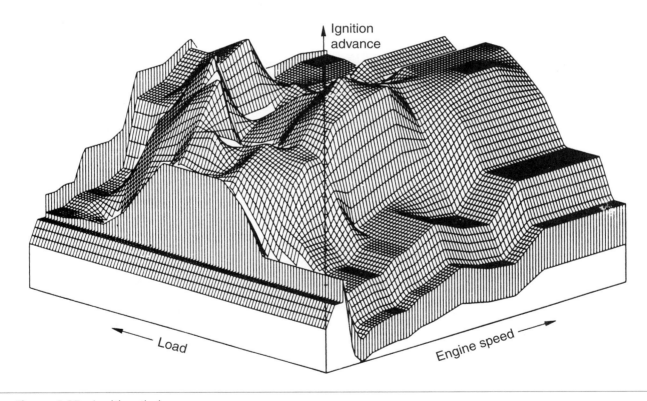

Figure 4-27 Ignition timing map.

Distributorless Electronic Ignition System Operation

Manufacturers have evolved their systems from the point-type, to solid state, to computer-controlled DI systems, and now to electronic ignition (EI). Electronic ignition (EI) systems (Figure 4-28) electronically perform the functions of a distributor. Distributorless systems use multiple coils and modules to provide and distribute high secondary voltages directly from the coil to the spark plug. Since the distributor is eliminated in EI systems, ignition timing remains more stable over the life of the engine.

Another advantage of the EI system is its higher voltage reserves. A specific amount of energy is available in the secondary ignition circuit. This energy is normally produced in the form of voltage required to start firing the spark plug, and then a certain amount of current flow across the spark plug electrodes. Electronic ignition systems are capable of producing much higher energy than distributor ignition systems. Since both the distributor ignition and the electronic ignition systems are firing spark plugs with approximately the same air gaps, the voltage required to start firing the spark plugs in both systems is nearly the same. However, since EI systems have higher voltage reserves, if the additional energy is not used to create the spark across the plug's air gap, it can be used to maintain the spark for a longer period of time. Since the additional energy is not produced in the form of voltage, it will be produced in the form of current flow for a longer time across the spark plug electrodes. The average current flow across the spark plug electrodes in an EI system is 1.5 milliseconds compared to approximately 1 millisecond in a DI system.

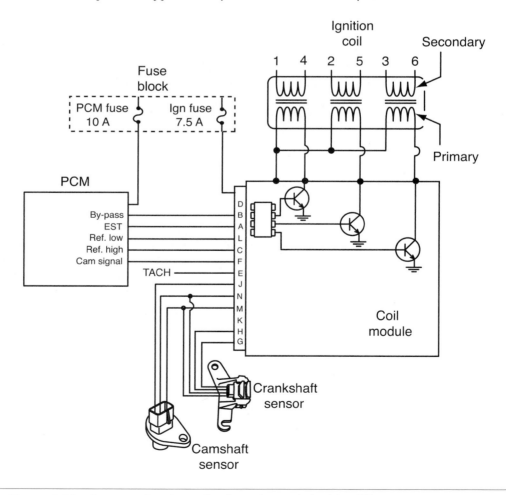

Figure 4-28 An example schematic of an electronic ignition (EI) system.

These extra 0.5 milliseconds of current flow duration may seem insignificant, but it is very important on today's engines with stricter fuel economy and emission regulations. Today's emission standards demand leaner air-fuel ratios. This additional spark duration on EI systems helps to prevent cylinder misfiring with leaner air-fuel ratios.

The ignition module uses crank/cam sensor data to control the timing of the primary circuit in the coils. Remember that there is usually more than one coil in a distributorless ignition system. The ignition module synchronizes the coils' firing sequence in relation to crankshaft position and firing order of the engine. Therefore, the ignition module takes the place of the distributor. Depending on the EI system, the ignition coils can be serviced as a complete unit or separately. The coil assembly is typically called a **coil pack** and consists of two or more individual coils.

Systems that use a coil for every two spark plugs use the **waste spark** method of spark distribution. Each end of the coil's secondary winding is attached to a spark plug. Each coil is connected to a pair of spark plugs in cylinders whose pistons rise and fall together. When the field collapses in the coil, voltage is sent to both spark plugs that are attached to the coil. In a V6 engine, the paired cylinders are 1 and 4, 2 and 5, and 3 and 6. In four cylinder engines, the pairing is 4 and 1, and 3 and 2. With this arrangement, one cylinder of each pair is on its compression stroke while the other is on the exhaust stroke. Both cylinders get spark simultaneously, but only one spark generates power while the other is wasted out the exhaust. During the next revolution, the roles are reversed.

Due to the way the secondary coils are wired, when the induced voltage cuts across the primary and secondary windings of the coil, one plug fires in the normal direction (positive center electrode to negative side electrode) and the other plug fires the reverse; side-to-center electrode (Figure 4-29). As shown in Figure 4-30, both plugs fire simultaneously, completing the series circuit. Remember, current must return to its source. In the case of the ignition coil, the source is the secondary winding. Current flows from the coil winding through the first spark plug, through the engine block and cylinder head(s) to the second spark plug, and back to the secondary winding to complete the circuit.

The coil is able to overcome the increased voltage requirements caused by reversed polarity and still fire two plugs simultaneously because each coil is capable of producing up to 100,000 volts. There is very little resistance across the plug gap on exhaust, so the plug requires very little voltage to fire, thereby providing its mate (the plug that is on compression) with plenty of available voltage.

This function of synchronizing the coils' firing sequence is called coil synchronizing.

Most EI systems have the function of the ignition module internal to the PCM. Some use a separate ignition module that receives signals from the PCM.

A coil assembly that contains two or more coils is referred to as a **coil pack**.

Waste spark is a spark that occurs during the exhaust stroke of a piston.

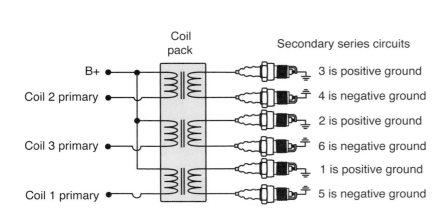

Figure 4-29 Spark plug firing for a six-cylinder engine with EI.

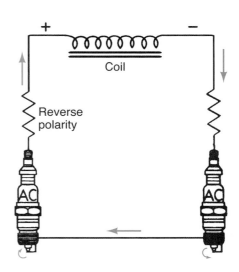

Figure 4-30 Complete circuit for spark plug firing in an EI system.

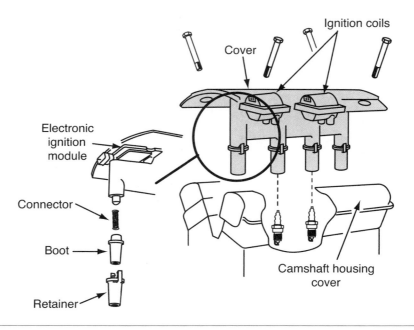

Figure 4-31   GM's Quad 4 with the ignition coils mounted directly over the spark plugs.

Figure 4-31 shows a waste spark system in which the coils are mounted directly over the spark plugs so no wiring between the coils and plugs is necessary. This type of system operates in the same way as other EI systems (Figure 4-32).

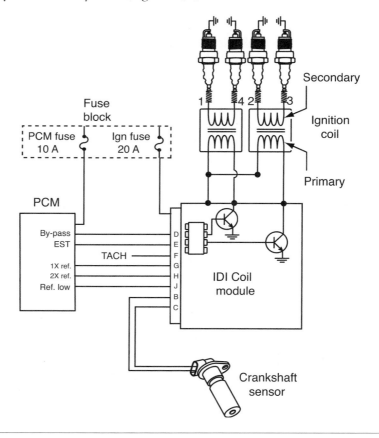

Figure 4-32   GM's Quad 4 ignition circuit.

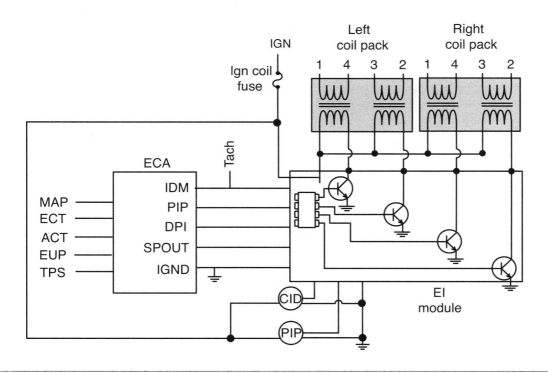

Figure 4-33 A dual plug EI system for a four-cylinder engine.

Some EI systems have two spark plugs per cylinder. These systems are called **dual plug systems** (Figure 4-33). One spark plug is located on the intake side of the combustion chamber, while the other is located on the exhaust side. The system illustrated in Figure 4-42 uses two coil packs (there are two coils in each pack). One coil pack operates the plugs on the intake side and the other operates the plugs on the exhaust side. While the engine is being started only one plug in each cylinder is fired. Once the engine is running, the system switches to dual plug mode. During dual plug operation, the two coil packs are synchronized so that each cylinder's two plugs fire at the same time. Therefore, on a four-cylinder engine, four spark plugs are fired at a time, two during the compression stroke of the cylinder and two during the exhaust stroke of the companion cylinder.

Dual plug systems
have two spark
plugs per cylinder.

Examples of EI System Operation

AUTHOR'S NOTE: The following is offered as examples of EI system operation. A look at systems used by DaimlerChrysler, General Motors, and Nissan will provide an understanding of how different systems operate. Also, you will see how similar they are.

DaimlerChrysler's V6 EI systems which are equipped with a single-board engine controller (SBEC), uses a Hall-effect-type crankshaft position sensor that is mounted in an opening in the transaxle bell housing. The inner end of this sensor is positioned near a series of notches and slots that are cut into the transaxle drive plate (refer to Figure 4-23).

A group of four slots is located on the transaxle drive plate for each pair of engine cylinders, and thus a total of 12 slots are positioned around the drive plate. The slots in each group are positioned 20 degrees apart. When the slots on the transaxle drive plate rotate past the crankshaft timing sensor, the voltage signal from the sensor alternates between 0 and 5 volts. This digital voltage signal informs the PCM regarding crankshaft position and speed, and the PCM

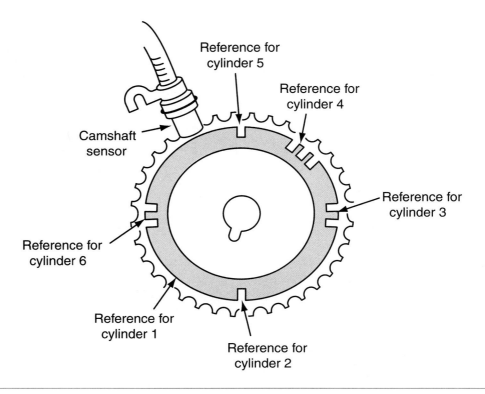

Figure 4-34   Notched camshaft drive sprocket.

calculates spark advance from this signal. The slots are spaced so the first rise from 0 to 5 volts is 69 degrees BTDC. The next is 49 degrees BTDC, then 29 degrees BTDC and the final slot is 9 degrees BTDC. The PCM also uses the crankshaft position sensor signal, along with other inputs, to determine fuel-injection pulse width. Base timing is determined by the signal from the 9-degree slot in each group of slots, and base timing adjustment is not possible.

The Hall-effect-type camshaft reference sensor is mounted in the top of the timing gear cover. A notched ring on the camshaft gear rotates past the end of the camshaft reference sensor. This ring contains two single slots, two double slots, a triple slot, and an area with no slots (Figure 4-34).

When a camshaft gear notch rotates past the camshaft reference sensor, the signal changes from 0 volts to 5 volts. The single, double, and triple notches provide different voltage signals from the camshaft reference sensor as they rotate past the sensor, and these signals are sent to the PCM. The PCM determines the exact camshaft and crankshaft position from the camshaft reference sensor signals, and the PCM uses these signals to sequence the coil primary windings and the activation of the injectors at the correct instant.

The PCM is able to make these calculations within one engine revolution during engine starting. While observing both the camshaft and crankshaft position sensors, the PCM determines which cylinder is approaching TDC compression stroke. When the PCM sees one group of slots from the crankshaft position sensor, followed by three slots (for example) from the camshaft position sensor, it knows cylinder number 4 is approaching TDC. If the PCM sees one group of

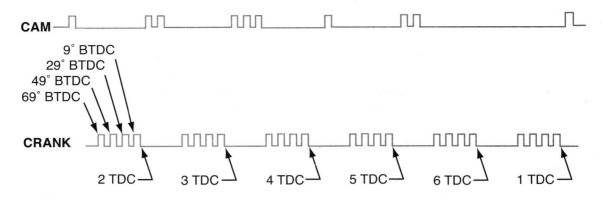

Figure 4-35 Coil firing events in response to camshaft and crankshaft position sensor input.

slots from the crankshaft position sensor, followed by no signal from the camshaft position sensor (for example), it knows cylinder number 1 is approaching TDC (Figure 4-35). The PCM is in synchronization once it recognizes number 1 or number 4 pistons approaching TDC.

The coil assembly contains three ignition coils. Two spark plug wires are connected from the secondary terminal on each coil to the spark plugs. In each coil, the ends of the secondary winding are connected to the two secondary terminals on that coil.

The closed contacts of the automatic shutdown (ASD) relay supplies 12 volts to the positive side of the primary windings in each coil when the ignition switch is turned on. The other end of each primary winding is connected to the PCM and onto individual low-side drivers.

When the engine is being started, the spark plugs fire and the injectors discharge fuel within one crankshaft revolution. The PCM determines when to sequence the coils and injectors from the camshaft reference sensor signals. If the camshaft reference sensor or the crankshaft timing sensor is defective, the engine will may not start. Later PCMs will start the engine if one of these inputs is missing, but only after an extended crank time. Each coil fires two spark plugs at the same instant, and the current flows down through one spark plug and up through the other spark plug. One of the cylinders in which a spark plug is firing is on the compression stroke, while the other cylinder is on the exhaust stroke. When the engine is cranking, all spark plug firings are at 9 degrees BTDC on the compression stroke. The PCM also fires all injectors one time. Once the PCM synchronizes the CMP and CKP sensors, it begins firing the appropriate coils.

The spark plug wires from coil number 1 are connected to cylinders 1 and 4, whereas the spark plug wires from coil number 2 go to cylinders 2 and 5, and the spark plug wires on coil number 3 are attached to cylinders 3 and 6. The cylinder firing order for the 3.3-L V6 engine is 1-2-3-4-5-6.

Once the engine is started, the PCM knows the exact crankshaft position and speed from the crankshaft sensor signals. Since the PCM synchronized the CMP and CKP sensors, it is now able to recognize the order from the CMP sensor.

With the engine running, the PCM determines the precise spark advance required when the next cylinder fires. This precise spark advance is provided when the PCM opens the primary circuit on the appropriate coil. Since DaimlerChrysler engines are equipped with sequential fuel injection (SFI), the PCM grounds each injector individually, but the proper injector sequencing is determined from the camshaft reference sensor signals.

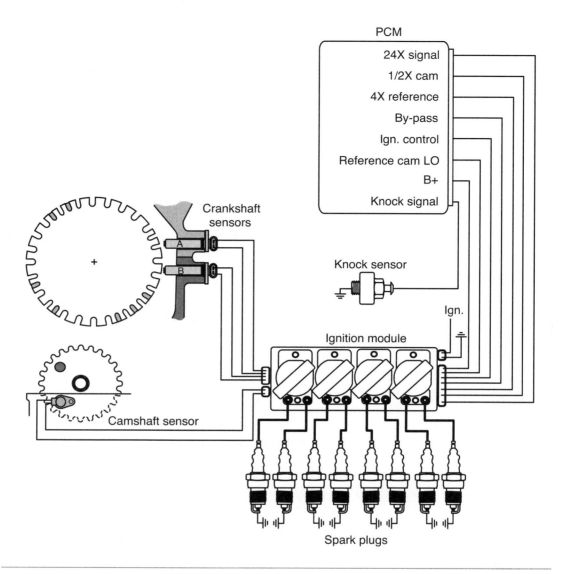

PCM

24X signal
1/2X cam
4X reference
By-pass
Ign. control
Reference cam LO
B+
Knock signal

Crankshaft sensors

Knock sensor

Ign.

Ignition module

Camshaft sensor

Spark plugs

Figure 4-36   General Motors' Northstar EI system.

The EI system used on the General Motors' Northstar (Figure 4-36) engine uses two crankshaft position sensors referred to as A and B sensors. A reluctor ring with 24 evenly spaced notches and 8 unevenly spaced notches is cast onto the crankshaft between the number 3 and 4 main bearing journals.

When the reluctor ring rotates past the magnetic-type A and B crankshaft sensors, each sensor produces 32 high= and low-voltage signals per crankshaft revolution. The A sensor is positioned in the upper crankcase, and the B sensor is positioned in the lower crankcase. Since the A sensor is above the B sensor, the signal from the A sensor occurs 27 degrees before the B sensor signal (Figure 4-37).

The signals from the A and B sensors are sent to the ignition control module (ICM). This module counts the number of B sensor signals between the A sensor signals to sequence the

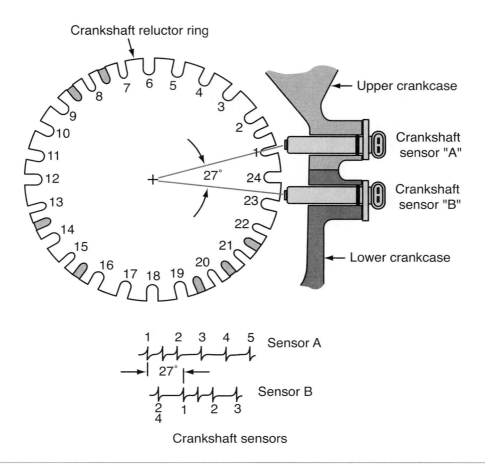

Figure 4-37 Dual piston sensors.

ignition coils properly. There can be zero, one, or two B sensor signals between the A sensor signals. When starting the engine, the ICM begins counting B sensor signals between A sensor signals as soon as the module senses zero B sensor signals between A sensor signals. After the ICM senses four B sensor signals, the module sequences the coils properly. This action allows the ignition system to begin firing the spark plugs within 180 degrees of crankshaft rotation while starting the engine.

Once the engine starts, the PCM switches from the bypass mode to the ignition control mode. The ICM uses the A and B crankshaft sensor signals to determine the exact crankshaft position and rpm. The ICM relays this information to the PCM with 4X reference and 24X voltage signals. The PCM scans all the other input sensor signals and calculates the precise spark advance required by the engine. The PCM sends an ignition control signal to the ICM, which informs the ICM to turn off the appropriate coil primary at the proper time to supply the correct spark advance. The reference cam low wire is a ground wire between the ICM and the PCM. If the engine detonates, the knock sensor signal informs the PCM to reduce the spark advance.

The camshaft position sensor is located in the rear cylinder bank in front of the exhaust camshaft sprocket. A reluctor pin in the sprocket rotates past the sensor. This sensor produces one high- and one low-voltage signal during every camshaft revolution or every two crankshaft revolutions. The PCM uses the camshaft position sensor signal to sequence the injectors properly.

Figure 4-38   Two spark plugs per cylinder are operated by a single coil set.

The EI system used by Mercedes-Benz on their 3.2 L V6 uses one coil set and two spark plugs per cylinder (Figure 4-38). The coils are mounted on the rocker cover and are connected to the spark plugs with short cables. The coil set contains two separate coils in one unit. The two primary windings share a common battery feed with the ignition switch in the RUN and START positions (Figure 4-39). Individual secondary windings are connected to the spark plugs. The spark plugs for each cylinder are identified as A and B plugs. Individual coils for each cylinder ensures performance at high rpms and allows individual cylinder spark control. Dual spark plugs provide more complete combustion, particularly near the cylinder walls and they decrease emissions. This ignition system enhances combustion efficiency when firing mixtures are diluted by EGR, and it improves overall efficiency and lowers emissions.

This is referred to as a "coil near plug" ignition system.

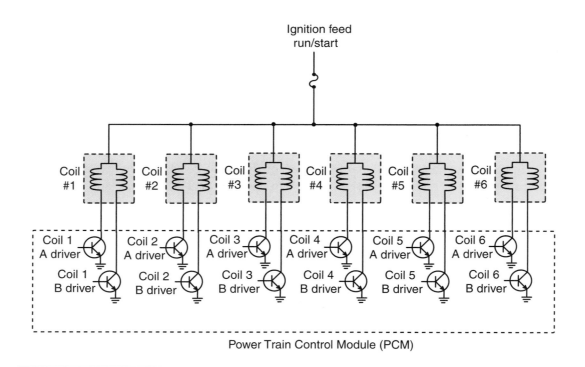

Figure 4-39   Simplified wiring schematic of the primary control circuit.

Under medium and high engine loads, the two spark plugs are fired slightly out of phase to prevent the cylinder pressures from rising too quickly, which could cause knocking. Offset between the plug firing varies from 0 to 10 degrees of crankshaft revolution. To prevent one spark plug from eroding more quickly than the other, the oils are "Phase-Shift" triggered (alternately lead each other). This means plug A fires, then B, then B, then A, then A, then B, and so on.

Under normal conditions, the timing is the same for all cylinders, but the timing can be delayed in individual cylinders if knocking is present in one or more of the cylinders. Highly sensitive knock sensors are used that can distinguish knocking conditions in individual cylinders and retard the ignition timing on only those cylinders that are experiencing knocking.

Coil-on-Plug Systems

Coil-on-plug (COP) is the next step in the evolution of the ignition system. These EI systems use one coil per spark plug. The advantage of this system is there are no spark plug cables; the coil sits directly on top of the spark plug. The electronic ignition module determines when each spark plug should fire and controls the on/off time of each plug's coil. The inputs for this system are the same as other EI systems. The ignition module will have individual drivers for each coil's primary circuit (Figure 4-40). Although not exclusive of COP systems, an adaptive dwell

Shop Manual
Chapter 4,
page 176

The **coil-on-plug (COP)** system generally uses one coil pack per spark plug and does not use spark plug cables since the coil is installed directly onto the spark plug.

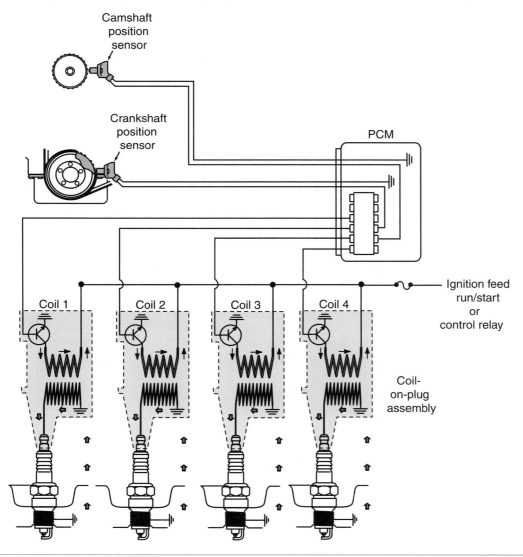

Figure 4-40 Simplified wiring diagram of a coil-on-plug (COP) system.

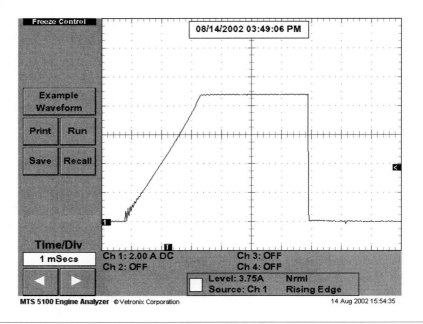

Figure 4-41 A current trace of the primary circuit with current limiting.

strategy can be used. The adaptive dwell is driven by sensing the current flow through the ignition coil drivers. Current flow is limited to about 8 amperes. The resistance of the primary winding is low enough to allow over 15 amperes of current. This excessive current may damage the control module.

The adaptive dwell strategy operates primary circuit dwell between 4 and 6 ms when engine speed is below 3,000 rpm. During engine cranking when the air-fuel mixture is rich, extra current is required to ignite the mixture. During this time, dwell can be as high as 200 ms. To limit the current in the primary windings, current in the circuit is measured as soon as dwell time starts. When the current-sensing device registers 8 amperes, the control module will regulate the current flow so it does not exceed this amount during the rest of the dwell time (Figure 4-41).

Figure 4-42 illustrates Ford's coil-on-plug system. This system fires each spark plug three times (per compression stoke) at idle to assure complete combustion. Once the engine is running in off idle mode the system reverts to firing the spark plug once. In Figure 4-42, notice that the secondary windings of the coil are not connected to the battery feed of the primary windings. The high secondary voltage is induced in the same way, but this method tends to have less RFI. Some COP system coils will have a high voltage diode in the secondary circuit (Figure 4-43). This diode is used for rapid cutoff of secondary ignition.

The 5.7L V8 Hemi engine produced by DaimlerChrysler uses a unique COP system. This system is a dual COP waste spark ignition, and it uses two spark plugs per cylinder. There are eight coils that are used to fire the sixteen spark plugs. This system is a combination of the waste spark EI system discussed earlier and the COP system. Each cylinder has a COP-fired plug and a

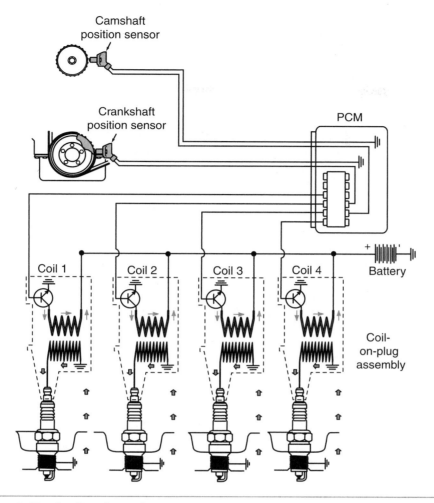

Figure 4-42 Ford's coil-on-plug system.

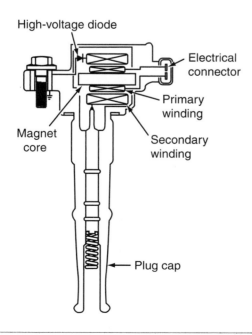

Figure 4-43 Some COP coils have a high-voltage diode in the secondary windings circuit.

Figure 4-44 The 5.7L Hemi has two spark plugs per cylinder and one coil per cylinder.

remote fired plug. The coil has two secondary terminals (Figure 4-44). One terminal is a COP terminal. The other connects to a spark plug cable and is run to a remote fired spark plug in the companion cylinder on the opposite bank.

The firing order for the engine is 1-8-4-3-6-5-7-2. This means the cylinder pairing is 1 and 6, 5 and 8, 4 and 7, 2 and 3. Every coil will fire once per engine revolution. To illustrate the firing sequence, assume piston number 1 is approaching TDC compression stroke. This means piston number 6 is approaching TDC on its exhaust stroke. The cylinder number 1 coil is fired by the PCM and the spark is delivered to the COP-fired spark plug underneath the coil. In addition, secondary voltage is also sent out the second terminal of the coil to cylinder number 6's remote-fired spark plug by means of a spark plug cable. At the same time as coil number 1 is being fired, coil number 6 is also being fired. This sends secondary voltage to cylinder number 6's COP-fired spark plug and also out the second terminal to cylinder number 1's remote-fired spark plug. Both spark plugs in a cylinder are fired at the same time, but by different coils. In actuality, every time there is a spark event, four spark plugs are fired.

Summary

❏ One of the many laws of physics utilized within the automotive engine is the law of thermodynamics. The driving force of the engine is the expansion of gases. Heat is generated by compressing the gasoline and air mixture within the engine. Igniting the compressed mixture causes the heat, pressure, and expansion to multiply.

❏ Most automotive and truck engines are four-stroke cycle engines. A stroke is the movement of the piston from one end to the other.

❏ The first stroke of the cycle is the intake stroke.

- ❏ The compression stroke begins as the piston starts its travel back to TDC.
- ❏ When the spark is introduced to the compressed mixture, the rapid burning causes the molecules to expand. This is the beginning of the power stroke.
- ❏ The exhaust stroke of the cycle begins with the upward movement of the piston back toward TDC and pushes out the exhaust gases from the cylinder past the exhaust valve and into the vehicle's exhaust system.
- ❏ The ignition system supplies high voltage to the spark plugs to ignite the air-fuel mixture in the combustion chambers.
- ❏ The arrival of the spark is timed to coincide with the compression stroke of the piston. This base timing can be advanced or retarded under certain conditions such as high engine rpm or extremely light or heavy engine loads.
- ❏ The ignition system has two interconnected electrical circuits: a primary circuit and a secondary circuit.
- ❏ The primary circuit supplies low voltage to the primary winding of the ignition coil. This creates a magnetic field in the coil.
- ❏ A switching device interrupts primary current flow, collapsing the magnetic field and creating a high-voltage surge in the ignition coil secondary winding.
- ❏ The secondary circuit carries high-voltage surges to the spark plugs. On some systems, the circuit runs from the ignition coil, through a distributor, to the spark plugs.
- ❏ Ignition timing is directly related to the position of the crankshaft. Magnetic pulse generators and Hall-effect sensors are the most widely used engine position sensors. They generate an electrical signal at certain times during crankshaft rotation. This signal triggers the electronic switching device to control ignition timing.
- ❏ EI systems provide longer spark duration at the spark plug electrodes than conventional electronic ignition systems. This helps to fire leaner air-fuel ratios in today's engines.
- ❏ Compared to electronic ignition systems with distributors, EI systems provide more stable control of ignition timing and spark advance, which reduces emissions and improves fuel economy.
- ❏ Computer-controlled ignition eliminates centrifugal and vacuum-timing mechanisms. The computer receives input from numerous sensors. Based on this data, the computer determines the optimum firing time and signals an ignition module to activate the secondary circuit at the precise time needed.
- ❏ Coil-on-plugs (COP) use one coil per spark plug. The advantage of this system is that there are no spark plug cables; the coil sits directly on top of the spark plug.
- ❏ Adaptive dwell strategy is driven by sensing the current flow through the ignition coil drivers.

Review Questions

Short Answer Essay

1. Describe how the basic laws of thermodynamics are used to operate the automotive engine.
2. List and describe the four strokes of the four-stroke cycle engine.
3. Describe the three major functions of an ignition system.
4. Name the two major electrical circuits used in ignition systems and their common components.

Terms-to-Know

Advanced timing

Base timing

Bottom dead center (BDC)

Coil-on-plug (COP)

Coil pack

Cycle

Distributor

Distributor cap

Distributor ignition (DI)

Dual plug systems

E coil

Efficiency

Electronic ignition (EI)

Firing order

Heat range

Horsepower

Ignition cables

Ignition timing

Induction reactance

Internal combustion engine

Ionize

Kinetic energy

Knock sensor

Mechanical efficiency

Mutual induction

Potential energy

Primary circuit

Primary coil windings

Pulse transformer

Reach

Reciprocating engines

Reserve voltage

Resistor plugs

Retarded timing

Rise time

Rotor

Saturation

Secondary circuit

5. Describe the difference between DI and EI systems

6. Under light loads, what must be done to complete air-fuel combustion in the combustion chamber by the time the piston reaches 23 degrees ATDC?

7. At high engine rpm, what must be done to complete air-fuel combustion in the combustion chamber by the time the piston reaches 23 degrees ATDC?

8. Explain the purpose of the camshaft position sensor in an EI system.

9. Describe how each pair of spark plugs is fired in an EI waste spark systems.

10. Describe how secondary ignition voltage is induced in the ignition coil.

Fill-in-the Blanks

1. Most automotive and truck engines are _____ cycle engines.

2. During the intake stroke, the intake valve is _____ while the exhaust valve is _____.

3. In the four-stroke engine, four strokes are required to complete one _____.

4. The arrival of the spark is timed to coincide with the _____ stroke of the piston.

5. In an EI system, the _____ _____ _____ identifies the next plug to be fired.

6. The ignition system has two interconnected electrical circuits: a _____ circuit and a _____ circuit.

7. The difference between the required voltage and the maximum available voltage is referred to as _____ _____ voltage.

8. The _____ _____ of the spark plug refers to its heat dissipation properties

9. In an EI system coil, two spark plug wires are connected to the ends of each _____ winding.

10. The calibrated or programmed information in the computer concerning spark advance is contained in what is called _____ _____ _____.

Multiple Choice

1. *Technician A* says the driving force of the engine is the expansion of gases.
   Technician B says compressing the gasoline and air mixture within the engine cools the air-fuel mixture.
   Who is correct?
   A. A only
   B. B only
   C. Both A and B
   D. Neither A nor B

2. *Technician A* says mechanical efficiency is a comparison between brake horsepower and indicated horsepower.
   Technician B says volumetric efficiency is a measurement of the amount of air-fuel mixture that actually enters the combustion chamber compared to the amount that could be drawn in.
   Who is correct?
   A. A only
   B. B only
   C. Both A and B
   D. Neither A nor B

3. Horsepower is defined as:
   A. the twisting effort applied to the crankshaft.
   B. the turning effort applied to the flywheel.
   C. the measure of the rate of work.
   D. the measure of torque.

4. What happens when the low-voltage current flow in the coil primary winding is interrupted by the switching device is being discussed. *Technician A* says the magnetic field collapses.
   Technician B says a high-voltage surge is induced in the coil secondary winding.
   Who is correct?
   A. A only   C. Both A and B
   B. B only   D. Neither A nor B

5. *Technician A* says the speed of the piston increases as it moves from its compression stroke to its power stroke.
   Technician B says the time needed for combustion is constantly varying.
   Who is correct?
   A. A only   C. Both A and B
   B. B only   D. Neither A nor B

6. Components of the primary circuit include a(n):
   A. Ignition switch.
   B. Distributor cap.
   C. Spark plug.
   D. All of the above.

7. *Technician A* says an ignition system must generate sufficient voltage to force a spark across the spark plug gap.
   Technician B says the ignition system must time the arrival of the spark to coincide with the movement of the engine's pistons and vary it according to the operating conditions of the engine.
   Who is correct?
   A. A only   C. Both A and B
   B. B only   D. Neither A nor B

8. The amount of voltage the coil will deliver to the spark plug depends on what factor(s)?
   A. Coil size
   B. Air-fuel ratio
   C. Reserve capacity
   D. Rotor blade width

9. The margin of coil secondary voltage that can be produced above that which is required to fire the spark plug is called?
   A. Heat range
   B. Saturation
   C. Dwell
   D. Reserve capacity

10. The heat range of a spark plug is determined by:
    A. Primary coil dwell.
    B. How far the center insulator goes into the shell around the center electrode.
    C. The thread diameter.
    D. The resistance used to reduce electromagnetic induction created within the ignition system.

Fuel Delivery Systems

Upon completion and review of this chapter, you should be able to:

❏ Explain the purpose of the fuel delivery system.

❏ List the commonly used components of the fuel delivery system.

❏ Detail the function of the individual components of the fuel delivery system.

❏ Explain how expansion, contraction, and overflow are controlled in the fuel tank.

❏ Explain the purpose of the check valves used in the fuel delivery system.

❏ Describe the function of the rollover valve.

❏ Detail the emission controls associated with the fuel delivery system.

❏ Describe the operation of the electric fuel pumps.

❏ Explain how the fuel pressure regulator operates.

❏ Explain the differences between return-type and returnless-type fuel systems.

❏ Describe the types of materials and fittings used in common fuel lines.

❏ Detail the function of the intake manifold.

❏ Explain typical fuel pump circuit operation.

Introduction

The fuel delivery system consists of all of the components responsible for storing and delivering fuel into the engine's combustion chambers. In addition, emissions associated with the storage of fuel must also be strictly controlled by this system. Since the fuel delivery system can have a major impact on engine performance and emissions, today's technician must possess knowledge of the components used and their function. This chapter details the common components of typical fuel delivery systems. Figure 5-1 shows a typical fuel delivery system used on fuel-injected vehicles. Fuel is drawn from the fuel tank by an in-tank or chassis-mounted electric fuel pump. Before it reaches the injectors, the fuel passes through a filter that removes dirt and impurities. A fuel line pressure regulator maintains a constant fuel line pressure in the system. This fuel pressure generates the spraying force needed to inject the fuel. On some systems, excess fuel not required by the engine returns to the fuel tank through a fuel return line. A typical fuel delivery includes the following components:

1. Fuel tank

2. Filler cap

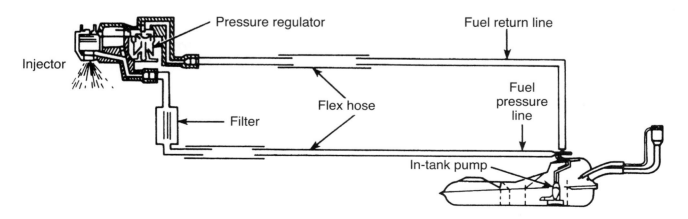

Figure 5-1   The typical components of the fuel delivery system.

3. Fuel pump
4. Fuel lines
5. Fuel filter
6. Check valves
7. Pressure regulator
8. Liquid-vapor separator
9. Pressure and vacuum controls
10. Fuel rail
11. Injectors
12. Intake manifold

Fuel Tank

Shop Manual
Chapter 5,
page 195

The **fuel tank** stores the liquid fuel until it is delivered to the engine.

The main function of the **fuel tank** is to provide a means for storing the fuel to be used by the engine. The expanded function of the fuel tank is to prevent gasoline vapors (HC) from escaping into the atmosphere. Although the fuel tank appears to be a simple device, in actuality there is much research and development that goes into the design of the fuel tank.

The fuel tank on most passenger cars and sport utility vehicles (SUVs) is located at the rear of the vehicle situated between the frame and above the rear suspension cross-member (Figure 5-2), or in front of the rear axle under the rear seat floor. On light-duty and medium-duty trucks, the fuel tank may be located under the bed and behind the cab. Straps or brackets support the fuel tank in the vehicle. Fuel tanks are normally mounted with insulators between the top of the tank and the chassis to protect the tank and prevent noise from transferring into the passenger compartment (Figure 5-3).

Fuel tanks can be made from corrosion-resistant stamped steel, aluminum, or high density polyethylene (HDPE) material. The tank must pass stringent testing for HC leakage and crash integrity prior to its use in a vehicle. Porosity, strength, size, and weight are all considerations that go into tank design and material selection.

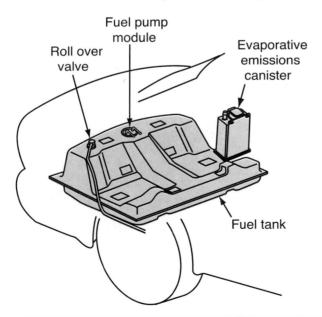

Figure 5-2 The fuel tank is usually located near the rear of the vehicle on most passenger cars and SUVs.

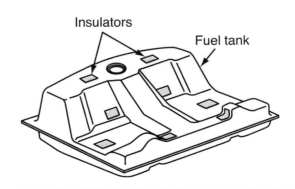

Figure 5-3 Most fuel tanks have insulators on the top surface to prevent noise transfer into the passenger compartment.

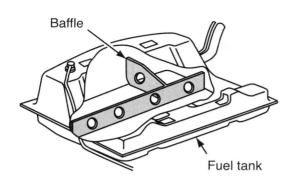

Figure 5-4   Internal baffles prevent fuel from rapidly moving from one side of the tank to the other.

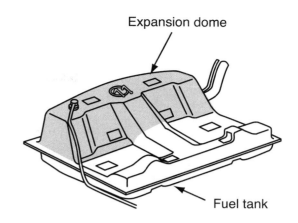

Figure 5-5   The expansion dome is an internal air chamber that allows for expansion of fuel vapors.

Usually the fuel tank will contain one or more baffles to control fuel slosh (Figure 5-4). The baffles will have a series of holes in them to allow the fuel to transfer evenly throughout the tank. However, if the vehicle experiences hard braking, hard acceleration, rough roads, or leans from cornering, the holes slow down the transfer of fuel from one section to another. This also assures that the fuel pickup tube to the fuel pump is always immersed in fuel.

Fuel vapors cannot be allowed to vent to the atmosphere, thus the fuel tank must be constructed to allow for expansion, contraction, and overflow. Most tanks will have an internal air expansion chamber dome (Figure 5-5), or an **expansion tank** (Figure 5-6) to allow for expansion of fuel as ambient temperature increases. Fuel tank overfill is controlled by the design of the tank. This can be accomplished by locating the **filler tube** lower in the tank (Figure 5-7) and/or

The **expansion tank** is a chamber of the fuel tank that allows for fuel expansion, resulting in temperature increases.

The **filler tube** is the pipe the fuel flows through to fill the tank.

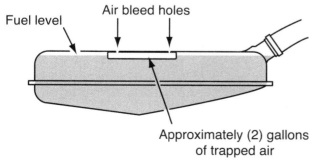

Figure 5-6   This internal expansion tank allows for expansion of vapors and also prevents overfilling of the tank.

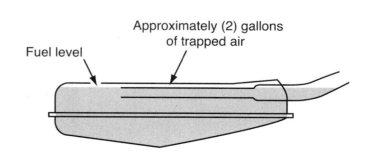

Figure 5-7   Locating the filler neck lower into the side of the tank prevents overfilling.

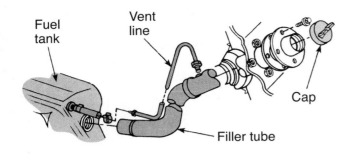

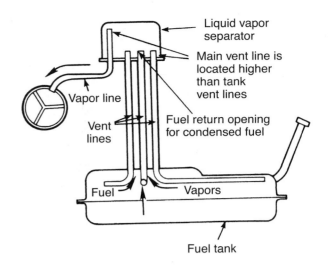

Figure 5-8 The use of vent lines controls fill volume of the tank.

Figure 5-9 The fuel/vapor separator allows fuel vapors to be condensed back to fuel and returned to the tank. It also prevents raw fuel from entering the evaporative control system.

The **fuel pump module** assembly incorporates the fuel pump, fuel filter, fuel gauge sending unit, and fuel temperature sensor (if used) in a single assembly.

by using vapor vent lines to the filler tube (Figure 5-8). These designs prevent the tank from being filled more than 90 percent of its total capacity. The 10 percent of empty tank capacity is used to contain fuel vapors and to allow for expansion. Some fuel tanks may also contain a check valve at the bottom of the filler tube that closes when the tank is full and prevents overfilling.

Some form of liquid vapor separator is incorporated into the system to prevent liquid fuel or bubbles from reaching the vapor storage canister or the engine crankcase. The separator may be located inside the tank, on the tank (Figure 5-9), in fuel vent lines, or near the fuel pump. Other forms of control include the use of two or three pressure relief/rollover valves, one or two at the top of the tank and the other in the fuel filler tube. These valves prevent fuel flow through the vent valve hose serving the evaporative canister.

Also, most of today's vehicles have the **fuel pump module** (Figure 5-10) attached to the top or side of the tank. The module may also contain a fuel level sending unit that includes a pickup tube and a float-operated fuel gauge (Figure 5-11). The fuel tank pickup tube is connected to the fuel pump by a fuel line. The pick up tube extends nearly, but not completely, to the bottom of the tank. This prevents rust, dirt, sediment, or water from being drawn into the fuel tank filter, which can cause it to clog. A ground wire is often attached to the fuel tank unit.

Figure 5-10 Fuel pump module.

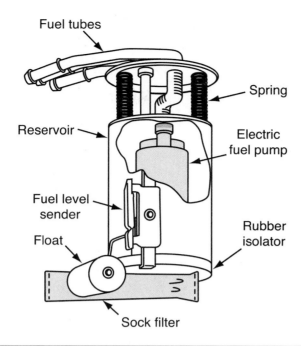

Fuel tubes

Spring

Reservoir

Electric fuel pump

Fuel level sender

Float

Rubber isolator

Sock filter

Figure 5-11 The fuel pump module includes the fuel pump, fuel gauge sending unit, and the pick up filter.

Fuel Filler Caps

The fuel filler tube is sealed with a special **fuel cap**. To prevent fuel vapors from being expelled into the atmosphere, the filler tube caps are nonventing (under normal operation). However, they will usually have some type of pressure vacuum relief valve arrangement (Figure 5-12) to prevent fuel tank damage. The cap allows excessive pressures to be vented and for fresh air to enter the fuel tank. If air could not enter the tank and replace the fuel, the fuel pump would create a vacuum in the tank that could cause it to collapse. However, as the temperature of the fuel increases, a pressure is created inside the tank. If the pressure surpasses the calibrated set point of the valve (approximately 1 psi, or 6.89 kPa), the valve will open. The fuel vapors are not allowed to vent into the atmosphere, rather they are vented to the vehicle's emission system. Once the pressure or vacuum has been relieved, the valve closes.

Shop Manual
Chapter 5,
page 198

The **fuel cap** seals the gas tank, but it must also provide for pressure release and fresh air intake.

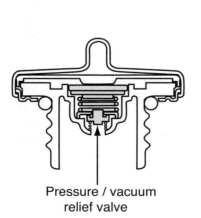

Pressure / vacuum relief valve

Figure 5-12 Filler cap with pressure/vacuum relief valve.

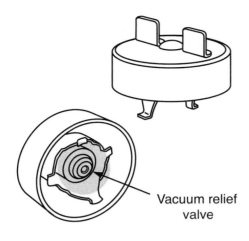

Figure 5-13   The four-tank fuel cap releases any tank pressure prior to the cap being removed.

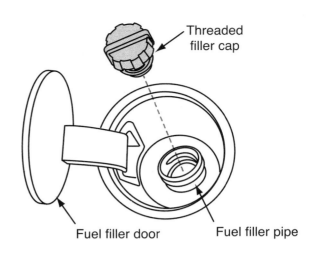

Figure 5-14   The threaded fuel cap releases pressure slowly as the cap is removed. This cap will also have a ratchet system to indicate it is properly tightened.

Manufacturers may also use a cap design that does not contain a pressure/vacuum valve. To control tank pressures a three-way valve is located in the vapor line between the tank and the carbon canister. The carbon canister stores the fuel vapors.

The cap must provide a positive seal between itself and the filler tube. Different designs of locking are used to accomplish this. Also, if there is pressure in the tank, it must be vented to prevent splash back of gasoline up the filler tube and onto the ground when the cap is removed. Many pressure caps have four anti-surge tangs that lock onto the filler neck to prevent the delivery system's pressure from pushing fuel out of the tank (Figure 5-13). By turning such a cap one-half turn, the tank pressure is relieved. Then, with another quarter turn, the cap can be removed. A threaded filler cap may be threaded into the upper end of the filler pipe. These caps require several turns counterclockwise to remove (Figure 5-14). The long threaded area on the cap is designed to allow any remaining fuel tank pressure to escape during cap removal. The cap and filler neck have a ratchet-type, torque-limiting design to prevent overtightening. When the cap is installed, a clicking noise will be heard as the ratchet releases indicating the cap is installed tight and fully seated. A third design is a variation of the threaded cap; however, it only requires one quarter turn to seal.

Rollover Valve

The **rollover check-valve** is a safety valve that closes the fuel flow in the event of a rollover accident. It may also be part of the evaporative emission controls.

The rollover valve is also referred to as the fuel cut-off valve.

Starting with the 1976 model year, a Federal Motor Vehicle Safety Standard (FMVSS 301) required a control system be installed to prevent gasoline leakage from passenger cars, certain light trucks, and buses after they were subjected to barrier impacts and rollovers. Tests conducted under these severe conditions showed the most common gasoline leak path was the gasoline supply line from the fuel tank to the carburetor. Although a carburetor system is no longer used, the manufacturers must still prevent leakage between the fuel tank and the evaporative charcoal canister. One method of doing this is to use a **rollover check-valve** located at the top of the fuel tank (Figure 5-15). The rollover valve can also be used to vent the fuel tank to the charcoal canister. During normal operation, the weight of the valve, combined with the vapor pressure, overpowers the force of the spring and opens the passage for fuel vapor. When the fuel level is high (or when driving around sharp corners) the liquid fuel causes the float to rise. This will cause the float to close the port and prevent liquid fuel from entering the evaporative canister line. If the vehicle overturns, gravitational pull on the float, along with spring pressure, will close the port. With the port closed, liquid fuel will not be able to escape through the evaporative canister line. In addition, if the float sticks, causing the valve to remain closed, there is a relief valve that provides a means through which fuel vapor can be routed to the evaporative canister.

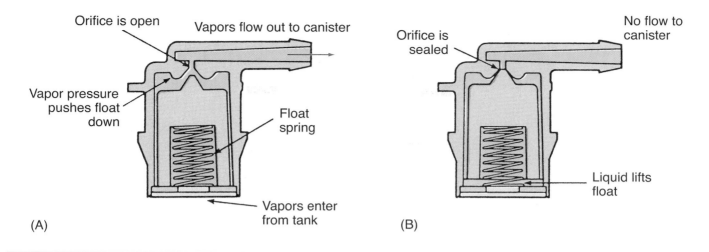

Figure 5-15 The rollerover check valve will close the passage to the canister if it floats to the top or if the vehicle is involved in a rollover accident. The valve can also be used to prevent raw fuel from entering the evaporative control system.

Onboard Refueling Vapor Recovery (ORVR)

Hydrocarbon vapors that occur within the fuel tank are prevented from entering the atmosphere by the vehicle's evaporative control system. This system is discussed in a later chapter, but basically it consists of a carbon-filled canister that stores the fuel vapors until the engine is started. Once the engine is running and other operating conditions are met, the vapor in the canister is allowed to enter the intake manifold and be burned as part of the air-fuel mixture. However, during the process of refueling, the vehicle's fuel tank hydrocarbons can escape into the atmosphere. Regulations require that the vehicle have a method of preventing these vapors from escaping. Phase in of this regulation began in the 1998 model year. This control system is commonly referred to as **onboard refueling vapor recovery (ORVR)**. The manufacturer can develop any means of refueling vapor control. The following are a couple of examples of these systems.

One system example uses a tapered filler tube, a one-way check valve at the base of the filler tube, a liquid-vapor separator, and a vapor control valve (Figure 5-16). As the fuel tank is

The purpose of the onboard refueling vapor recovery (ORVR) system is to prevent hydrocarbon vapors from escaping to the atmosphere during refueling by drawing them into the fuel tank.

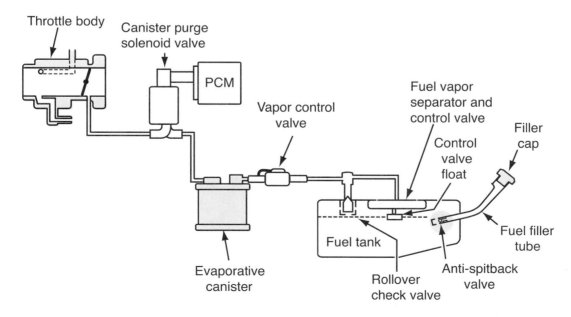

Figure 5-16 Onboard refueling vapor recovery system.

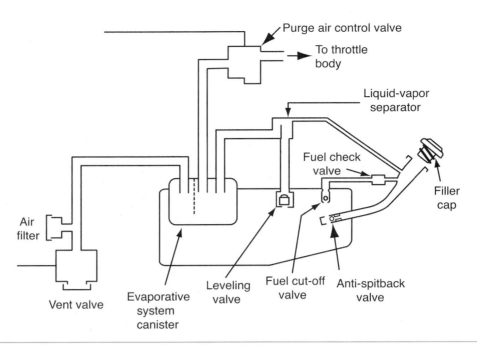

Figure 5-17   Another example of onboard refueling vapor recovery.

The one-way check valve is also referred to as the "anti-spitback valve."

being filled, the fuel from the nozzle travels down the filler tube and opens the one-way check valve. This allows the fuel to enter the fuel tank. The filler tube is tapered to create a venturi vacuum as the fuel flows down it. This vacuum draws the hydrocarbon vapors that would escape to the atmosphere down the tube with the fuel and to the top of the tank. The vapors then flow through the vapor control valve, the liquid-vapor separator, and into the charcoal canister for storage. This action vents the tank. As the fuel tank is filled, the float of the vapor control valve will rise. Once the tank is full, this float closes the passage to the canister. Since the tank is no longer venting, tank pressure increases and turns off the fuel nozzle. With the stop of fuel flow down the filler tube, the one-way check valve closes to prevent a sudden surge of fuel up the filler tube.

Another system example uses several check valves and additional tubing to prevent the escape of vapors (Figure 5-17). This system consists of a venturi filler tube (Figure 5-18), a one-way shut-off valve near the top of the filler tube (Figure 5-19), a liquid-vapor separator (Figure 5-20), a two-way check valve (Figure 5-21), fuel shut-off valve at the base of the filler tube (Figure 5-22), a fuel cut-off valve (Figure 5-23), and a leveling valve. In addition, the system also uses a canister vent solenoid valve assembly (Figure 5-24).

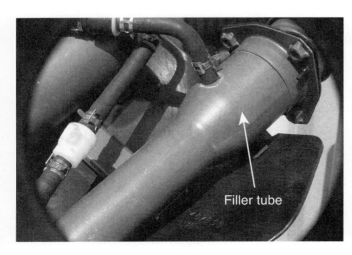

Figure 5-18   Venturi-shaped filler tube.

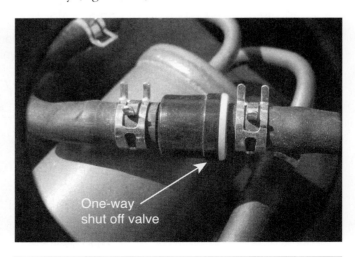

Figure 5-19   One-way shut-off valve.

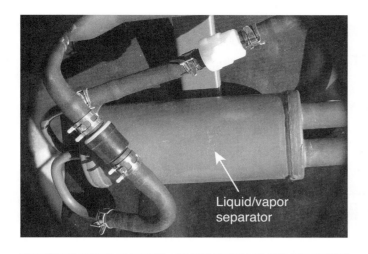

Figure 5-20 Liquid/vapor separator.

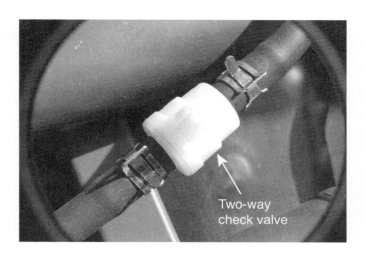

Figure 5-21 Two-way check valve.

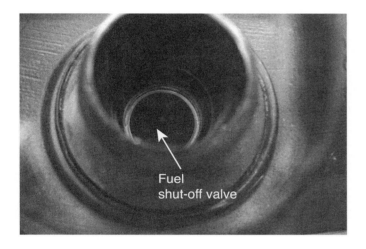

Figure 5-22 Fuel shut-off valve located at the bottom of the filler tube.

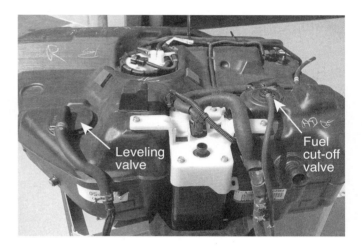

Figure 5-23 Fuel cut-off and leveling valve location.

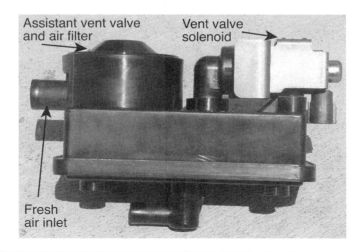

Figure 5-24 Vent solenoid valve assembly.

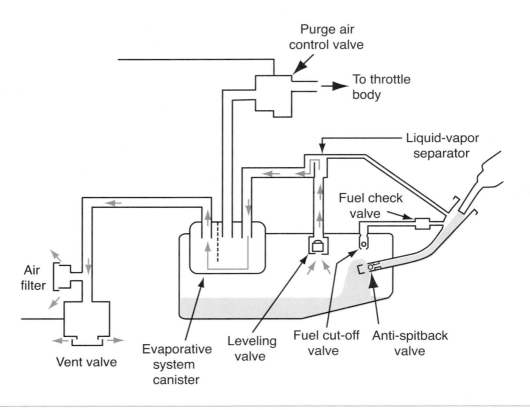

Figure 5-25 Vapor flow while tank is being filled.

As fuel is dispensed into the tank from the filling nozzle, the fuel flowing down the filler tube creates a venturi vacuum to draw the vapors into the tank. The fuel shut-off valve is open to allow the fuel and vapors to enter the tank (Figure 5-25). The vapors will flow past the leveling valve and into the liquid-vapor separator. Liquid fuel will return to the tank past the leveling valve while vapors are routed to the canister for storage. Once the tank is filled, the leveling valve closes the passage to the canister and causes tank pressure to increase and shut off the fuel nozzle.

With the vehicle parked, pressures inside the tank can increase due to an increase of ambient temperature. Depending on fuel level, the vapors in the tank can flow past the fuel cutoff valve or the leveling valve into the canister. Vapors passing the fuel cutoff valve enter the top of the filler tube through the two-way fuel check valve then flow through the separator and into the canister. Vapors that flow past the leveling valve enter directly into the separator and onto the canister. The canister vent solenoid valve and the assist vent valve are turned off to allow vapors in the canister to be cleaned by the filters and vent the tank (Figure 5-26).

While the vehicle is being driven, the level of the tank will slowly lower. The space that was once occupied by vapors is now replaced with fresh air (Figure 5-27). The canister vent solenoid valve and the assist vent valve are turned on. Fresh air passes the air filter and enters the canister. The atmospheric pressure is used to "push" the vapors from the canister into the intake manifold when the purge air control valve allows this. Also fresh air passes through the canister, backward through the separator, past the leveling valve, and into the tank. The atmospheric pressure in the tank is used to "push" fuel into the fuel pump module and to prevent a vacuum on the tank.

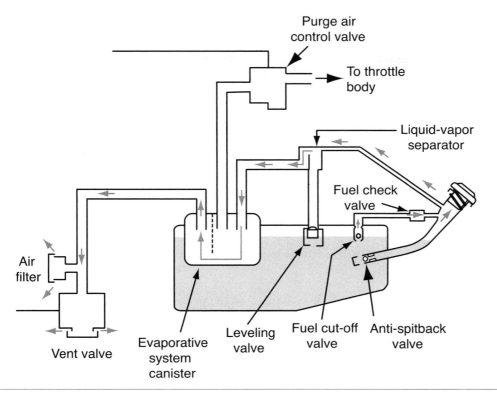

Figure 5-26 Vapor flow with vehicle parked.

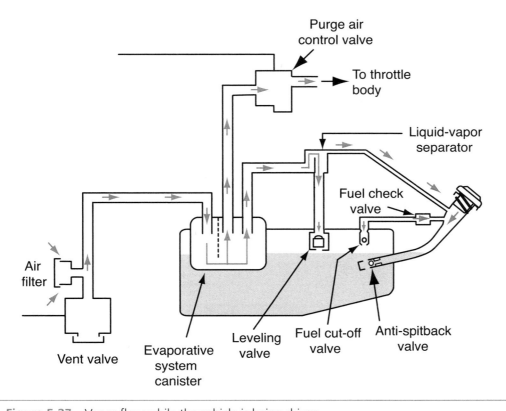

Figure 5-27 Vapor flow while the vehicle is being driven.

Fuel Pumps

During the early years of the automobile, many cars did not have a fuel pump. On some cars, like the Model A Ford, the fuel tank was mounted in the front of the instrument panel. The filler cap was positioned in front of the windshield. Since the gasoline tank was mounted higher than the carburetor, the fuel flowed by gravity from the tank to the carburetor. Other cars in those years had a vacuum tank, which used engine vacuum to move the fuel from the tank to the carburetor. Since gravity was used to fill the carburetor bowl, it was not uncommon to see motorists driving up a hill backwards so fuel from the tank would still flow to the carburetor. The first successful mechanical fuel pump was introduced in 1927 by the AC Spark Plug Company.

The **fuel pump** is the device that draws the fuel from the fuel tank through the fuel lines to the engine's carburetor or injectors.

An **electric fuel pump** uses a DC motor to drive a pump gear or roller set to draw fuel.

Positive displacement means that the same volume of fuel is delivered every rotation of the pump, regardless of speed.

The **fuel pump** supplies pressurized fuel to the carburetor or injectors. Early carbureted engines used a mechanical pump that was run off the engine's camshaft. Electronic fuel-injection systems utilize an **electric fuel pump**.

Electric fuel pumps offer several advantages over mechanical fuel pumps. First, they maintain constant fuel pressure, which aids in starting and reduces the potential for vapor lock. In addition, an electric fuel pump can be located either inside or outside the fuel tank. The reasons for locating the fuel pump in the fuel tank include: (1) to keep the fuel pump cool while it is operating and (2) to keep the entire fuel line pressurized to prevent premature fuel evaporation.

The in-tank pump is usually a **positive displacement** design pump. It can be either a roller vane type design with a permanent magnet electric motor, a two-stage gerotor design pump, or a design that was an impeller connected to the end of the armature shaft. The pump and its electric motor are fitted in a common housing in the fuel tank and are continuously surrounded by fuel. The fuel pump assembly consists of the pump, noise dampener, fuel house, intake strainer (filter), and mounting plates.

The impeller design in-tank electric fuel pump contains a small direct current (DC) electric motor with an impeller mounted on the end of the motor shaft (Figure 5-28). A pump cover containing the inlet and discharge ports is mounted over the impeller. When the armature and impeller rotate, fuel is moved from the tank to the inlet port, and the impeller grooves pick up the fuel and force it around the impeller cover and out the discharge port. Fuel moves from the discharge port through the inside of the motor and out the check valve and outlet connection, that is connected to the fuel line.

AUTHOR'S NOTE: Although it is dangerous to have a spark near gasoline, the in-tank fuel pump is safe because there is no oxygen to support combustion in the tank.

General Motors introduced the in-tank electric fuel pump on the 1969 Buick Riviera and then again in 1971 on the Chevrolet Vega.

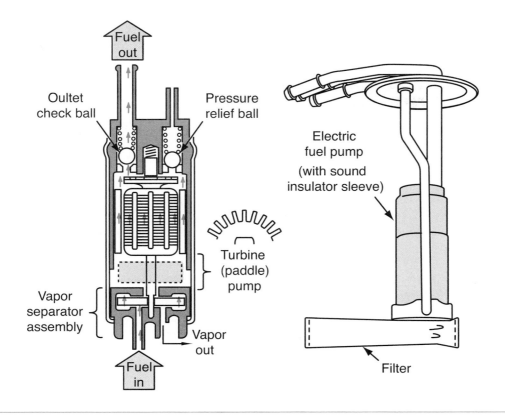

Figure 5-28 Electric fuel pump components.

The fuel pump/sending unit assembly is installed inside the fuel tank through a large flange opening on top of the fuel tank. This flange also provides the necessary electrical connections and fuel line connections. When installed, the flange seats against the synthetic rubber seal and is held in position by a lock ring.

Reservoir Chamber. The purpose of the reservoir chamber is to provide fuel at the pump inlet during all operating conditions, especially when the fuel level is low. The reservoir chamber prevents the pump from losing prime while the vehicle is going around corners, or when the vehicle experiences hard braking or acceleration with low-fuel levels.

Pressure Relief Valve. The fuel pump usually has two internal check valves. One check valve is located on the inlet side of the pump. This valve is used to regulate maximum fuel pump output. The **pressure relief valve** opens if the fuel supply line is restricted and pump pressure becomes too high. When the relief valve opens, fuel is returned through this valve to the pump inlet. This action protects fuel system components from excessive fuel pressure. Different fuel-injection systems will have varying specifications to the relief pressure. Many fuel systems used today have a relief pressure of approximately 120 psi (8 bar).

The **pressure relief valve** is a safety valve that opens to prevent damage to the fuel delivery system if fuel pump pressure becomes excessive.

Shop Manual
Chapter 5,
page 203

The **check-valve** is
used to prevent the
fuel in the lines from
returning to the tank
after the fuel pump
shuts off. It also
prevents fuel from
entering the fuel
lines with the pump
off in the event of an
accident.

Shop Manual
Chapter 5,
page 213

Fuel filters are
elements made from
pleated paper,
ceramic, or bronze
material used to
remove
contamination in the
fuel delivery system.

One-Way Check-Valve. The second **check-valve** is located on the outlet side of the fuel pump and prevents any movement of fuel in either direction when the pump is not operating. This traps a volume of fluid between the pump and injectors to help assure faster engine starting.

> **AUTHOR'S NOTE:** Remember the purpose of the check valve is to maintain a volume of fluid in the supply line. It does not maintain pressure over long periods of time. It is normal for fuel pressures to drop as the fuel cools. However, pressure should not drop off rapidly after the engine is shut off.

Return Line Check-Valve. The return line check-valve is hot staked in the fuel tank swirl reservoir assembly. The valve's only purpose is to prevent fuel flow in the reverse direction. Whenever the fuel tank is rotated from the normal position (rollover accident for example), the pressure in the fuel tank is greater than the pressure in the return line, so the valve closes to prevent fuel from draining out of the tank into the fuel line.

Fuel Filters

The fuel delivery system usually consists of two filters and a screen. The first filter is located in the fuel tank on the fuel pump inlet. This filter is actually a strainer located on the fuel pickup tube ahead of the fuel pump. The strainer is made of a finely woven fabric. The purpose of this strainer is to prevent large contaminant particles from entering the fuel system where they could cause excessive fuel pump wear or plug fuel-metering devices. It also helps to prevent passage of any water that might be present in the tank. The second filter is the main fuel filter, which is discussed next. The screen is installed into each injector's inlet port.

There are two different types of **fuel filters**. One type is contained in the same housing as the fuel pressure regulator and attaches to the fuel pump module (Figure 5-29). The second is mounted outside the tank in the fuel supply line between the fuel tank and the engine (Figure 5-30). Most fuel filters contain a pleated paper element mounted in the filter housing, which may be made from metal or plastic. Fuel filters on fuel injected vehicles usually have a metal case. On many fuel filters, the inlet and outlet fittings are identified, and the filter must be installed properly. An arrow on some filter housings indicates the direction of fuel flow through the filter.

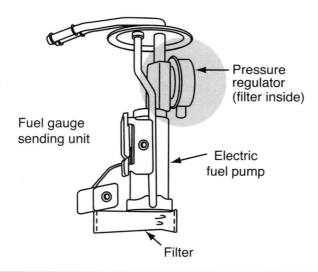

Figure 5-29 Combination fuel filter and pressure regulator.

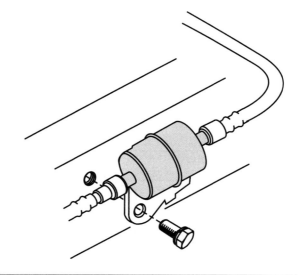

Figure 5-30 EFI fuel filter mounted on the frame rail.

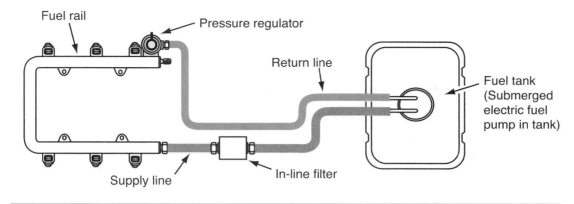

Figure 5-31 The fuel lines connect the fuel tank to the engine.

Fuel Lines

The fuel delivery system is the link between the fuel tank and the injector(s) (Figure 5-31). Depending on system design, the fuel lines are part of the high-pressure and low-pressure side of the system. The high-pressure side (fuel supply side) delivers a constant flow of fuel, under pressure, from the tank to the injector(s). On fuel injected engines, since the fuel pump is usually located in the fuel tank, the entire delivery side is pressurized so there is less likelihood of vapor bubbles forming. Those that may form are readily purged through the return circuit.

Some fuel systems contain a fuel return arrangement (low-pressure side) that aids in keeping the gasoline cool, thus reducing chances of vapor lock. Since the engine only requires a small portion of the delivered fuel, the remainder of the fuel may be returned to the fuel tank through the fuel return circuit. The fuel return line generally runs next to the fuel supply line. The fuel return system allows a metered amount of cool fuel to circulate through the tank and fuel pump, thus reducing vapor bubbles caused by overheated fuel in the tank. The return circuit is not pressurized.

Fuel lines can be made of metal tubing, flexible nylon, or synthetic rubber hose. Usually the fuel lines will be constructed using all of these materials in different locations.

Metal lines are generally used where the supply lines and return lines follow the frame along the under chassis of the vehicle. These metal lines extend from near the tank to a point near the fuel injector rail. Clips retain the fuel lines to the chassis to prevent line movement and damage (Figure 5-32). To absorb vibrations, the metal line is joined to the fuel pump and fuel rail by lengths of high-pressure flexible hose.

Shop Manual
Chapter 5,
page 215

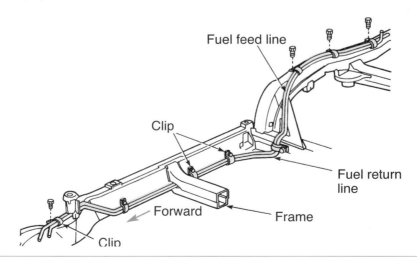

Figure 5-32 The fuel line must be properly supported along the frame.

The flexible high-pressure line, unlike filler neck or vent hoses, must work under pressure. Because of this, the flexible synthetic hoses must be stronger. This is especially true for the hoses on fuel-injection systems, where pressures reach 58 psi (400 kPa) or more. For this reason, flexible fuel line hose must also have special resistance properties.

Any synthetic rubber hose used in the fuel delivery or evaporative control systems must be able to resist gasoline. It must also be nonpermeable, so gas and gas vapors cannot evaporate through the hose. Ordinary rubber hoses, such as those used for vacuum lines, deteriorates when exposed to gasoline. Only hoses made for fuel systems should be used for replacement. Similarly, vapor vent lines must be made of materials that resist attack by fuel vapors. Replacement vent hoses are usually marked with the designation "EVAP" to indicate their intended use.

Fuel Line Connections

There are several methods used for making connections to the fuel lines. If the fuel filter is located in the fuel line, it is usually attached by means of a **banjo fitting** (Figure 5-33). The banjo fitting provides a positive connection by using two metal gaskets compressed between the two ends of the connection.

Sections of fuel line are assembled together by special fittings. A common fitting is the **quick connect** design (Figure 5-34). Leakage through the connection is prevented by the dual O-ring arrangement and the plastic spacer.

The fitting can also be a threaded fitting. The two most common threaded-type tube fittings are the **compression fitting** and the **double flare fitting** (Figure 5-35). The most common of these two is the double flare. The double flare is made with a special tool that has an anvil and a cone (Figure 5-36).

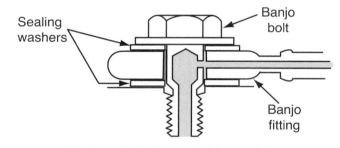

Figure 5-33   The banjo fitting uses a hollow bolt and two sealing washers.

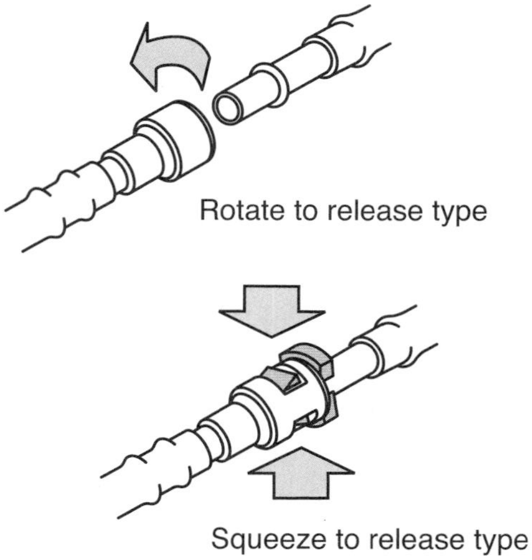

Figure 5-34   Example of quick-connect fittings.

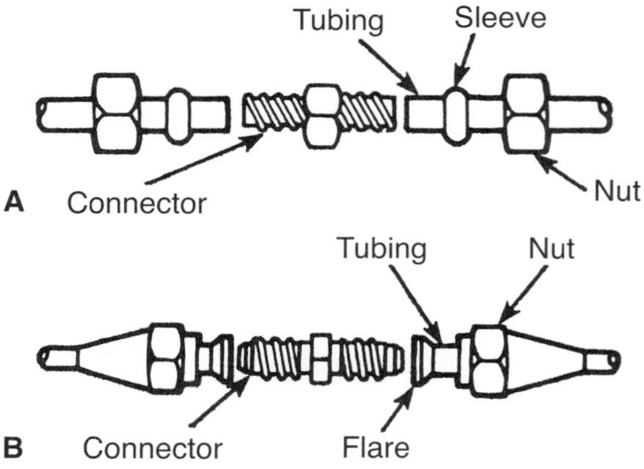

Figure 5-35   (A) Compression fitting, (B) Compression fitting.

Shop Manual
Chapter 5,
page 224

The **banjo fitting** is a metal connection using a circular filling with a hollow bolt and two steel sealing rings.

The **quick connect** fuel coupler consists of a plastic retainer, two O-rings, and a plastic spacer. The retainer engages and locks into position on a raised bead on the fuel line.

The **compression fitting** compresses the end of a sleeve to provide a seal.

The **double flare fitting** uses a procedure that folds the ends of the tube over itself, which doubles the thickness of the tube end and creates two sealing surfaces.

Shop Manual
Chapter 5,
page 217

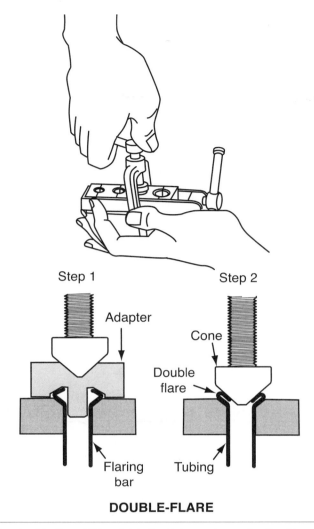

Step 1

Adapter

Step 2

Cone

Double
flare

Flaring
bar

Tubing

DOUBLE-FLARE

Figure 5-36   The double flare fitting folds the tube end over itself with a special tool.

Some fuel lines have threaded fittings with an O-ring seal to prevent fuel leaks (Figure 5-37). These O-ring seals are usually made from Viton, which resists deterioration from gasoline. On return lines that have no pressure, the fuel hose may be clamped to the steel line.

To control the rate of vapor flow from the fuel tank to the vapor storage tank, a plastic or metal restrictor may be placed in either the end of the vent pipe or in the vapor/vent hose itself.

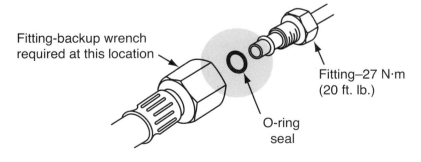

Fitting-backup wrench
required at this location

Fitting—27 N·m
(20 ft. lb.)

O-ring
seal

Figure 5-37   Some fittings use an O-ring seal.

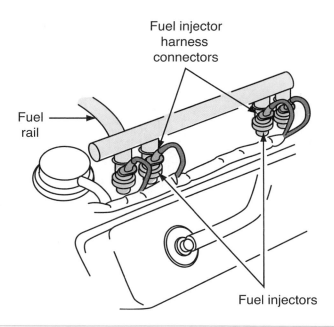

Fuel injector
harness
connectors

Fuel
rail

Fuel injectors

Figure 5-38   The fuel rail connects the fuel lines to the injectors.

Fuel Rail

Shop Manual
Chapter 5,
page 225

The **fuel rail** can be mounted on the intake manifold (Figure 5-38) or be an integral part of the intake manifold. The fuel rail can be constructed of metal or composite materials. The injectors mount to the fuel rail with push-on retaining clips, and use O-rings to prevent leakage between the injectors and the fuel rail.

Fuel Pressure Regulation

The **fuel rail** is a manifold system used to supply fuel to the injectors.

Since the fuel pump is a positive displacement design, a **pressure regulator** is needed to assure a constant pressure difference at the injector(s). The same pressure difference is programmed into the computer because the fuel pump produces more pressure and volume than the engine needs. A pressure regulator is needed to lower the pressure to the value programmed into the computer.

There are two basic types of regulation used depending on fuel delivery system classification: return type and returnless type. Although regulation works similarly in both systems, there are some major differences.

Shop Manual
Chapter 5,
page 203

Return-Type Systems

The **pressure regulator** maintains the proper fuel pressure at all times.

On most **return-type fuel systems**, pressure is controlled by a fuel pressure regulator mounted on the fuel rail (Figure 5-39). The pressure regulator is a mechanical device that is not controlled by the powertrain control module (PCM). The regulator contains a calibrated spring and a diaphragm that actuates the regulator valve. Fuel pressure operates on one side of the diaphragm, while spring pressure operates on the other side. The diaphragm opens the valve to the return port, allowing fuel to be dumped back into the fuel tank. System fuel pressure reflects the amount of fuel pressure required to open the port. The spring on the opposite side of the diaphragm attempts to close the valve, causing an increase of pressure on the fuel as it travels to the fuel rail.

On **return-type fuel systems**, all fuel is routed through the fuel rail, then what is not used by the engine is returned to the fuel tank.

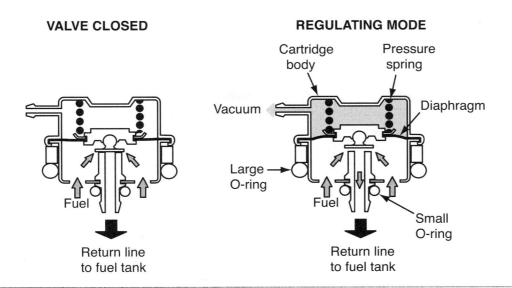

VALVE CLOSED

REGULATING MODE

Cartridge body

Pressure spring

Vacuum

Diaphragm

Large O-ring

Fuel

Fuel

Small O-ring

Return line to fuel tank

Return line to fuel tank

Figure 5-39 Fuel pressure regulator operation.

On port fuel-injection systems, there is a need to assure a constant pressure difference between the tip of the injector and the intake manifold under all driving conditions. To accomplish this, manifold vacuum is applied to the spring side of the diaphragm. The holding pressure of the spring will vary according to the amount of assist it receives from the intake manifold vacuum. At idle, vacuum being applied to the regulator lessens the effective spring rate, allowing fuel to be returned to the tank when pressure reaches desired specifications (approximately 38 psi (262 kPa), for example). Because a lesser volume of fuel is being used at idle, a greater amount of fuel must be bypassed to the tank to maintain the correct fuel pressure differential. Also, there is less fuel pressure at idle because high manifold vacuum helps pull fuel from the injector.

With low vacuum, such as full throttle, very little vacuum assist is available and the full force of the spring is exerted to seal the outlet, thus raising positive fuel pressure Under high speed and load conditions a greater volume of fuel is required. With limited vacuum assist on the diaphragm, the flow back to the tank is restricted to prevent a drop in pressure and volume from occurring. The fuel pressure must increase since there is less manifold vacuum pulling the fuel from the injector. With a reduction of vacuum on the injector, less fuel is delivered. The increase in fuel pressure will overcome this.

In short, the pressure regulator maintains a constant pressure differential by controlling the amount of fuel entering the fuel return line. As intake manifold vacuum changes, the differential pressure between the positive pressure in the fuel rail and the negative pressure (vacuum) in the intake manifold also changes. The following is offered as an example of this function. When intake manifold vacuum is a steady 20 inches Hg (–10 psi), the fuel pressure regulator must maintain a positive 29 psi fuel pressure. Under these conditions the fuel system (regulated by the pressure regulator) is pushing fuel through the injector and into the intake manifold at 29 psi. In addition, 20 inches Hg (–10 psi) of intake manifold vacuum is assisting the fuel supply system by pulling the fuel into the manifold at a negative pressure of –10 psi. Therefore, with the positive fuel pressure pushing at 29 psi and the negative intake manifold pressure (vacuum) pulling at –10 psi, the correct fuel pressure difference of 39 psi in the intake manifold is established.

Returnless-Type Fuel Systems

The **returnless-type fuel system** does not have a return line routed from the fuel rail to the fuel tank.

In recent years, many manufacturers have changed over to a **returnless-type fuel system**. There are three major advantages to this system over the return type system. The first advantage is lower fuel temperatures since all fuel is not being routed through the hot environment of the engine compartment and then returned to the tank. This reduces evaporative emissions, resulting in less evaporative canister purging. Second, since fuel only passes through the fuel filter one time before it is used by the engine, the fuel filter lasts longer. The third advantage is reduced cost since fewer components are used.

Like the return type system, the pressure regulator on a returnless-type system is a mechanical device containing a calibrated spring and a diaphragm that actuates the regulator valve (Figure 5-40). Fuel pressure operates on one side of the diaphragm, while spring pressure operates on the other side. The diaphragm opens the valve to the return port, allowing fuel to be dumped back into the fuel tank. System fuel pressure reflects the amount of fuel pressure required to open the port. The spring on the opposite side of the diaphragm attempts to close the valve, causing an increase of pressure on the fuel as it travels to the fuel rail. Usually the pressure regulator is part of the fuel pump module (Figure 5-41). It is part of a filter/regulator assembly on some vehicles and a separate component on others (Figure 5-42).

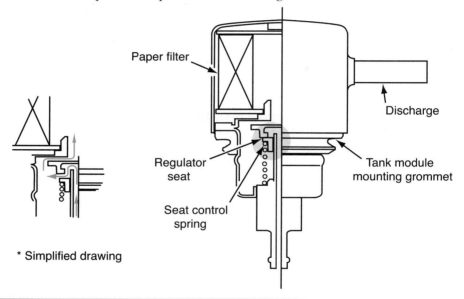

Figure 5-40   Returnless fuel system pressure regulator.

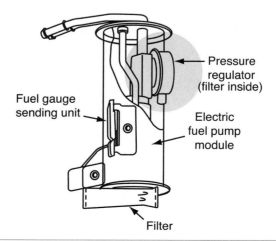

Figure 5-41   Returnless fuel system pressure regulator attached to the fuel pump module.

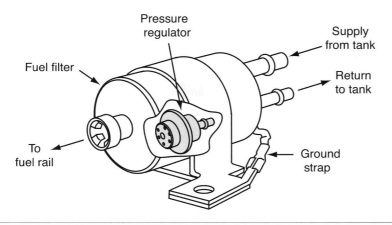

Figure 5-42   Returnless fuel system with remote mounted pressure regulator.

Returnless systems do not use engine vacuum like the return-type system. With the regulator mounted at the tank, a constant fuel pressure is always supplied to the injectors. The PCM uses a special formula that calculates the pressure differential across the injector and then adjusts injector pulse width accordingly.

Intake Manifolds

The most common materials used for **intake manifold** construction was cast iron and cast aluminum. In recent years the use of composites has expanded as well. On throttle body injected engines, the intake manifold will carry the air-fuel mixture to the combustion chamber (Figure 5-43). This means the intake charge will have droplets of fuel in it. The fuel droplets will stay in the charge only if the charge flows at high velocity. If the velocity drops below 50 feet (15.24 m) per second, the droplets separate from the charge. Unfortunately, an idling engine

The **intake manifold** directs the air or air-fuel mixture into the cylinders.

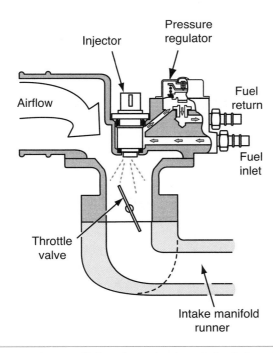

Figure 5-43   The air mixture will flow from the throttle body into the intake manifold.

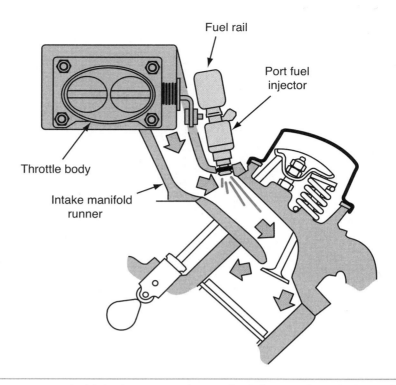

Fuel rail

Port fuel
injector

Throttle body

Intake manifold
runner

Figure 5-44 Port fuel-injection systems have only air flowing through the intake manifold. The fuel is added to the air just above the intake valve.

may not supply the velocity needed to keep the fuel droplets suspended in the charge. To overcome this, extra fuel must be supplied so a combustible mixture is delivered to the combustion chamber. Another way to overcome this tendency is to incorporate port fuel injection. Port injection uses an injector for each cylinder. The fuel is delivered on the manifold side of the intake valve, so the intake manifold only carries air on these engines (Figure 5-44).

Manifold design is a factor in velocity. The cross section of the runners must be large enough to supply a sufficient amount of charge, yet small enough to keep velocity high. Original equipment intake manifolds represent a compromise between these two concerns. A round runner shape provides the greatest cross-sectional area. However, many passenger vehicle engines are equipped with intake manifolds with rectangle-shaped runners. The flat floors allow for any fuel droplets that separate from the charge to be spread in a thin layer on the floor. The thin layer of fuel is quicker to evaporate than a puddle that would be developed from a round runner. In addition, the rectangular shape of the runner creates **eddy currents** in the corners. As the charge moves through the runner, turbulence is set in motion. The turbulence is a result of the runner design, bends, and interior surface. Normal turbulence of the charge is in a clockwise direction, however, the corners of the rectangle will cause a small turbulence in a counterclockwise direction. These eddy currents work to pick up any separated fuel from the floor and place them back into the charge.

To prevent fuel droplets from accumulating, the floor of the runner is designed so it will be level when the engine is installed in the vehicle. In addition, sharp bends in the runners will cause the fuel droplets to separate. Since air has less mass than fuel, it is capable of making the sharp turns easily. However, the fuel droplets cannot turn as easily and are thrown out of the charge.

Manufacturers design the manifold so an equal amount of air or air-fuel mixture is delivered to each cylinder. This is done by making each runner of the manifold carry the same volume. This is accomplished by making the runners the same length and equal cross section size. Most V-type engines use a dual plane intake manifold. This manifold has runners on two levels.

Eddy currents are currents that flow in reverse of the main current.

Tuned manifolds are designed to take advantage of the pressure wave that occurs naturally in a gas column. The length of the runner is determined so the pressure wave will reach the combustion chamber the instant the intake valve opens. This provides a ram effect and increases the efficiency of the engine.

Manifold design varies depending on engine types. Some manifold designs utilize engine coolant flow through internal passages. This design allows the warmed coolant to heat the air-fuel mixture.

All of the efforts of intake manifold design cannot compensate for a plugged air intake system. If the air filter becomes dirty, it will block the flow of air into the intake manifold. Without the proper amount of air, the engine will burn a rich air-fuel mixture, resulting in poor performance, low fuel economy, and increased emissions.

Electric Fuel Pump Circuits

There are several methods that manufacturers use to control the operation of electric fuel pumps used on fuel injected engines. This section will detail some of the most common circuits.

Fuel Pump Relay

Most electric fuel pump systems use a relay that is controlled by the PCM. Since the PCM controls when the relay will be energized, it in effect controls the fuel pump activation. The fuel pump relay is located inside the Power Distribution Center (PDC) or junction box on most vehicles. The fuel pump relay is energized under the following conditions to provide power to operate the fuel pump for approximately 0.7–1.5 seconds during the initial key-on cycle, and while the PCM is receiving an rpm signal that exceeds a predetermined value.

Ignition voltage is provided to the fuel pump relay's coil any time the key is in the RUN/START position (Figure 5-45). The PCM provides the ground control to energize the relay. The relay is energized when the key is cycled to RUN in order to prime the fuel rail with liquid fuel, allowing for a quick startup.

Any time the PCM receives an rpm signal that exceeds a predetermined value; the relay is energized to ensure proper fuel pressure and volume during engine cranking and running conditions. When the contacts in the relay close, battery voltage is supplied to the fuel pump. If the rpm signal is lost (engine has been shut off or the sensor indicates no rpm), the fuel pump relay is deenergized.

Shop Manual
Chapter 5,
page 204

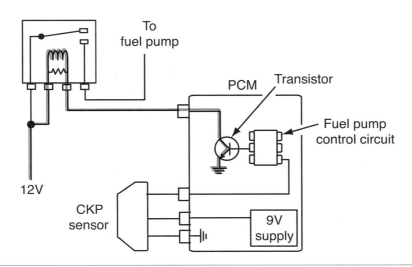

Figure 5-45   Fuel pump relay control circuit.

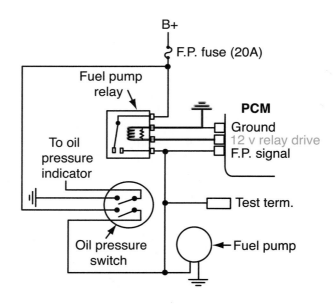

Figure 5-46 Fuel pump relay control using high-side switching.

Some manufacturers will supply voltage to the winding of the fuel pump relay instead of providing ground (Figure 5-46). Except for the use of high-side switching, the operation of the fuel pump is similar to that discussed above.

Inertia Switch

The **inertia switch** is used to shut off the fuel pump in the event of an accident that may rupture the fuel line and cause the engine to shut off after an accident.

On Ford products, an **inertia switch** is connected in series in the fuel pump circuit. The inertia switch consists of a steel ball inside a conical ramp, a target plate, and a set of electrical contacts. The magnet holds the steel ball in the bottom of the conical ramp. In the event of a collision, the inertia of the ball causes it to break away from the magnet, roll up the conical ramp, and strike the target plate (Figure 5-47). The force of the ball striking the plate causes the electrical contacts

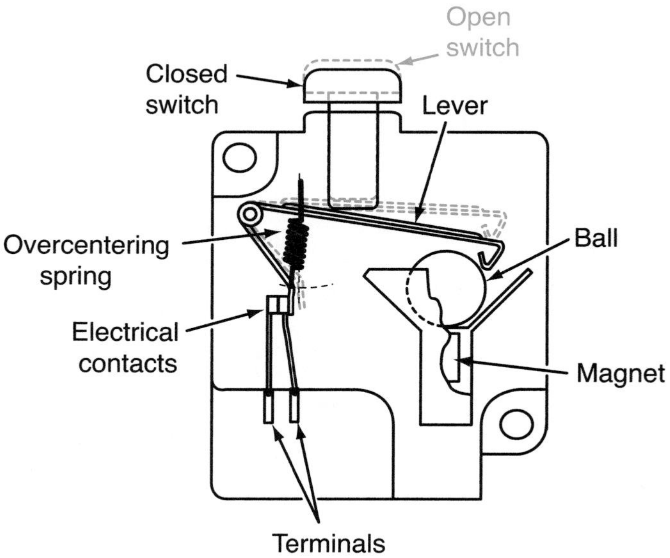

Figure 5-47 The inertial switch uses a weighted ball to trip the circuit contacts open if the vehicle is involved in an accident. The reset button closes the contacts.

in the inertia switch to open, removing voltage to the fuel pump. If tripped, a reset button on top of the inertia switch must be pressed to close the switch and restore fuel pump operation.

Circuit-Opening Relay

On many Toyota vehicles, the fuel pump relay is called a **circuit-opening relay**. This relay has dual windings (Figure 5-48). One of the windings is connected in series with the starter relay contacts and ground. The second relay winding is connected from the battery positive terminal to the PCM. When the engine is cranking and the starter relay contacts are closed, current flows through the starter relay contacts to the circuit opening relay winding and on to ground. This current flow creates a magnetic field around the circuit-opening relay winding that closes the relay contacts. When these contacts are closed, current flows to the fuel pump.

Once the engine starts, the starter relay is no longer energized, and current stops flowing through these relay contacts and to the first winding of the circuit-opening relay. However, the PCM grounds the other winding of the circuit-opening relay whenever the engine is running. This action keeps the relay contacts closed while the engine is running.

The **circuit-opening relay** is a dual-winding relay used to send current to the fuel pump while the engine is cranking and after it starts.

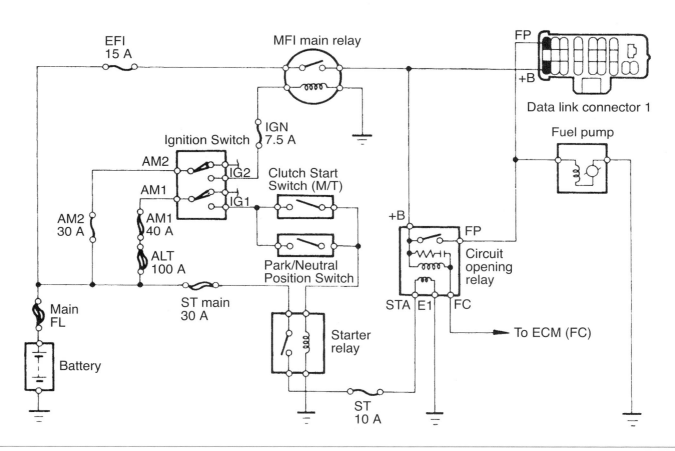

Figure 5-48   Circuit diagram of a circuit-opening relay fuel pump system.

Summary

❏ The main function of the fuel tank is to provide a means for storing the fuel to be used by the engine. The expanded function of the fuel tank is to prevent gasoline vapors (HC) from escaping into the atmosphere.

❏ Fuel tanks can be made from corrosion-resistant stamped steel, aluminum, or high density polyethylene (HDPE) material. The tank must pass stringent testing for HC leakage and crash integrity prior to its use in a vehicle.

❏ The fuel cap allows excessive pressures to be vented and for fresh air to enter the fuel tank. If air could not enter the tank and replace the fuel, the fuel pump would create a vacuum in the tank that could cause it to collapse. The cap must provide a positive seal between itself and the filler tube.

❏ The rollover check-valve, located at the top of the fuel tank, prevents fuel leakage from the tank in the event of a rollover accident.

❏ The onboard refueling vapor recovery (ORVR) system prevents hydrocarbon vapor from escaping into the atmosphere while the fuel tank is being filled.

❏ The fuel pump supplies pressurized fuel to the carburetor or injectors. Early carburetor-equipped engines used a mechanical pump while electronic fuel injection systems utilize an electric pump.

❏ The electric in-tank pump is a positive displacement design pump. It can be either a roller vane type design with a permanent magnet electric motor, a two-stage gerotor design pump, or a design that has an impeller connected to the end of the armature shaft.

❏ The pressure relief valve opens if the fuel supply line is restricted and pump pressure becomes too high.

❏ The fuel pump check-valve is located on the outlet side of the fuel pump, and it prevents any movement of fuel in either direction when the pump is not operating.

❏ The fuel delivery system usually consists of two filters and a screen.

❏ The fuel lines carry fuel from the tank to the fuel pump, fuel filter, and fuel injectors. They are made of either metal tubing or flexible nylon or synthetic rubber hose.

❏ The fuel rail connects the fuel lines to the individual injectors.

❏ Because the fuel pump is a positive displacement design, a pressure regulator is needed to control system pressures.

❏ There are two basic types of regulation used depending on fuel delivery system classification: return type and returnless type.

❏ The regulator contains a calibrated spring and a diaphragm that actuates the regulator valve. Fuel pressure operates on one side of the diaphragm, while spring pressure operates on the other side. The diaphragm opens the valve to the return port, allowing fuel to be dumped back into the fuel tank. System fuel pressure reflects the amount of fuel pressure required to open the port. The spring on the opposite side of the diaphragm attempts to close the valve, causing an increase of pressure on the fuel as it travels to the fuel rail.

❏ Intake manifold design is a factor in velocity. The cross section of the runners must be large enough to supply a sufficient amount of charge, yet small enough to keep velocity high.

❏ Manufacturers design the manifold so an equal amount of air or air-fuel mixture is delivered to each cylinder. This is done by making each runner of the manifold carry the same volume.

❏ Most electric fuel pump systems use a relay that is controlled by the PCM. Since the PCM controls when the relay will be energized, it in effect controls the fuel pump activation.

❏ An inertia switch in the fuel pump circuit opens the fuel pump circuit immediately if the vehicle is involved in a collision.

❏ If tripped, a reset button on top of the inertia switch must be pressed to close the switch and restore fuel pump operation.

❏ The circuit opening relay has dual windings. One winding is connected in series with the starter relay contacts and ground. The second relay winding is connected from the battery positive terminal to the PCM, which provides the ground.

Review Questions

Short Answer Essays

1. Name the components of a typical fuel delivery system.
2. Explain how expansion, contraction and overflow are controlled in the fuel tank.
3. Describe the function of the rollover valve.
4. Describe the purpose on the onboard refueling vapor recovery (ORVR) system.
5. Explain how the vacuum assisted fuel pressure regulator operates.
6. Explain the differences between return-type and returnless-type fuel systems.
7. Explain the purpose of the inertia switch.
8. Explain the purpose of the check-valves used in an electric fuel pump.
9. Describe the function of the intake manifold.
10. Explain the operation of the circuit opening relay.

Fill-in-the-Blanks

1. Some fuel tank filler caps contain a pressure valve and a _____ valve.
2. In an electric fuel pump, the relief valve opens if the fuel line pressure becomes _____.
3. The main function of the fuel tank is to provide a means for storing the fuel to be used by the engine. The expanded function of the fuel tank is to prevent _____ _____ from escaping into the atmosphere.
4. The _____ check valve, located at the top of the fuel tank, prevents fuel leakage from the tank in the event the vehicle overturns in an accident.
5. The on-board refueling vapor recovery (ORVR) system prevents _____ vapor from escaping into the atmosphere while the fuel tank is being filled.
6. The electric in-tank pump is a _____ displacement design pump.
7. The _____ _____ connects the fuel lines to the individual injectors.
8. The fuel pressure regulator contains a _____ _____ and a _____ that actuates the regulator valve.
9. Most electric fuel pump systems use a relay that is controlled by the _____ .
10. An _____ switch in the fuel pump circuit opens the fuel pump circuit immediately if the vehicle is involved in a collision.

Multiple Choice Questions

1. *Technician A* says the threaded filler cap should be tightened until it clicks.
   Technician B says the location of the filler tube to the tank is used to control fill level.
   Who is correct?
   - **A.** A only
   - **B.** B only
   - **C.** Both A and B
   - **D.** Neither A nor B

2. *Technician A* says the one-way check-valve prevents fuel flow from the underhood fuel system components into the fuel pump and tank when the engine is shut off.
   Technician B says the one-way check-valve prevents fuel flow from the pump to the fuel filter and fuel system if the engine stalls and the ignition switch is on.
   Who is correct?
   - **A.** A only
   - **B.** B only
   - **C.** Both A and B
   - **D.** Neither A nor B

3. *Technician A* says some electric fuel pumps are combined in one unit with the gauge sending unit.
   Technician B says on a vehicle with circuit open relays, low engine oil pressure may cause the engine to stop running.
   Who is correct?
   - **A.** A only
   - **B.** B only
   - **C.** Both A and B
   - **D.** Neither A nor B

4. Most onboard refueling vapor recovery (ORVR) systems work using:
   - **A.** the fuel pump to draw vapors down the filler tube.
   - **B.** venturi vacuum and a series of check valves.
   - **C.** a vacuum internal to the filling nozzle.
   - **D.** positive pressure to move the vapors back into the filling nozzle.

5. A port fuel injected engine with wet intake manifold refers to:
   - **A.** an air-fuel mixture flowing through the intake manifold to the intake valves.
   - **B.** a special glycol coating on the inside of the intake manifold.
   - **C.** a layer of moisture that lines the runners of the intake manifold.
   - **D.** air only flowing in the intake manifold to the injectors.

6. *Technician A* says in a circuit-opening relay system, one winding is connected in series with the starter relay contacts and ground.
   Technician B says the second relay winding is connected from the battery positive terminal to the fuel pump.
   Who is correct?
   - **A.** A only
   - **B.** B only
   - **C.** Both A and B
   - **D.** Neither A nor B

7. The function of the fuel tank is:
   - **A.** to provide a means for storing the fuel to be used by the engine.
   - **B.** to prevent gasoline vapors (HC) from escaping into the atmosphere.
   - **C.** Both A and B.
   - **D.** Neither A nor B.

8. Which of the following filters may be found within the fuel-delivery system?
   - **A.** Pick-up strainer
   - **B.** Screen in the injector
   - **C.** Filter in the supply circuit
   - **D.** All of the above

9. *Technician A* says banjo fittings may be used to connect the fuel filter to the supply circuit.
   Technician B says a quick connect fitting may require special tools.
   Who is correct?
   - **A.** A only
   - **B.** B only
   - **C.** Both A and B
   - **D.** Neither A nor B

10. *Technician A* says that return-type fuel systems usually use a fuel pressure regulator that does not sense engine vacuum.
    Technician B says that returnless-type fuel systems usually use pressure regulators that have a vacuum line connected to them to control pressures under different operating conditions.
    Who is correct?
    - **A.** A only
    - **B.** B only
    - **C.** Both A and B
    - **D.** Neither A nor B

Fuel System Principles

Upon completion and review of this chapter, you should be able to:

❏ Describe the importance of the air-fuel ratio to engine performance and emission levels.

❏ Describe the importance of stoichiometric as related to emissions and performance.

❏ Explain the difference between open loop and closed loop fuel strategies.

❏ Explain the operation of the speed density electronic fuel injection (EFI) system.

❏ Explain the operation of the mass air flow (MAF) electronic fuel injection (EFI) system.

❏ Explain the operation of the mass air flow sensor.

❏ Describe the function of adaptive memory.

❏ Explain the fuel strategies used during different modes of operation.

❏ Describe the operation of the throttle body fuel injection (TBI) system.

❏ Describe the difference between the sequential fuel injection (SFI) system and the multipoint fuel injection (MPI) system.

❏ Describe the meaning of pulse width.

❏ Describe the function of the central port injection (CPI).

Introduction

Although some European manufacturers used fuel injection for many years, it was not until the 1980s that U.S. automotive engineers started to change engines from carburetors or computer-controlled carburetors and then to **electronic fuel injection (EFI)** systems. This action was taken to improve fuel economy, performance, and emission levels. Automotive manufacturers had to meet increasingly stringent **corporate average fuel economy (CAFE)** regulations and comply with stricter emission standards at the same time.

✐ **AUTHOR'S NOTE:** The term "electronic fuel injection" was used by some manufacturers to describe their fuel system. Today, this term is no longer used; it has been replaced with more specific descriptions. In this book, the term EFI is applied to any electronic or computer-controlled fuel injection system.

Many of the EFI systems of the 1980s were **throttle body injection (TBI)** systems in which the fuel was injected above the throttles. The TBI systems have been gradually changed to port fuel injection (PFI) systems with the injectors located in the intake ports. On throttle body injected engines, some intake manifold heating was required to prevent fuel condensation on the intake manifold passages. When the injectors are positioned in the intake ports, intake manifold heating is not required. This provided engineers with increased intake manifold design flexibility. Intake manifolds could now be designed with longer, curved air passages, which increased air flow and improved torque and horsepower.

Since intake manifolds no longer require heating, they can now be made from plastic materials such as fiberglass reinforced nylon resin. This material is considerably lighter than cast iron or even aluminum. Saving weight means an improvement in fuel economy. The plastic-type intake manifold does not transfer heat to the air and fuel vapor in the intake passages; this improves fuel economy and hot start performance.

This chapter discusses the importance of the proper air-fuel ratio to engine performance and emissions. We will also cover the principles of operation of typical types of EFI systems.

The Importance of the Air-Fuel Ratio

The internal combustion engine converts the energy found in fuel into heat energy, which is changed to the mechanical energy that is used to move a vehicle. The conversion of energy

---

Electronic fuel-injection (EFI) refers to any fuel-injection system that is electronically controlled including TBI, MFI and SFI systems.

Corporate average fuel economy (CAFE) standards are federally imposed regulations requiring vehicle manufacturers to meet a specific average fleet fuel mileage for all their vehicles.

Throttle body injection (TBI) refers to a system that has the injector(s) located in a throttle body assembly and above the throttle plates.

results from the mechanical design of the engine and through the burning process of combustion. During combustion, a chemical reaction takes place between the air and the fuel that entered the engine's cylinders. During this chemical reaction, heat energy is released. The release of heat results in an increase in the pressure of the air in the cylinders. This high pressure pushes on the engine's piston, forcing it to move downward. The movement of the piston is converted to rotational movement in order to move the vehicle.

The energy resulting from the movement of the piston is dependent upon the amount of pressure there is on the piston. The amount of pressure on the piston depends on the amount of heat generated by the combustion process. Complete combustion of the air and fuel will provide for the maximum amount of heat from that amount of air and fuel.

In order to have complete combustion, four things must be present: (1) the correct amount of air must be mixed with (2) the correct amount of fuel in (3) a sealed container, and this mixture must be (4) shocked by the correct amount of heat at the correct time. Although there are other factors that can affect combustion, these are the most important ones.

Although no engine can achieve complete combustion during all operating conditions, total combustion is strived for in the design of engine and fuel systems. The result of complete combustion is the generation of large amounts of heat energy, and the conversion of all of the cylinder's air and fuel into water and oxygen. When complete combustion does not take place, full conversion of the air fuel and also does not take place. This results in the release of pollutants and reduced power from the engine.

The correct **air-fuel ratio** is required for an engine to provide good driveability, to reduce emissions, and to prevent internal damage to engine parts. The desirable air-fuel mixture is about 15 parts air to one part fuel (expressed as 15:1). However, the requirements change based on driving conditions. A 15:1 ratio would not produce enough energy to satisfactorily power the vehicle up a hill or to accelerate away from a stop sign. A normal air-fuel mixture for high power is about 12 parts of air to 1 part of fuel (Figure 6-1). The ratio for maximum economy is 15:1–16:1 while maximum power occurs from an air-fuel ratio of 12:1–12.5:1 (Figure 6-2).

The **air-fuel ratio** of the engine is measured by weight in pounds. The desirable air-fuel mixture is about 15 pounds of air to one pound of fuel. A 15:1 air-fuel ratio is 9,000 gallons of air to 1 gallon of fuel.

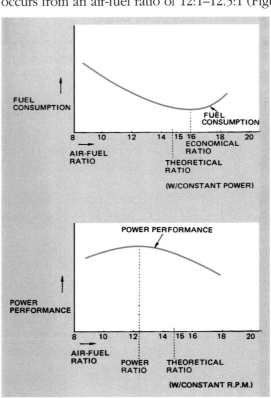

Figure 6-1 The air-fuel ratio must change to meet the requirements of various operating conditions of the engine.

ENGINE OPERATING CONDITION	AIR/FUEL RATIO (AIR:FUEL)
Starting (Air temperature approx. 0°C)	Approx. 1 : 1
Starting (Air temperature approx. 20°C)	Approx. 5 : 1
Idling	Approx. 11 : 1
Running slow	12 – 13 : 1
Accelerating	Approx. 8 : 1
Max. output (full load)	12 – 13 : 1
Running at medium (economical) speed	16 – 18 : 1

Figure 6-2   Different air-fuel ratios for different operating conditions.

Not only is the air-fuel ratio a major factor in engine performance, it is a common element in the formation of each of the three main pollutant gases: hydrocarbons (HC), carbon monoxide (CO), and oxides of nitrogen (NOx). Each of these emissions can be caused by improper air-fuel ratio during the combustion process (Figure 6-3). Maintaining the correct air-fuel ratio has there-

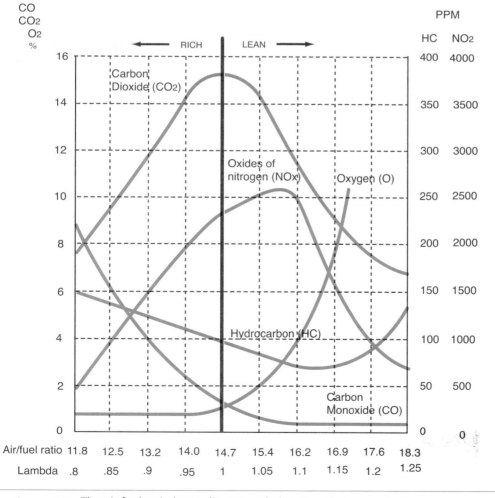

Figure 6-3   The air-fuel ratio has a direct correlation to emission production.

fore become one of the main goals of emissions control system design, and it has ultimately resulted in today's computer-controlled fuel-injection systems.

An air-fuel mixture that is too **lean** can result in excess levels of HCs and NO_x (depending on other circumstances). An excessively lean mixture can also cause a **lean-burn misfire**. Since there is too little fuel in the mixture to ignite, what is in the combustion chamber passes through the exhaust system unburned. This results in a cylinder that produces reduced power output and high HC levels. In addition, when the fuel mixture is lean, but not too lean to burn, the excess oxygen produces such a hot flame that combustion temperatures become excessive. These high temperatures cause nitrogen in the intake air to combine with oxygen to form harmful oxides of nitrogen (NO_x).

A **rich** mixture may be desirable under certain operating conditions. When the engine is under load, the mixture must be enriched (12:1 ratio for example) in order to produce enough energy to move the vehicle. In addition, when the engine is started, it requires a richer air-fuel mixture. The colder the engine temperature, the richer the mixture must be.

On EFI systems, the mixture is enriched by increasing the injector pulse width. The amount of pulse width increase is based on the input from coolant temperature sensor signal to the computer. The PCM supplies the proper air-fuel ratio and engine rpm when starting a cold engine. This also eliminates the need for the driver to depress the accelerator pedal while starting the engine.

AUTHOR'S NOTE: When an EFI-equipped engine is cold, the computer provides a very rich air-fuel ratio for faster starting. However, if the engine does not start because of an ignition defect, the engine becomes flooded quickly. Under this condition, excessive fuel may run past the piston rings into the crankcase. Therefore, when a cold EFI engine does not start, periods of long cranking should be avoided.

If the driver suspects that the air-fuel ratio is extremely rich, the driver may depress the accelerator pedal to the wide-open position while cranking the engine. Under these conditions, the computer program provides a very lean air-fuel ratio. Most manufacturers have programmed their PCMs to shut down injector operation altogether during this time. This mode is referred to as a **clear flood**. However, under normal conditions, the driver should not push on the accelerator pedal at any time when starting an engine with EFI. When the engine is decelerated, the computer reduces injector pulse width to provide a lean air-fuel ratio, which reduces emissions and improves fuel economy.

Although there are circumstances that require a richer air-fuel mixture, a mixture that is too rich (9:1 for example) can lead to reduced fuel economy and increased levels of HCs and CO, but NO_x emissions will generally be low.

Interestingly, HC emissions can be caused both by an excessively lean or an excessively rich mixture. In the case of a lean mixture, there is not enough gasoline to ignite, and it passes unburned through the exhaust system. In a very rich mixture, there may not be enough oxygen present to support combustion. The end result is the same unburned HCs go out the tailpipe.

High CO levels are a pretty sure sign of a rich fuel mixture, especially if high HC levels are also present. In a rich mixture, there is insufficient oxygen to combine with the carbon atoms from the gasoline to form harmless carbon dioxide (CO_2). Instead, each carbon atom combines with only one oxygen atom to form poisonous carbon monoxide (CO).

NO_x levels will usually be low in a rich mixture situation because the shortage of oxygen results in lower combustion temperatures. Remember, NO_x is formed when nitrogen from the air combines with oxygen at high temperatures.

A **lean** mixture is one in which there is an excess of oxygen mixed with the gasoline.

The **lean-burn misfire** results when a given volume of fuel mixture simply does not contain enough gasoline to ignite.

A **rich** air-fuel ratio contains excessive fuel in relation to the amount of air entering the cylinders.

Clear flood mode of operation is accomplished by pressing the accelerator pedal to the wide-open throttle position while cranking the engine. The combination of these inputs has the PCM turn off the injectors.

Starting a fuel injected engine without depressing the accelerator pedal may be called no-touch starting.

The Stoichiometric Ratio

At this point it is obvious that there must be an air-fuel mixture that would be considered optimum. The optimum air-fuel ratio is called the **stoichiometric** ratio (pronounced stoy-kee-o-MET-ric). In a gasoline fuel system, the stoichiometric ratio is an air-fuel mixture is 14.7:1. This indicates that for every 14.7 pounds of air entering the air intake, the fuel injectors supply 1 pound of fuel. At the stoichiometric air-fuel ratio, combustion is most efficient, and nearly all the oxygen in the air is mixed with fuel and burned in the combustion chambers. Computer-controlled fuel injection systems maintain the air-fuel ratio at stoichiometric under most operating conditions. As the air-fuel ratio cycles from slightly rich or slightly lean from the stoichiometric ratio, the oxygen sensor (O_2S) voltage cycles from high to low. The computer **feedback** fuel system tailors the air-fuel mixture to adapt to various engine conditions.

The stoichiometric ratio is the point at which the three pollutant gases are at controllable levels. At this ratio, there are exactly enough oxygen atoms to combine with the fuel molecules in the combustion chamber.

In working with air-fuel ratios, you may occasionally see the term **lambda**. This is simply a number that compares the actual air-fuel ratio to the stoichiometric ratio:

$$Lambda = Actual\ ratio/Stoichiometric\ ratio$$

When the actual air-fuel ratio is at stoichiometric, lambda = 1. When the actual mixture is rich, lambda is less than one. When the actual mixture is lean, lambda is greater than one.

As the air-fuel mixture moves away from stoichiometric (either too lean or too rich), a number of conditions could occur, including emissions test failures and engine problems (Table 6-1).

TABLE 6-1 EFFECTS OF AIR-FUEL RATIO ON EMISSIONS AND DRIVEABILITY

Air-Fuel Mixture	Effect on Vehicle
Too Lean	Increased NO_x emissions Poor engine power Misfiring at cruising speeds Burned valves Burned pistons Scored cylinders Spark knock or ping
Slightly Lean	Low exhaust emissions High gas mileage Reduced engine power Slight tendency to knock or ping
Stoichiometric	Best all-around performance and emissions levels
Slightly Rich	Increased CO emissions Increased HC emissions Maximum engine power Higher fuel consumption Less tendency to knock or ping
Too Rich	Increased CO emissions Increased HC emissions Poor fuel mileage Misfiring Oil contamination Black exhaust

The name stoichiometric comes from Greek words meaning "measured element." Mixing air and fuel at the stoichiometric ratio of 14.7:1 is the single most important technique that is used to control emissions levels. When the air-fuel ratio is at the stoichiometric ratio, every fuel molecule combines with every oxygen molecule, with nothing left over (theoretically, at least).

Feedback, as it relates to fuel systems, means the system uses an oxygen sensor to provide feedback information to the computer concerning combustion quality.

The air-fuel mixture is expressed either as the ratio of air-to-fuel vapor or as a lambda value. The lambda value is derived from the stoichiometric air-fuel ratio. The stoichiometric ratio is 14.7:1 when expressed as an air-fuel ratio, or 1 when expressed as a lambda value.

The quest for a reliable method of achieving stoichiometry has led to the evolution of fuel delivery systems from simple carburetion to today's computer-controlled multi-port fuel injection. Along the way, there have been interim developments such as electronic feedback carburetors and mechanical fuel injection. However, conditions inside the engine's combustion chamber are not ideal. Even with a stoichiometric ratio, the engine's exhaust gases contain a certain percentage of unburned fuel in the form of hydrocarbons (HC) and carbon monoxide (CO). In addition, at higher combustion temperatures, some of the free oxygen and nitrogen gases combine, forming various oxides of nitrogen (NO_x). Obtaining stoichiometric efficiency makes the vehicle's emission system more efficient at controlling these left over emissions.

Open Loop vs Closed Loop

In a computer-controlled fuel system, **open loop** occurs when the engine coolant is cold or the O_2 sensor signal is ignored by the PCM. When the engine is first started, the oxygen sensor (O_2S) is too cold to produce a satisfactory signal, and the computer program controls the air-fuel ratio without the O_2S input.

Open loop is also the computer strategy used for some driving conditions. The engine will always use open loop fuel strategy when it is first started (regardless of coolant temperature). Open loop strategy is also used during wide-open throttle operation to provide a rich air-fuel mixture to meet the load requirements. Also, open loop strategy is used when the throttle is closed quickly. Under this condition, the injectors may be turned off briefly and cause a very lean air-fuel mixture to prevent unburned fuel from exiting the exhaust.

Once the O_2 sensor warms sufficiently from the heat of the exhaust gases (and by an internal heating element) that it is able to provide an accurate voltage signal, the computer enters the **closed loop** mode. In closed loop, the computer uses the (O_2S) signal to control the air-fuel ratio. Determination of closed loop operation varies between manufacturers. Some base closed loop operation on O_2 sensor activity along with throttle position sensor (TPS) and manifold absolute pressure (MAP) or mass air flow (MAF) inputs. Some manufacturers use a closed loop timer that will put the PCM into closed loop after a certain amount of engine run time has elapsed. The length of the time is based on the engine coolant temperature (ECT) sensor value at engine start-up. In addition, a few systems do not enter closed loop until the (ECT) sensor signal informs the computer that the coolant temperature is approximately 175°F (79.4°C) or hotter and the O_2S signal is valid.

Therefore, the ECT sensor signal is very important because it determines the open or closed loop status. For example, if the engine thermostat is defective and the coolant temperature never reaches 175°F (79.4°C), the computer-controlled fuel system never enters closed loop. When this open loop condition occurs, the air-fuel ratio is continually rich, fuel economy is reduced, and emissions are increased. Some systems go back into open loop during prolonged periods of idle operation when the O_2S cools down. Also, most systems revert to open loop at or near wide-open throttle to provide a richer air-fuel ratio.

To keep the computer in closed loop during idle, and to get the O_2S hot faster, most (since 1993) O_2S are heated electrically (Figure 6-4). Battery voltage is sent to the heater element when the ignition switch is in the RUN position. The heater element is a positive temperature coefficient (PTC) resistor that operates as a current regulator. When the O_2S is hot, the PTC resistance increases, thus it causes a reduction in current flow through the heater element.

Open loop means the oxygen sensor is out of the fuel strategy loop. The oxygen sensor is ignored during open-loop operation.

Closed loop means the oxygen sensor is involved in the fuel strategy loop. The voltage value of the oxygen sensor is used as feedback.

Shop Manual
Chapter 6,
page 253

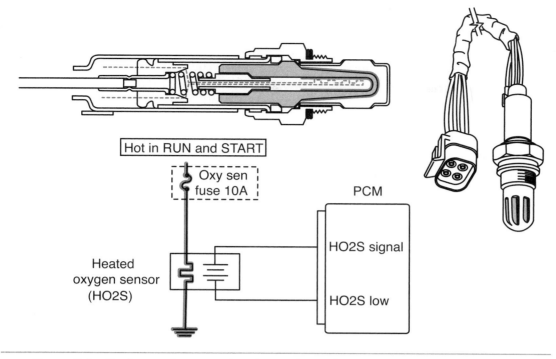

Figure 6-4 The heater element makes the O₂S hotter faster and keeps the PCM in closed loop during extended idle.

Computer-Controlled Air-Fuel Ratio Strategy

In an EFI system, the computer must know the amount (mass) of air entering the combustion chamber so it can determine the proper amount of fuel required to maintain the stoichiometric air-fuel ratio. Since the amount of air entering the combustion chambers is a constantly moving value, fast responding onboard computer systems are needed. The computer will use an oxygen sensor (O₂S) to measure the oxygen content of the exhaust. This provides information concerning how well the computer did in calculating the correct air-fuel mixture for the previous burn. The computer then works to maintain driveability while maintaining the 14.7 to 1 ratio as closely as possible.

Injector **pulse width** determines the amount of fuel that will be delivered into the combustion chamber. Under most closed-loop operating conditions, the computer provides the correct injector pulse width to maintain the stoichiometric air-fuel ratio. For example, the computer might ground the injector for 2 milliseconds at idle speed and 7 milliseconds at part throttle to provide the stoichiometric air-fuel ratio. There are three basic strategies, or methods, manufacturers use to maintain the stoichiometric ratio: speed density, mass airflow, and density speed. Regardless of the system used, the computer must know the amount of air entering the engine in order to provide the proper pulse width.

Shop Manual
Chapter 6,
page 264

The length of time the computer energizes the injector is referred to as **pulse width**.

Speed density
refers to the fuel
injection system that
used engine speed
and MAP inputs to
infer the mass of air
entering the
combustion chamber
and responded
based on
programmed look-
up tables in the
PCM's memory
chips.

The **manifold
absolute pressure
(MAP) sensor** is a
piezoresistive or
capacitance
discharged sensing
device used to
determine the load
on the engine by
sensing the
difference between
atmospheric pressure
and engine vacuum.

Speed Density Systems

The **speed density** fuel-injection system uses the rpm input as the most important sensor. Without the rpm input, the engine cannot run. The **manifold absolute pressure (MAP) sensor** provides a varying voltage or frequency with engine load. The MAP sensor is the second most important sensor in this type of system. These two inputs are used to calculate air density entering the combustion chamber. This calculated value is applied to a formula to determine the amount of fuel to be injected.

In addition, the rpm input determines the injection frequency (how many times the injector is fired), while the MAP sensor will determine the major amount of fuel (pulse width). So the rpm signal "says" how often the engine gets fuel and the MAP sensor "says" how much fuel the engine gets. The other sensors/switches will also have an effect on pulse width, but the MAP sensor determines the major amount (base pulse width). This type of EFI system is referred to as a speed density system, because the computer calculates the air intake flow from the engine rpm input, and the density of the intake charge through the manifold vacuum input. Therefore, the computer must have accurate signals from these inputs to maintain the correct air-fuel ratio. The other inputs are used by the computer to "fine tune" the air-fuel ratio. For example, if the TPS input indicates sudden acceleration, the computer momentarily supplies a richer air-fuel ratio. Below is the formula for a typical speed density pulse width calculation. The computer will assign a multiplicative value to the input signal and then determine the pulse width based on these values. The following is the pulse width calculation used by the PCM:

Load		Base PW Calculation		Adaptives		
RPM MAX RPM (X)	MAP BARO	(X) TPS (X) ECT (X) IAT (X) Sensed B+ (X) LT*	(X) O2	Short (X)Term	Long (X)Term	Pulse = Width

*After long-term adaptive information is stored in memory, it becomes part of the Base PW Equation, and is used under ALL operating conditions; hot or cold, open or closed loop.

The PCM requires crankshaft and camshaft position sensor inputs to determine when to fire the injectors. The crankshaft position sensor supplies engine speed information to the PCM. This information is required to determine if the engine is starting or in the run mode. In addition, the PCM calculates the volume of fuel needed from the rpm signal (and other sensors). The rpm signal is used to determine how much air can enter the combustion chamber under that particular engine speed. All EFI systems use some method of determining the volume of airflow. Most systems will not start if the crankshaft speed sensor input is missing. However, some systems will start and run the engine as long as either a crankshaft or camshaft position sensor signal is received.

Most MAP sensors are a form of a Wheatstone bridge stress gauge. The manifold absolute pressure (MAP) sensor is usually mounted in the engine compartment. A hose may be connected from the intake manifold to a vacuum inlet on the MAP sensor, and three wires are connected from the sensor to the computer. Some systems insert the map sensor directly into the intake manifold, thus eliminating the need for a hose. The computer supplies a constant 5-volt reference voltage to the sensor; the other wires are a signal wire and a ground wire.

The computer uses the MAP sensor signal to determine the engine load and volume of air entering the engine. When the MAP sensor signal indicates wide-open throttle and heavy load conditions, the computer provides a richer air-fuel ratio. The computer supplies a leaner air-fuel ratio if the MAP sensor signal indicates a light load, and moderate cruising speed conditions.

Many MAP sensors will also perform the function of a barometric pressure (BARO) sensor. When the ignition switch is turned on, but before the engine is started, the MAP sensor signal informs the computer regarding atmospheric pressure. Atmospheric pressures vary in relation to altitude and atmospheric conditions such as humidity. Some applications have a separate barometric pressure sensor.

Many manufacturers use MAP sensors that contain a silicon diaphragm. When the manifold vacuum stresses the silicon diaphragm, an analog voltage signal is sent from the sensor to the computer in relation to the amount of vacuum. On a typical MAP sensor, the signal voltage changes from between 1 volt and 1.5 volts at idle to 4.5 volts at wide-open throttle.

Another variation of the MAP sensor is a capacitance discharge MAP. Instead of using a silicon diaphragm, this system uses a variable capacitor. In the capacitor capsule-type MAP sensor, two flexible alumina plates are separated by an insulating washer (Figure 6-5). A film electrode is deposited on the inside surface of each plate and a connecting lead is extended for external connections. The result of this construction is a parallel plate capacitor. This capsule is placed inside a sealed housing that is connected to intake manifold pressure. As manifold pressure increases (goes toward atmospheric), the alumina plates deflect inward, resulting in a decrease in the distance between the electrodes. Since the formula for capacitance is:

$$C = \frac{\in_0 A}{d}$$

Where,

$\quad$ $\in_0$ is the dielectric constant of air

$\quad$ A is the area of the film electrodes

$\quad$ d is the distance between electrodes

Since every value in the formula is constant except for the distance between the electrodes, a measure of capacitance constitutes a measurement of pressure.

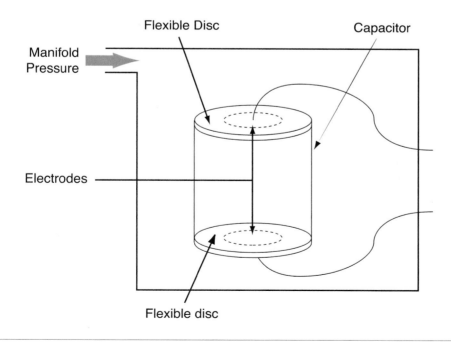

Figure 6-5   Capacitance discharge MAP sensor.

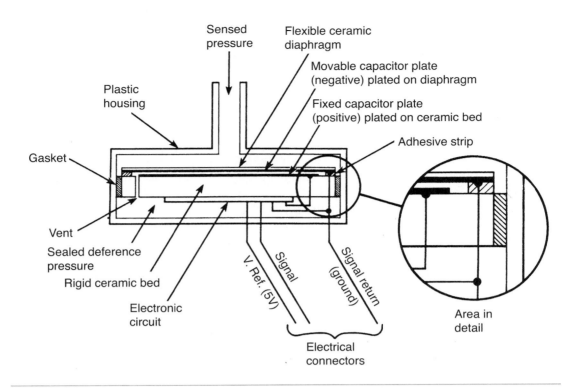

Ford refers to this condition as low MAP because there is a greater difference between atmospheric pressure and manifold vacuum.

This condition may be called high MAP because the manifold vacuum is closer to atmospheric pressure.

The **mass air flow (MAF) sensor** is located in the air intake system to directly measure the mass of air entering the engine.

Shop Manual
Chapter 6,
page 239

The **mass air flow (MAF)** systems use a MAF sensor that provides a direct input regarding the amount of air entering the engine.

The mass air flow system is also referred to as an "air density system."

Figure 6-6 Typical variable capacitor sensor construction.

Ford MAP sensors contain a pressure-sensitive disk capacitor (Figure 6-6). This type of MAP sensor changes manifold pressure to a varying frequency digital voltage signal. The MAP sensor actually senses the difference between atmospheric pressure and manifold vacuum. When the engine is idling, high intake manifold vacuum is about 18 inches of mercury (Hg). Under this condition, the MAP sensor signal is approximately 95 hertz.

If the engine is operating at, or near, wide-open throttle, the manifold vacuum may be 2 inches Hg, and the MAP sensor hertz will be approximately 160.

Most manufacturers will use a limp-in value if the MAP sensor signal is considered invalid by the PCM. Limp-in values are used to keep the engine running. However, engine performance suffers greatly under these conditions. Usually the predetermined value is based on throttle position sensor and engine speed inputs.

Mass Air Flow Systems

AUTHOR'S NOTE: Each manufacturer's engine package is unique. The values provided in the following explanation of inputs are offered as an example based on the actual strategies that are used. The amount of influence an input has on the air-fuel mixture and ignition timing will vary. Like the speed density system, the MAF system uses the rpm input, and it is still the most important sensor. The MAF is the second most important sensor in this system.

EFI systems that use a **mass air flow (MAF) sensor** to determine air volume are referred to as **mass air flow (MAF)** systems. The MAF sensor measures the temperature and mass of air entering the engine, as well as altitude for the PCM. The mass of a given amount of air is calculated by multiplying its volume by its density and is expressed as grams per second. The denser the air, the more oxygen it contains. From a measurement of mass, the PCM modifies injector pulse width for the oxygen content in a given volume of air. The following is a list of typical sensors used in the air density system and there order of importance:

1. *Crank Angle Sensor (RPM)*. In most cases, the engine cannot run without this sensor. First, this sensor is used for an rpm input. With that information, the PCM will determine ignition coil dwell and timing. Second, the PCM will use this sensor to start the injector firing.

2. *Mass Air Flow (MAF) Sensor*. This sensor is the load sensor of this system. The MAF sensor will "count" the air going into the engine. The PCM will then add fuel based on the MAF sensor's output. A higher airflow will result in more fuel, while a lower airflow will result in less fuel. The MAF sensor affects the major amount of fuel (base pulse width). It is important to note that the MAF sensor's output also has a major influence on ignition timing. If the load is high (a large amount of airflow), the PCM will give the engine less advance than if there is a light load on the engine. Load is based on MAF, throttle position sensor (TPS), and rpm.

Shop Manual
Chapter 6,
page 239

3. *Throttle Position Sensor (TPS)*. Based on the TPS input, the PCM can increase fuel by 500 percent (quickly increasing TPS voltage) or decrease fuel by 70 percent (during deceleration). Also, based on the TPS input, the PCM can lower the timing advance as the TPS voltage increases. If the TPS voltage is steady, it will have no effect on fuel modification.

Shop Manual
Chapter 6,
page 243

4. *Engine Coolant Temperature (ECT) Sensor*. Based on the ECT input, the PCM can increase fuel by 60 percent at –22°F (–30°C) and have no effect on fuel at 176°F (80°C). The ETC input can increase the timing by 14° at -22°F (–30°C) and will have no effect on timing at 95°F (35°C).

Shop Manual
Chapter 6,
page 247

5. *Barometric Pressure Sensor (BARO)*. Based on the BARO input, the PCM can increase fuel by $12^1/_2$ percent (below Sea level) or decrease fuel by 50 percent when above 8,000 feet (2.4 km) altitude. The BARO sensor input can also add 7 degrees of timing when above 7,000 feet (2.1 km) altitude and will have no effect on timing below 1,500 feet (.45 km) altitude.

6. *Intake Air Temperature (IAT) Sensor*. Based on the IAT input, the PCM can increase fuel by 23 percent at 26°F (–3°C) air temperature, or decrease fuel by 17 percent at 185°F (85°C) air temperature. This sensor will have no effect on fuel at 73°F (23°C) air temperature. This sensor has no effect on timing.

Shop Manual
Chapter 6,
page 247

7. *Oxygen Sensor (O₂S)*. The O_2S can increase fuel by 17 percent (lean condition/too much oxygen) or decrease fuel by 17 percent (rich condition/lack of oxygen). This sensor has no effect on timing. This sensor's output will only be used when the engine is in closed loop.

Shop Manual
Chapter 6,
page 249

8. *Adaptive memory*. This value can add or take away fuel, and it is always used in the pulse width formula. It comes from the oxygen sensor switching activity over a period of time.

Shop Manual
Chapter 6,
page 261

The following is a typical formula used by the computer to calculate pulse width in an air density system:

Base Pulsewidth = $\dfrac{\text{Airflow Hz}}{\text{Engine RPM}}$ × TPS (decel) × Coolant × Baro × Intake Air Temp × O_2 (Closed Loop)

x Air-Fuel Compensator (Open Loop) × Adaptive Memory × Detonation Sensor (Turbo)

Total Pulsewidth = Base Pulsewidth × TPS (accel) + Injector Lag Time

There are some manufacturers that use both MAF and MAP sensors. In this case, the MAP sensor is used mainly as a backup if the MAF sensor fails to determine faults. In some 4-cylinder applications using the MAF strategy, the TPS sensor input is unique in that it does not affect the fuel injector base pulse width. Instead, the PCM (based on TPS input) will add four additional injector firings based on the rate of TPS increase. If the TPS sensor input identifies a rapid opening of the throttle plates, the PCM will add four additional injector pulses (one to each cylinder). The extra pulses occur every 10 milliseconds as long as the TPS input increases at a rapid rate. The actual amount of fuel delivered by the extra pulses is determined by ECT inputs. However, the maximum pulse width of these pulses is 4 milliseconds. The effect of TPS inputs on fuel strategy is listed below:

TPS Voltage	Multiplicative Value
Steady	1.0 (no effect)
Rapidly increasing	5.0 (increase pulse width by 500%)
Rapidly decreasing	0.3 (decrease pulse width by 70%)

Usually the barometric pressure (BARO) sensor input has no effect on fuel strategies at sea level. The range of authority of a typical BARO sensor on fuel ranges from reducing pulse width by about 50 percent to $12\frac{1}{2}$ percent increase. This means at higher altitudes this sensor input can reduce the calculated base pulse width in half. If the engine is operating below sea level, more fuel will be added.

Shop Manual
Chapter 6,
page 239

Some vane-type MAF sensors are referred to as volume airflow meters.

Mass Air Flow Sensors.   The MAF sensors may be classified as vane, heated grid, hot wire, pressure, or ultrasonic. The MAF sensor is usually mounted in the hose between the air cleaner and the throttle body so that all the intake air must flow through the sensor.

In a vane-type MAF sensor (Figure 6-7), a pivoted air-measuring plate is lightly spring loaded in the closed position. As the intake air flows through the sensor, the air-measuring plate moves toward the open position.

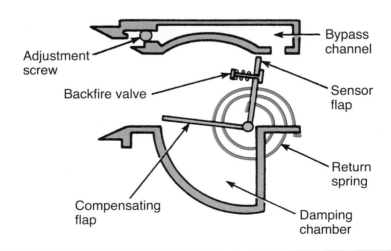

Figure 6-7   A vane-type mass air flow sensor used to measure the volume of air entering the engine.

A movable pointer is attached to the measuring plate shaft. The pointer contacts a resistor to form a potentiometer that sends a voltage signal to the PCM in relation to intake airflow. A thermistor in the airflow meter sends a signal to the PCM in relation to air intake temperature.

The heated resistor-type MAF sensor has a heated resistor mounted in the center of the air passage. In addition, an electronic module is mounted on the side of the sensor. When the ignition switch is turned on, voltage is supplied to the module. The module sends current through the resistor to maintain a specific resistor temperature.

Shop Manual
Chapter 6,
page 241

If an engine is accelerated suddenly, the rush of air tries to cool the resistor. Under this condition, the module supplies more current to maintain the resistor temperature. The module sends the increasing current signal (proportional to the airflow entering the engine) to the PCM. When the PCM receives this increasing current flow signal, it increases the amount of fuel delivery to go with the additional airflow entering the engine. The MAF module reacts in a few milliseconds to maintain the resistor temperature. Some MAF sensors have an electric grid in place of the resistor.

In a hot wire-type MAF sensor, a hot wire is positioned in the air stream through the sensor, and an ambient temperature sensor wire is located beside the hot wire. The ambient temperature wire senses intake air temperature and may be referred to as a cold wire (Figure 6-8).

When the ignition switch is turned on, the MAF module sends enough current through the hot wire to maintain the temperature of this wire at 392°F (200°C) above the ambient temperature sensed by the cold wire. If the engine is accelerated suddenly, the rush of air tries to cool the hot wire. The module immediately sends more current through the wire to maintain the temperature of the wire. The module sends the increasing current signal to the PCM that is directly proportional to the intake airflow. When this signal is received, the PCM supplies more fuel to go with the increased intake airflow. Some MAF sensors have a burn-off relay and related circuit. When the ignition switch is turned off, after the engine has been running for a specific length of time, the computer closes the burn-off relay. This relay activates a burn-off circuit in the MAF that heats the hot wire to a very high temperature (1,000°C) for a short time period. This action burns contaminants off the hot wire.

Another version of the hot wire MAF is the hot film. A nickel foil sensor is maintained at 75°C above ambient temperature. This type of sensor does not require a burn-off period.

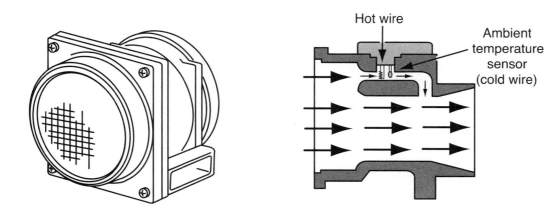

Figure 6-8 Hot-wire type MAF sensor.

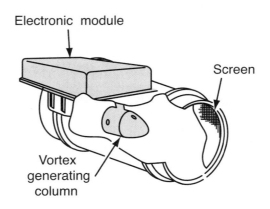

Figure 6-9 Pressure-type MAF sensor.

The pressure type MAF sensor uses the Karman vortex phenomenon to determine the volume of air entering the engine. A vortex-generating column is located in the path of airflow (Figure 6-9). Vortexes are generated in proportion to airflow speed (air volume).

A pressure inlet is positioned downstream from the vortex-generating column (Figure 6-10). The pressure inlet is connected to a pressure sensor (stress gauge). The vortexes generated by the column result in pressure variations that are detected by the pressure sensor. As the airflow through the MAF sensor increases, so does the number of pressure variations. The voltage changes resulting from the pressure variations are changed to a pulsed digital signal that is proportional to intake airflow. The digital signal is then sent to the PCM.

The ultrasonic MAF sensor also uses the Karman vortex phenomenon. The difference is this sensor measures airflow with ultrasonic waves. An ultrasonic transmitter is positioned downstream of the vortex-generating column. The transmitter is located across from an ultrasonic receiver (Figure 6-11). The sensor determines the volume of intake air by measuring the length of time required for the ultrasonic waves from the transmitter to reach the receiver.

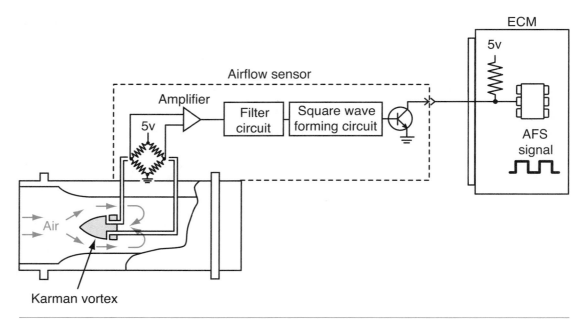

Figure 6-10 Pressure-type MAF sensor operation.

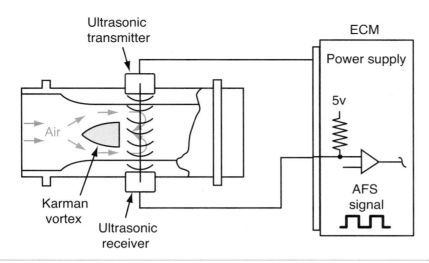

Figure 6-11   Ultrasonic mass airflow sensor.

When the ignition switch passes through the RUN position on the way to the START position, there is no airflow over the sensor. During this phase, a fixed reference time for the ultrasonic transmission is established. A clockwise rotating vortex speeds up the ultrasonic waves, and they reach the receiver quicker than the reference time. A counterclockwise vortex tends to slow down the ultrasonic waves. The resulting signal is then converted into digital pulses and sent to the PCM. The greater number of pulses indicates a greater volume of air.

Barometric Pressure Sensors.   Many systems use a barometric pressure (BARO) sensor to determine the amount of fuel delivery needed to start the engine and for spark timing strategies. In some applications, the BARO sensor is part of the MAF sensor. As its name implies, this sensor measures atmospheric pressure. At lower pressures, the engine requires more air for efficient combustion. At higher pressures, it requires less air. Basically, the BARO sensor operates the same as the MAP sensor discussed earlier. The main difference is there is no vacuum source attached it.

Idle Air Control Motors.   Many TBI systems will use a DC reversible motor to control idle speed. In this system, the pintle of the idle speed control (ISC) motor is located against a lever that is attached to the throttle plates. As the motor moves the pintle in and out, the throttle plates move. Moving the plate allows more or less air to enter the engine. For example, if idle speed needs to be increased, the PCM would activate the motor to move the pintle so the throttle plates would open slightly. At this time, more air is allowed to enter the engine. The O_2S senses the high oxygen content and the PCM will increase the pulse width to maintain the stoichiometric ratio. With the increase of air and fuel, engine speed increases. The opposite will occur if idle speed needs to be decreased.

Newer EFI systems control idle speed by regulating airflow around the throttle plates. The idle air control (IAC) stepper motor is used to control the amount of air that bypasses the closed throttle plates.

Most IAC stepper motors use permanent magnet motors with two to six electromagnets. Each command from the PCM results in one step movement. By switching which circuit supplies the current and which supplies ground (through the use of a H driver), the PCM can control the

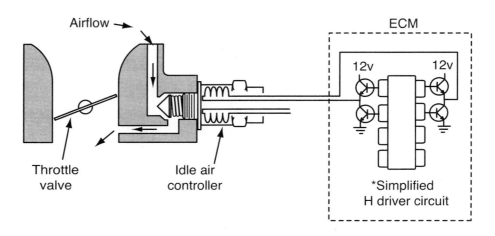

Figure 6-12 The H driver controls the direction of current, thus the direction of motor rotation.

direction of motor rotation (Figure 6-12). To decrease engine speed, the controller will step the motor in to close off the air bypass. If engine speed needs to be increased, the PCM drives the stepper motor pintle out to increase the size of the bypass.

Many manufacturers now use a solenoid to perform the function of idle air control. The solenoid is pulse width modulated to control air bypass. The use of the solenoid provides faster reaction time to obtain the desired results.

Density Speed System

The **density speed** system is similar to the air-density system except the engine will not run if either the rpm or MAP sensor inputs are missing.

In most **density speed** systems, the most important input is from the MAP sensor. In fact, the engine will not run if this input is missing. The second important input is a combination TDC/Crankshaft position/Engine speed sensor (all three are the same sensor, but they may be called by any one of these three names). If this input is missing, the engine will not run. In the density speed system, rpm input determines the frequency of injection (how many times the injector is fired), while the MAP sensor will determine the major amount of fuel (pulse width). Other sensors will affect the pulse width, but again, the MAP sensor will determine the base pulse width.

EFI Modes of Operation

The following are typical modes of operation for most EFI systems. Variations of these modes are also used. For this reason, the correct service manual should be referenced when diagnosing a fuel system problem.

Key ON

Typically, when the ignition switch is turned to the ON or RUN position, the PCM receives a key "on" signal. The PCM will then gather the needed information in order to prepare the engine for starting. Usually the information required is:

- Barometric pressure
- Engine coolant temperature
- Throttle position
- Intake air temperature

Also during this time, the fuel pump is turned on for a short time to prime the system. This mode is known as initialization.

Crank

When the ignition switch is located in the START position and the engine is cranking, the PCM receives crankshaft and camshaft position sensor signals. The PCM will then turn on the fuel pump. Depending on the system, the PCM may also energize all of the injectors at the same time to provide the initial fuel to start the engine. This may be the only firing of the injectors until the rpm signal indicates the engine has started. The amount of fuel injected during this phase is based upon the ECT sensor input. Other systems will fire the injectors in order (sequentially) while the engine is being cranked. During the time the engine is being cranked, the PCM ignores the O_2S input.

The TPS, MAP/MAF, and BARO sensors allows the PCM to monitor the amount of air delivered to the engine. Some EFI systems also receive an injector-sync signal at this point to inform the PCM which cylinder is approaching the compression stroke for proper fuel injector sequencing.

Open Loop

When the engine rpm increases to about 450 rpm, the PCM switches to "run" mode. Based on the input from the MAP or MAF sensor and engine speed inputs, the PCM determines base pulse width and generates ignition timing signals. Initially, the O_2S input is ignored since the sensor is cold and a rich mixture was delivered during cranking to start the engine. The PCM delivers a predetermined air-fuel ratio and a higher idle speed to bring the engine up to operating temperature faster.

Closed Loop

Once the PCM is satisfied that all required criteria has been met and has determined the O_2S is warm enough to send valid readings, it switches to closed loop operation. The PCM will now adjust the air-fuel mixture based on O_2S inputs.

Acceleration

The rate of TPS voltage change is monitored to determine how much acceleration the driver is requesting. Based on the TPS input, along with MAP or MAF, the injection pulse width is increased to prevent hesitation and stumble. Based on the rate of acceleration, and how far the accelerator pedal is depressed, the system will stay in closed-loop operation.

Wide-Open Throttle

The wide-open throttle (WOT) signal is received by the PCM from the TPS. MAP or MAF sensor inputs also identify a WOT condition. Upon recognition of a WOT condition, the PCM switches back to open-loop operation and ignores the O_2S input. The main priority of the PCM during this mode is to increase the injector pulse width.

Deceleration

Based on the response of the TPS when the accelerator pedal is released and on MAP or MAF signals, the PCM may enter deceleration mode. Depending on the conditions and programming, the PCM may turn off the injectors very briefly. The system will go into open loop during this very lean condition.

Adaptive Memory

The PCM's goal is to use the input information and control outputs to provide a constant stoichiometric ratio. The operating modes just discussed represent decision categories for the PCM. Within each of the modes of operation, smaller and more finite decisions are made.

Shop Manual
Chapter 6,
page 261

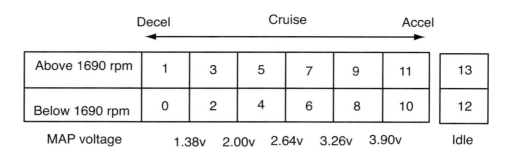

	Decel		Cruise			Accel	
Above 1690 rpm	1	3	5	7	9	11	13
Below 1690 rpm	0	2	4	6	8	10	12
MAP voltage		1.38v	2.00v	2.64v	3.26v	3.90v	Idle

Figure 6-13   An example of memory cell construction based on engine rpm and MAP inputs.

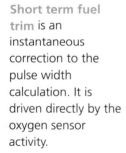

Adaptive memory is a function of the PCM to make corrections to the pulse width calculation based on oxygen sensor activity.

Adaptive memory is a component of the air-fuel ratio calculation. Within each operating cell, fuel pulse width calculations occur. Usually, a cell is defined by the MAP/MAF and rpm signals (Figure 6-13). As discussed, a set formula determines pulse width. If the O_2S signals a rich or lean condition in a cell, the cell will update to aid in fuel control. There are different methods used by manufacturers to express the amount of correction occurring. A common method is a percentage value change to the formula. For example, assume that the engine is operating in closed loop (adaptive memory only updates in closed loop) and the fuel pressure is a little low. The O_2S reading will indicate high oxygen content in the exhaust stream until the PCM increases the injector pulse width till the air-fuel mixture is returned to 14.7 to 1. The percentage of pulse width increase would be recorded by the PCM for that cell. Any time the operating conditions reenter that cell, the PCM will use the percent of correction in its formula to determine pulse width.

Chrysler Adaptive Strategies

When the fuel system enters closed loop operation, there are two adaptive memory systems that begin to operate. The first system that becomes operational is **short term fuel trim**. Short term fuel trim corrects fuel delivery in direct proportion to the voltage signals from the O_2S (Figure 6-14). As the O_2S voltage switches in response to the air-fuel ratio changes, the PCM will adjust the amount of fuel delivered until the O_2S reaches its switch point. When the switch point is reached, short term adaptive begins with a quick kick. Then the adaptive memory slowly ramps until the O_2S's output voltage indicates the switch point in the opposite direction. Short term adaptive will continue to increase and decrease the amount of fuel delivered based upon the O_2S input. For example, if the O_2S output voltage goes toward 0 volt (high oxygen), short term adap-

Short term fuel trim is an instantaneous correction to the pulse width calculation. It is driven directly by the oxygen sensor activity.

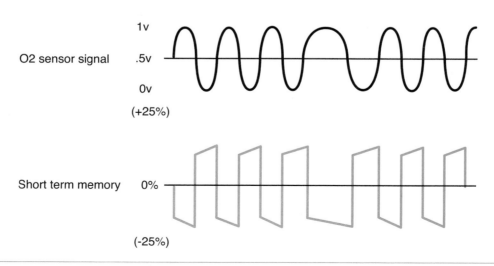

Figure 6-14   Short-term fuel trim is driven by O_2S inputs.

	Open Throttle						Idle	Decel
Above X rpm	1	3	5	7	9	11	13 "D"	15
Below X rpm	0	2	4	6	8	10	12 "N"	14
MAP (inches)	20	17	13	9	5	0		

Figure 6-15 Long-term adaptive memory cells.

tive memory will start to add additional fuel until the O_2S begins switching again. The maximum range of authority for short term adaptive memory is either ±25 percent or ±33 percent of base pulse width, depending on the PCM model.

Short term adaptive resets to a value of 0 when the engine is turned off. It will not update until the PCM enters closed loop. Normal short term adaptive activity is for the values to jump back and forth, crossing over 0.

The second system is called **long term adaptive memory**. This memory uses a cell structure to better maintain correct emission levels throughout all operating ranges (Figure 6-15). In this illustration, there are 16 cells; however, different engines may have fewer or more cells. Two of the cells are used only in idle, based on TPS and PARK/NEUTRAL switch inputs. Two cells are used for deceleration, based on TPS, rpm, and vehicle speed. The other 12 cells are based on rpm and load inputs.

As the engine enters a cell, the PCM monitors the short term adaptive value. The goal of long term adaptive is to keep short term at 0. For example, if short term is correcting for low oxygen content based on O_2S inputs, short term will begin to correct for it. After short term reaches a predetermined threshold, long term adaptive will start to correct in the same direction to bring short term back to 0 (Figure 6-16).

Long term adaptive memory is a correction to the pulse width calculation based on short term activity over a period of time.

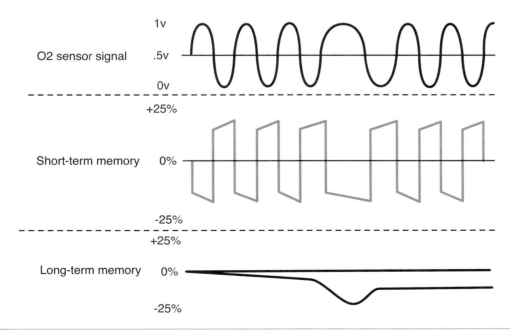

Figure 6-16 Short-term fuel trim and long-term adaptive memory activity based on O_2S inputs. Long-term attempts to keep short term at zero.

Long term is retained in memory even after the engine is turned off. However, long term does not update until the engine reaches about 170°F (77°C). Additionally, long term adaptive memory correction values are used in open loop operation. Long term can alter pulse width ±25 percent or ±33 percent from base, depending on the PCM model. Combined, short term and long term can correct pulse width 50 percent (or 66 percent) on either side of zero.

General Motors Integrator and Block Learn

Shop Manual
Chapter 6,
page 267

Integrator
represents a short-term correction to the pulse width.

Block learn makes long-term corrections to the injector pulse width.

On GM vehicles with OBD-II, integrator is called short term fuel trim and block learn is called long term fuel trim.

Pre-OBD-II General Motors learning ability is called **integrator** and **block learn**. Integrator is the first correction. As a rich or lean condition is sensed by the O₂S, integrator will attempt to correct for it by adjusting the pulse width. If integrator is being observed on a scan tool, the value is displayed by a number. The number displayed by the scan tool is a base ten conversion of a binary code. If there is no correction, then the number is 128 (half way between 0 and 255). A number larger than 128 represents increased pulse width over base. A number lower than 128 means pulse width is being reduced.

Block learn represents pulse width adjustments that have become a trend over an extended period of time. Once the engine has been running in closed loop long enough, the PCM will store the corrective factor (as a number) in memory. That number will be stored until additional corrections need to be made.

Block learn numbers are stored in 16 different cells (Figure 6-17). Each cell represents a different combination of engine speed and load. Integrator and block learn work together to correct pulse width. For example, if block learn is above 128, but the integrator is at 128, the system has a lean tendency that is being corrected by block learn, but the integrator is not having to make any additional changes. If block learn is at 128 but integrator is above 128, the system is attempting to run lean, but it has not been doing so long enough for block learn to make any corrections.

Adaptive memory strategies for a General Motors OBD-II certified vehicle is similar to that of the Chrysler system discussed earlier.

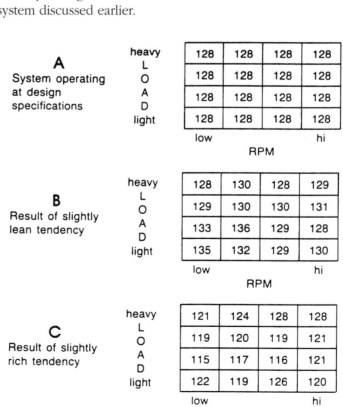

Figure 6-17 Block learn information.

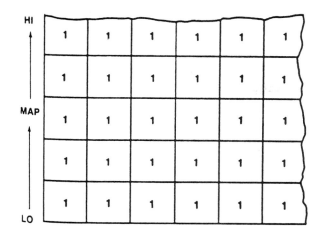

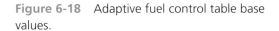

Figure 6-18   Adaptive fuel control table base values.

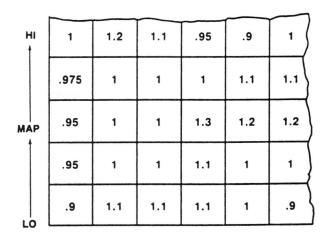

Figure 6-19   Adaptive fuel control table with adapted cell values.

Ford Adaptive Strategies

Inside of the PCM's ROM chip is a table that represents cells based on engine speed and load (Figure 6-18). Each cell is assigned a base value of 1. Based on rpm and engine load, the PCM will determine which cell it is operating in. The value in that cell is then used to correct pulse width. The value is the multiplier used in pulse width calculations. A value of 1 will not change pulse width over base.

Notice this table is located in the computer's ROM chip. Since the values in ROM cannot be changed by the microprocessor, there is a second copy of the table in volatile RAM. It is the second copy the computer uses for corrections (Figure 6-19). The adaptive control will add or subtract from 1. A number larger than 1 indicates an increase in pulse width over base; a number less than 1 represents a decrease in pulse width.

Import Adaptive Memory Strategy

Many import manufacturers will use a variation of the strategies just discussed; however, some do not report the adaptive value to the scan tool for the technician to review. The following is an example of the Mitsubishi adaptive memory.

The PCM recognizes the air-fuel mixture as being rich or lean based on which side of the O_2S switch point (usually around 0.5 volts) the input voltage is. If the voltage is above the switching point, the system is running rich (low oxygen). A lean mixture (high oxygen) is recognized when the voltage is below the switching point. Once the system is in closed loop, the upstream oxygen sensor(s) is used to update the short term fuel trim. If the oxygen sensor stops switching, the PCM either lengthens or shortens the injector pulse width until the sensor begins switching again.

The only time emission requirements are achieved is when the O_2S is switching back and forth. The short-term fuel trim is used to make instantaneous corrections to the pulse width and is only updated in closed loop. This correction factor is not stored in memory and is reset each time the ignition key is cycled to OFF.

The short term fuel trim in turn moves the correction factor to a long term fuel cell, where it becomes part of the pulse width equation. This memory location represents the amount of correction required by the vehicle under specific load to maintain oxygen sensor switching. The information stored in a long-term memory is nonvolatile, and it is not lost during a key cycle.

Short-term fuel trim range correction can add 25 percent to the base pulse width for lean conditions. For rich conditions, the short-term trim value will have a maximum authority of –30.5 percent. Long-term fuel trim uses two cells with a range of authority of ±17 percent.

Throttle Body Injection Systems

Throttle body injection (TBI) systems spray the fuel directly above the throttle plates. The **throttle body** assembly is mounted on top of the intake manifold. Four-cylinder engines usually have a single throttle body assembly with one throttle, whereas V6 and V8 engines may be equipped with dual throttle bodies with two throttles on a common throttle shaft (Figure 6-20). When the engine is cranking or running, fuel is supplied from the fuel pump through the lines and filter to the throttle body assembly. A fuel return line connected from the throttle body to the fuel tank returns excess fuel to the fuel tank.

When fuel enters the throttle body fuel inlet, the fuel surrounds the injector (or injectors) at all times. Each injector is sealed into the throttle body with O-ring seals, which prevent fuel leakage around the injector at the top or bottom. Fuel is supplied from the injector through a passage to the pressure regulator. A diaphragm and valve assembly is mounted in this regulator, and a diaphragm spring holds the valve closed. At a specific fuel pressure, the regulator diaphragm is forced upward to open the valve and some excess fuel is returned to the fuel tank.

In most TBI systems, the fuel pressure regulator controls fuel pressure at 10 psi to 25 psi (70 kPa to 172 kPa). The fuel pressure must be high enough to prevent fuel boiling in the TBI assembly. When the pressure on a liquid is increased, the boiling point is raised proportionally. If fuel boiling occurs in the TBI assembly, vapor and fuel are discharged from the injectors.

AUTHOR'S NOTE: The fuel pressure is regulated at 14.5 psi (100 kPa) in Chrysler TBI assemblies with a temperature sensor. These systems were used from 1986 through 1990, and they are referred to as low-pressure TBI (LPTBI). A black plastic rivet is located on the top of the pressure regulator in a LPTBI system. In 1991, Chrysler installed a pressure regulator on their TBI systems that controls the fuel pressure at 39.2 psi (270 kPa). These systems are called high-pressure TBI (HPTBI). A white plastic rivet is located on top of these pressure regulators, and the TBI temperature sensor is no longer required with the higher fuel pressure.

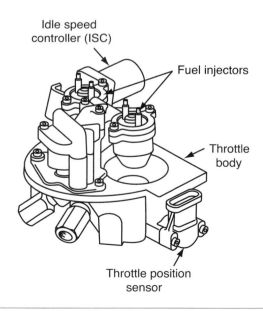

Figure 6-20   Dual throttle body assembly from a V8 engine.

In the four-cylinder TBI system, there is a single injector event for each individual cylinder event. This means the injector pulses twice for each engine revolution. When the engine is being started, double injector pulses occur to enrich the mixture. Double pulsing at cranking occurs only for a programmed time interval to avoid flooding. Pulse width during cranking is based on engine coolant temperature.

AUTHOR'S NOTE: In many throttle body injection (TBI) systems, the computer grounds an injector each time a signal is received from the distributor pickup. This provides an injector firing for each intake valve opening. This type of TBI system is referred to as a synchronized system because the injector pulses are synchronized with the pickup signals. In a dual injector throttle body assembly, the computer grounds the injectors alternately under most operating conditions.

Once the engine is started, the PCM provides the pulse width calculation based on coolant temperature and MAP inputs. Once the engine warms to allow the mode to change to closed loop, the PCM will use the feedback formula. Some systems will pulse the injector every 12.5 milliseconds, regardless of reference pulses, when the TPS signals it is near full throttle.

AUTHOR'S NOTE: Some General Motors TBI systems have a semi-closed loop mode (highway mode). During extended cruise conditions at highway speeds, the PCM sets the air-fuel ratio as lean as 16.5 to 1. This is done to increase fuel economy.

Port Fuel-Injection Systems

BIT OF HISTORY

Since the 1970s, fuel system technology has developed quickly from carburetors to computer-controlled carburetors and then to throttle body injection systems. Many of the throttle body injection systems have been replaced with multipoint fuel-injection systems, and a significant number of multipoint injection systems have been changed to sequential fuel-injection systems.

There are two basic types of **port fuel-injection (PFI)** systems: **multipoint fuel-injection (MPI)** and **sequential fuel injection (SFI)**. There are many similarities in the MPI and SFI systems supplied by the various domestic and import vehicle manufacturers. For example, both SFI and MPI systems have injectors installed in the intake ports near the intake valve, and many of these systems share similar inputs and outputs. One of the major differences in MPI and SFI systems is the method of connecting the injectors to the computer. In SFI systems, each injector is connected individually to the computer, so the computer energizes one injector at a time. In MPI systems, the injectors are grouped together in pairs or groups. Each group of injectors shares a common wire to the computer so they are energized together. For example, on some four-cylinder engines, the injectors are connected in pairs, and each pair of injectors has a common connection to the computer. On some V6 engines, each group of three injectors has a common ground wire connected to the computer. Groups of four injectors share a common ground connection to the computer on some V8 engines.

The computer is programmed to ground the injectors well ahead of the actual intake valve openings so the intake ports are filled with fuel vapor before the intake valves open

Firing the injector once for each time a cylinder fires is referred to as the synchronized mode.

Port fuel injection (PFI) is a term that may be applied to any fuel-injection system with the injectors located in the intake ports.

The multipoint fuel-injection (MPI) system has one injector per cylinder. The injectors are fired in pairs or in groups of three or four. Usually half of the fuel is delivered on each crankshaft revolution.

In a sequential fuel-injection (SFI) system, each injector is fired individually in ignition firing order prior to the intake valve opening.

A multipoint fuel-injection system is also called multiport fuel injection.

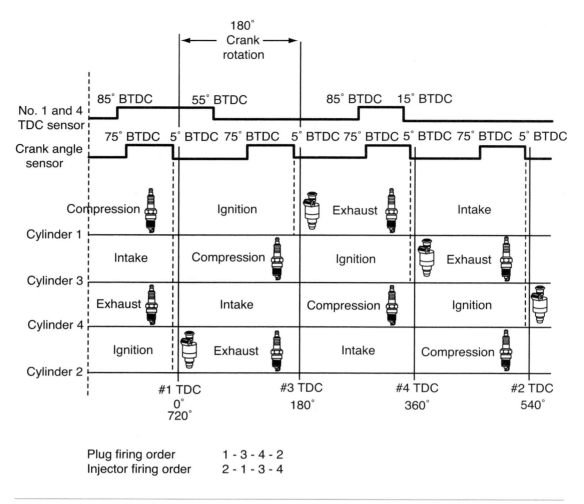

Figure 6-21 Fuel-injection timing strategy.

(Figure 6-21). In both SFI and MPI systems, the computer supplies the correct injector pulse width to provide the stoichiometric air-fuel ratio. The computer increases the injector pulse width to provide air-fuel ratio enrichment while starting a cold engine. A clear flood mode is also available in the computer in MPI and SFI systems. On many MPI and SFI systems, the computer decreases injector pulse width while the engine is decelerating to provide improved emission levels and fuel economy. On some of these systems, the computer stops operating the injectors while the engine is decelerating in a certain rpm range.

Injector Internal Design and Electrical Connections

The fuel injector is simply a fast-acting solenoid. Whenever the injector is opened, it sprays a constant amount of fuel for a given amount of time. Because pressure drop across the injector is fixed, the fuel flow through the injector is constant. With the injector connected to a pressurized fuel supply, atomized fuel sprays from the injector nozzle behind the intake valve.

Inside the injector, the plunger and valve seat are held downward by a spring. In this position, the seat closes the metering orifices in the end of the injector (Figure 6-22). Openings in the sides of the injector allow fuel to enter the cavity surrounding the injector tip. A mesh screen filter inside the injector openings prevents dirt particles in the fuel from entering the metering orifices. In some injectors, a diaphragm is located between the valve seat and the housing. The tip of the injector may contain one to six metering orifices. Injector design varies depending on the manufacturer.

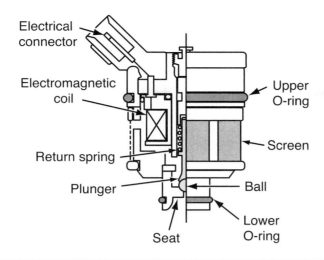

Figure 6-22 Throttle body injector design.

Each injector contains two terminals, and an internal coil is connected across these terminals. The plunger is positioned in the center of the coil, and the lower end of the plunger has a tapered valve seat. When the ignition switch is turned on, 12 volts are supplied to one of the injector terminals. The other injector terminal is connected to the PCM that controls ground. When the PCM grounds this terminal, current flows through the injector coil to ground. When this action occurs, the injector coil magnetism moves the plunger and valve seat upward. Fuel, under pressure, sprays out the injector orifices into the air stream.

Injector Deposits.   In some EFI systems, there have been problems with deposits on injector tips resulting from small quantities of gum present in gasoline. Injector deposits usually occur when this gum bakes onto the injector tips after a hot engine is shut off. Most oil companies have added detergents to their gasoline to help prevent injector tip deposits. Vehicle manufacturers and auto parts stores sell detergent additives to mix in the fuel tank to clean injector tips. Also, some vehicle manufacturers and parts suppliers have designed deposit-resistant injectors. These injectors have several different pintle tip and orifice designs to help prevent deposits. On one type of deposit-resistant injector, the pintle seat opens outward away from the injector body and more clearance is provided between the pintle and the body. Another type of deposit-resistant injector has four orifices in a metering plate rather than a single orifice. Some deposit-resistant injectors may be recognized by the injector body color. For example, conventional injectors supplied by Ford Motor Company are painted black, whereas their deposit-resistant injectors have tan or yellow bodies.

Shop Manual
Chapter 6,
page 259

Sequential Fuel-Injection System

It would be impossible to cover all SFI systems; however, most operate very similar to each other. The main differences are the inputs used between speed density and air density systems. The process of delivering the fuel to the combustion chamber is very similar between manufacturers. The following is offered as an example of an SFI system.

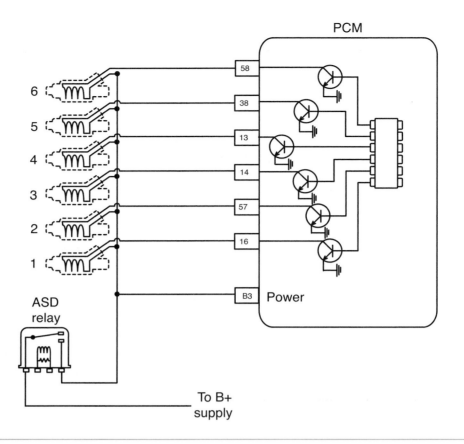

Figure 6-23 Sequential fuel injection uses individual connections to the PCM to control each injector.

The Chrysler SFI system used on a 3.5 L engine is wired so each injector has a separate ground wire connected into the PCM (Figure 6-23). Voltage to the injectors is supplied through the automatic shut down (ASD) relay contacts when the ignition switch is in the START position and an rpm signal is received. If the rpm signal is lost, the ASD relay is turned off, shutting off the injectors. This engine is equipped with an electronic ignition (EI) system and the crankshaft and camshaft position sensors are inputs for this system.

The crankshaft position sensor informs the PCM of engine speed and of two pistons approaching TDC. The camshaft position sensor informs the PCM which piston is approaching TDC on the compression stroke. The two sensors provide alternating high and low voltage signals to provide input for pulse width and ignition strategies (Figure 6-24).

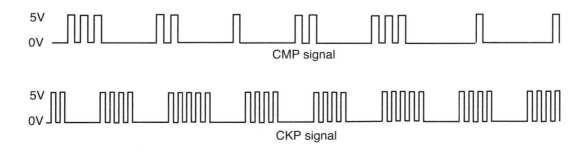

Figure 6-24 Cam and crank sensor pattern of the 3.5L engine.

The different slots on the crankshaft position sensor inform the PCM of needed functions as follows:

- 69 degrees BTDC
  MAP reading is updated and a value is stored in RAM.
  Ignition period is determined by the time between two consecutive 69-degree edges.
  Detonation is mainly controlled on the 69-degree edge.
  Synchronization is mainly conducted on this edge along with camshaft interrupt.
  Ignition coil firing for the current cylinder, as well as for the next cylinder, is generated.
  Injection firing may occur based on operating conditions.
- 49 degrees BTDC
  Ignition coil firing adjustment may occur if the engine speed is below a programmed value.
  Injector firing may occur based on operating conditions.
- 29 degrees BTDC
  Ignition coil firing adjustment may occur if the engine speed is below a programmed value.
  Injector firing may occur based on operating conditions.
  Adaptive knock is updated.
- 9 degrees BTDC
  If the engine is being started, the coil firing is done on this edge.
  Cylinder indication is updated.
  Adaptive dwell is updated.
  MAP is read. The value is averaged between the reading at 69 degrees BTDC and this reading.

Shop Manual
Chapter 6,
page 243

Minimum TPS is the lowest voltage value the PCM received from the TPS during that key cycle.

The PCM determines idle mode based upon inputs from the TPS. When the ignition switch is turned to the RUN position during engine starting, the PCM is programmed to monitor the TPS input. Once the engine is started, the PCM assumes that the lowest value it receives from the TPS must be where the throttle plate is fully closed. Normally this voltage is between 0.5 and 1.0 volts. At the low voltage position, the PCM records the signal as idle or **minimum TPS**. If the sensed voltage from the TPS goes lower than this value, the PCM will update its memory. However, the PCM will not make any corrections for an increase in voltage values. The PCM uses minimum TPS to determine other modes of operation.

If the throttle is opened and the TPS moves from its minimum TPS value by approximately 0.06 volt, the PCM enters off-idle strategies. At this time, spark advance is no longer used to control idle speed. Also, the idle air-control motor is positioned to act as a dashpot if the throttle is released suddenly.

If the PCM receives a rapidly increasing voltage signal from the TPS, the PCM will enter acceleration mode. During this time, the injector pulse width is increased based on the rate of TPS voltage increase. The PCM may activate three injectors at the same time: the injector for the cylinder whose intake valve just closed, the injector for the cylinder whose intake valve is being opened, and the injector for the cylinder whose intake valve will be opened next.

Wide-open throttle (WOT) is determined by an increase in TPS volts of 2.608 volts above the recorded minimum TPS. The PCM enters open loop and increases the injector pulse width. If a rapid deceleration is sensed, the PCM will lean out the air-fuel ratio. In some instances, the injectors are turned off. This action reduces the amount of emissions emitted.

If the throttle is at WOT during engine cranking, the injectors are turned off. This provides a clear flood mode. This program only occurs during cranking and when the TPS voltage exceeds 2.608 volts above minimum TPS. This function is not dependent upon engine coolant temperature.

The progress leading up to fuel injection systems began as early as 1883 with Edward Butler, Duetz, and other pioneers. During World War II, Germany pursued it further by bringing the Robert Bosch Company into developing fuel injection for the aviation field. During that time, Great Britain and the United States combined efforts to build a system to use in the Patton tank. Electronic fuel injection had its beginnings in Italy when an engineer named Ottavio Fuscaldo incorporated an electrical solenoid as a means to control fuel flow. In 1949, an Indy race featured a fuel-injected Offenhauser. The system was developed by Stuart Hillborn and featured an indirect injection system. Later, Chevrolet introduced the Rochester Ramjet in 1957. It was also used in the 1957 Pontiac Bonneville. This system used a lot of the features designed by Hillborn. The system was not popular with the general public and it was dropped after 1959, except for the Corvette which used it as an option until 1965. In 1975, General Motors introduced the first mass produced domestic fuel injection system on the 1976 model Cadillac Seville. It was a system based on the Bosch "D" Jetronic system used by European manufacturers since 1968. The system consisted of a throttle body, eight fuel injectors mounted on a fuel rail directing fuel into the intake, a crude analog computer and various sensors, all on a modified intake using an Oldsmobile 350 V8. It had been developed by Bendix, Bosch, and General Motors.

Typical Import Sequential Fuel Injection System

The Nissan electronic concentrated engine control system (ECCS) is an SFI system that has many of the same inputs and outputs as the other systems already discussed (Figure 6-25). As with most systems, this system has evolved over the years.

The Nissan systems incorporate both sequential and multipoint (simultaneous) injection operation. The sequential mode injects fuel once every complete engine cycle (two crankshaft revolutions) during the corresponding cylinder's intake stroke. The simultaneous injection mode sprays all the injectors at once, twice per complete engine cycle. The simultaneous injection mode is used when the engine is starting, if the engine is in fail-safe mode, or if the camshaft position sensor signal is lost and the engine is running on the crankshaft position sensor signal alone.

Early systems used a vane-type mass airflow sensor. The throttle valve switch is mounted in the throttle chamber, and the switch contacts are closed when the throttle is in the idle position (Figure 6-26). When the throttle is opened from the idle position, the switch contacts open.

Later systems use a hot film air mass sensor located in the intake air duct between the air filter and the throttle. The sensor will indicate a signal voltage less than 1 volt with the ignition in the RUN position and the engine OFF. At idle, the voltage will change to between 1.0 and 1.7 volts. At higher engine speeds and loads, the signal voltage can rise to as much as 4.0 volts.

Some systems use a barometric pressure sensor that is located in the PCM. This sensor sends a signal to the PCM in relation to barometric pressure as it varies with altitude and climatic conditions. Newer systems use an absolute pressure sensor. The Nissan absolute pressure sensor connects to the MAP/BARO switch solenoid through a hose. It detects both ambient pressure and intake manifold pressure. As the pressure rises, the voltage signal rises.

Early systems located the crank angle sensor in the distributor. Later systems have two crankshaft sensors. One of these sensors, referred to as the REF senor, is located next to the crankshaft accessory belt pulley. The REF signals the computer when the TDC position for individual cylinders occurs. This sensor is a coil-and-magnet inductive signal generator producing a signal that varies in voltage with engine speed. The second crankshaft position senor is referred to as the POS sensor. This sensor is located at the rear of the engine block and reads pulses from a trigger system on the flywheel or flexplate. The POS signals 1 degree change in crankshaft position.

Later systems also use a camshaft position sensor (PHASE) located in the front engine cover facing the camshaft sprocket. The PHASE sensor provides information that distinguishes which cylinder is at the firing position. The camshaft position sensor is a pulse generator consisting of a

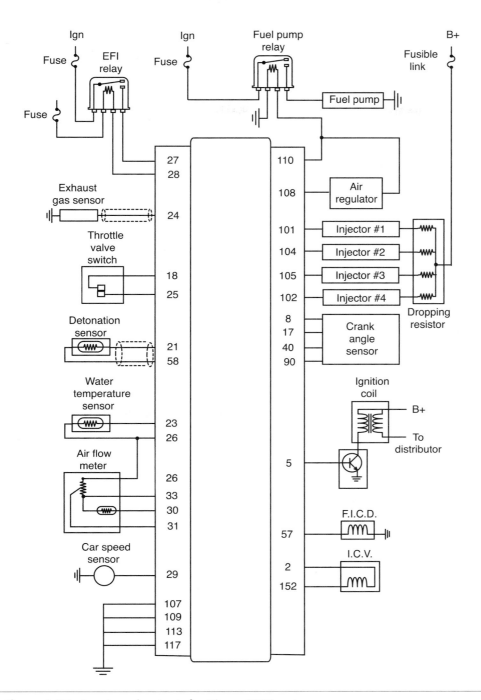

Figure 6-25 Inputs and outputs for Nissan ECCS system.

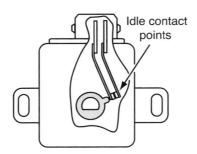

Figure 6-26 Typical idle switch.

permanent magnet wound in an electrical coil. When the gap between the camshaft sprocket and the magnet changes, the magnetic field changes and generates a voltage signal in the sensor output circuit. The computer uses the camshaft position signal to sequence the fuel injectors. At the beginning of the start-up cranking period, the camshaft position signal has not yet occurred, and all the injectors are simultaneously pulsed for engine start-up.

The dropping resistor assembly contains a resistor connected in series with each injector. These resistors protect the injectors from sudden voltage changes and provide constant injector operation.

Multipoint Fuel-Injection Systems

The following is offered as an example of a multipoint fuel-injection system. Like SFI systems, the MPI systems are very similar among manufacturers.

In many Ford MPI systems on V8 engines, the injectors are connected in groups of four on the ground side. Each group shares a common ground wire to the powertrain control module (PCM). On a Ford 5.8 L V8 engine, injectors 1, 4, 5, and 8 are connected to PCM terminal 58 with a common ground wire. The ground sides of injectors 2, 3, 6, and 7 are connected to PCM terminal 59 (Figure 6-27). When the injectors are connected in groups of four on the ground side, a group of injectors is grounded by the computer every crankshaft revolution.

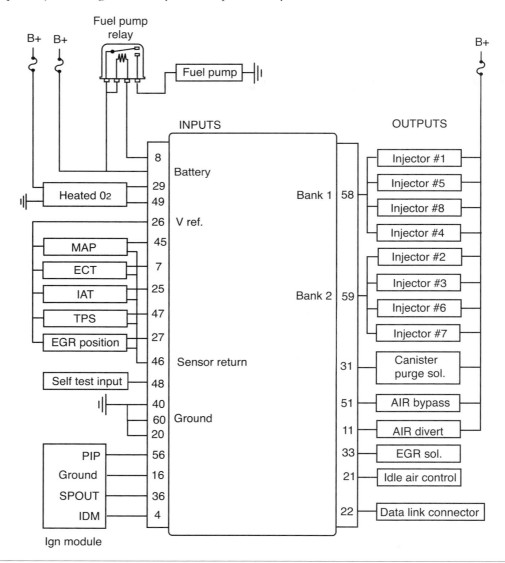

Figure 6-27 Inputs and outputs for the 5.8L multipoint fuel-injection system.

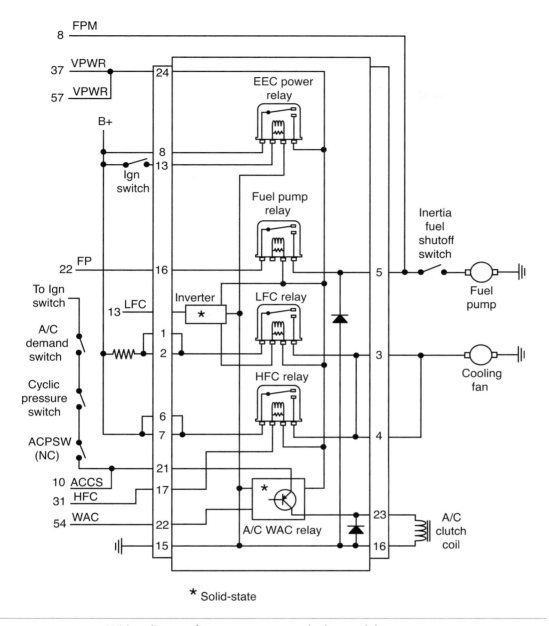

Figure 6-28 Wiring diagram for a constant control relay module.

Some Ford products are equipped with a **constant control relay module** (Figure 6-28). These relays perform the same function whether they are mounted separately or located in the constant control relay module. When the ignition switch is turned on, voltage is supplied to the power relay winding. This action closes the power relay points that supply voltage to the fuel pump relay winding, PCM terminals 37 and 57, and the electric drive fan (EDF) relay winding.

The **constant control relay module** contains several relays within its housing.

Central Port Injection

Another type of fuel injection system is the central port injection (CPI) system. This system uses a central port injector assembly mounted in the lower half of the intake manifold. Fuel inlet and return lines are connected from the rear of the intake to the CPI assembly. A retaining clip attaches

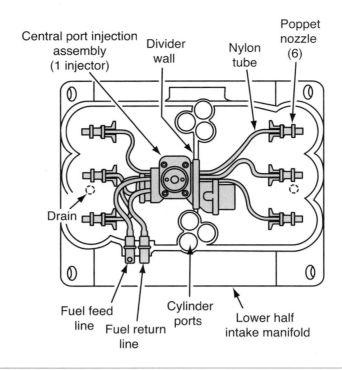

Figure 6-29 Central port injection components in the lower half of the intake manifold.

these lines to the CPI assembly. Small poppet nozzles are positioned in each intake port in the lower half of the intake. Nylon fuel lines connect these nozzles to the CPI assembly (Figure 6-29).

The pressure regulator is mounted with the central injector. Since this regulator is mounted inside the intake manifold, vacuum from the intake is supplied through an opening in the regulator cover to the regulator diaphragm. The regulator spring pushes downward on the diaphragm and closes the valve. Fuel pressure from the in-tank fuel pump pushes the diaphragm upward and opens the valve, which allows fuel to flow through this valve and the return line to the fuel tank (Figure 6-30).

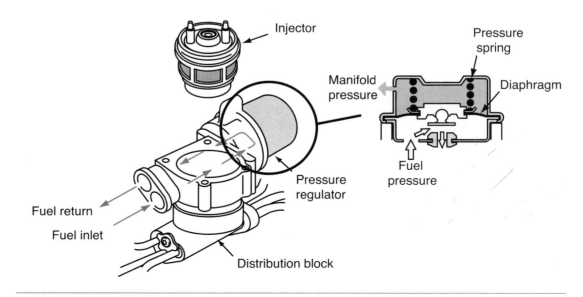

Figure 6-30 CPI system pressure regulator.

The pressure regulator is designed to regulate fuel pressure to 54 to 64 psi (370 to 440 kPa). High pressure is required in the CPI system to prevent fuel vaporization from the extra heat encountered with the CPI assembly, poppet nozzles, and lines mounted inside the intake manifold.

The central port injector contains a winding with two terminals extending from the ends of the winding through the top of the injector. When the ignition switch is on, voltage is supplied from the ignition switch to one of the injector terminals. The other injector terminal is connected to the PCM, which will control ground.

A pivoted armature is mounted under the injector winding in the central port injector. The lower side of this armature acts as a valve that covers the six outlet ports to the nylon tubes and poppet nozzles. A supply of fuel at a constant pressure surrounds the injector armature while the ignition switch is on. Each time the PCM grounds the injector winding, the armature is lifted upward, opening the injector ports. Under this condition, fuel is forced from the nylon tubes to the poppet nozzles (Figure 6-31).

The amount of fuel delivered by the central injector is determined by the length of time the PCM keeps the injector winding grounded. When the PCM opens the injector ground circuit, the injector spring pushes the armature downward and closes the injector ports. The injector winding has 1.5 ohms resistance, and the PCM operates the injector with a peak-and-hold current. When the PCM grounds the injector winding, the current flow in this circuit increases rapidly to 4 amperes. When the current flow reaches this value, a current-limiting circuit in the PCM limits the current flow to 1 ampere for the remainder of the injector pulse width. The peak-and-hold function provides faster injector armature opening and closing.

The poppet nozzles are snapped into openings in the lower half of the intake manifold, and the tip of each nozzle directs fuel into an intake port. Each poppet nozzle contains a valve

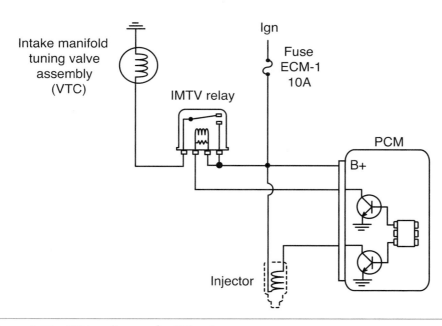

Figure 6-31 Wiring diagram for CPI system.

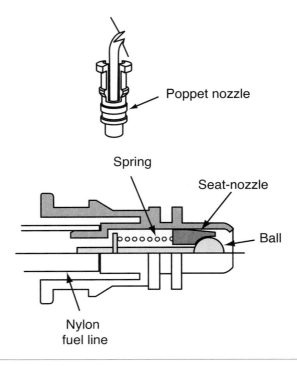

Figure 6-32 CPI injector operation.

with a check ball seat in the tip of the nozzle (Figure 6-32). A spring holds the valve and check ball seat in the closed position. When fuel pressure is applied from the central injector through the nylon lines to the poppet nozzles, this pressure forces the valve and check ball seat open against spring pressure. The poppet nozzles open when the fuel pressure exceeds 37 to 43 psi (254 to 296 kPa) to spray fuel from these nozzles into the intake ports.

When the fuel pressure drops, the poppet nozzles close. Under this condition, approximately 40 psi (276 kPa) fuel pressure remains in the nylon lines and poppet nozzles. This pressure prevents fuel vaporization in the nylon lines and nozzles during hot engine operation or hot soak periods. If a leak occurs in a nylon line or other CPI component, fuel drains from the bottom of the intake manifold through two drain holes to the center cylinder intake ports. The in-tank fuel pump, fuel filter, lines, and fuel pump circuit used with the CPI system are similar to those used with SFI and MFI systems.

Summary

Terms to Know

Adaptive memory

Air-fuel ratio

Block learn

Clear flood

Closed loop

Constant control
relay module

Corporate average
fuel economy
(CAFE)

❏ TBI, MPI, and SFI systems provide improved fuel economy, engine performance, and emission levels compared to carbureted engines.

❏ In a typical speed density TBI, MPI, or SFI system, the computer uses the MAP and engine rpm inputs to calculate the amount of air entering the engine. The computer then calculates the required amount of fuel to go with the air entering the engine.

❏ In a typical mass air flow (MAF) system the PCM uses the MAF and engine speed to determine the pulse width.

❏ In any TBI, MPI, or SFI system, the fuel pressure must be high enough to prevent the fuel from boiling.

❏ In a TBI, MPI, or SFI system, the computer supplies the proper air-fuel ratio by controlling injector pulse width.

❏ In a TBI, MPI, or SFI system, the computer increases injector pulse width to provide air-fuel ratio enrichment while starting a cold engine.

❏ Most computers provide a clear flood mode if a cold engine becomes flooded. This mode is activated by pressing the gas pedal to the floor while cranking the cold engine.

❏ In an SFI system, each injector has an individual ground wire connected to the computer.

❏ In an MPI system, the injectors are connected together in pairs or groups on the ground side.

❏ A central port injection system has one central injector and a poppet nozzle in each intake port. The central injector is operated by the PCM, and the poppet nozzles are operated by fuel pressure.

Review Questions

Short Answer Essays

1. Describe the importance of the air-fuel ratio to engine performance and emission levels.
2. What is meant by stoichiometric?
3. Explain the difference between open loop and closed loop fuel strategies.
4. Explain the operation of the speed density fuel system.
5. Explain the operation of the mass air flow (MAF) electronic fuel-injection (EFI) system.
6. Describe the difference between the sequential fuel-injection (SFI) system and the multipoint fuel-injection (MPI) system.
7. Describe the meaning of pulse width.
8. Describe the operation of the central injector and poppet nozzles in a central port injection (CPI) system.
9. Explain the purpose of the TPS input in a speed density fuel-injection system.
10. Describe the difference between short term and long term adaptive memory.

Fill-in-the-Blanks

1. The computer determines the air entering the engine from the _____ and _____ input signals in a speed density system.
2. The length of time the computer grounds the injector is referred to as _____ _____ .
3. In TBI, MPI, and SFI systems, the fuel pressure must be high enough to prevent _____ _____ .
4. On an SFI system, each injector has an individual _____ _____ connected to the computer.
5. If the injector pulse width is increased, the air-fuel ratio becomes _____.
6. _____ loop means the O_2S input is ignored by the PCM.
7. Not only is the air-fuel ratio a major factor in engine performance, it is a common element in the formation of _____, _____, and _____.
8. The optimum air-fuel ratio is called _____.
9. Short term adaptive memory is directly driven by the _____.
10. In a central port fuel-injection (CPI) system, the air-fuel ratio is determined by the pulse width on the _____ _____ _____ .

Multiple Choice Questions

1. *Technician A* says the computer uses the TPS and ECT signals to determine the mass of air entering the engine in a speed density system.
   Technician B says the computer uses the TPS and oxygen sensor signals to determine the mass of air entering the engine in a MAF system.
   Who is correct?
   A. A only
   B. B only
   C. Both A and B
   D. Neither A nor B

2. The stoichiometric ratio of a gasoline engine is:
   A. 9.3:1.
   B. 12.6:1.
   C. 14.7:1.
   D. None of the above.

3. *Technician A* says that in an EFI system, higher-than-normal fuel pressure causes a lean air-fuel ratio.
   Technician B says that in these systems lower-than-normal fuel pressure causes a rich air-fuel ratio.
   Who is correct?
   A. A only
   B. B only
   C. Both A and B
   D. Neither A nor B

4. While discussing TBI, MPI, and SFI systems:
   Technician A says that in EFI systems the PCM provides the proper air-fuel ratio by controlling fuel pressure.
   Technician B says the PCM provides the proper air-fuel ratio by controlling injector pulse width.
   Who is correct?
   A. A only
   B. B only
   C. Both A and B
   D. Neither A nor B

5. Open loop refers to:
   A. The O_2S being used as feedback to the PCM.
   B. The injectors operating in a nonsynchronized mode.
   C. All injectors being pulsed at the same time.
   D. The O_2S input is being ignored by the PCM.

6. *Technician A* says that in a central port injection (CPI) system the poppet nozzles are opened by the computer during engine starting.
   Technician B says the poppet nozzles are opened by fuel pressure when the engine is running.
   Who is correct?
   A. A only
   B. B only
   C. Both A and B
   D. Neither A nor B

7. *Technician A* says in a MAF system, the mass air flow (MAF) sensor is used to measure the vacuum in the intake manifold to determine engine load.
   Technician B says the rpm input is needed for the engine to start.
   Who is correct?
   A. A only
   B. B only
   C. Both A and B
   D. Neither A nor B

8. *Technician A* says short-term fuel trim corrects fuel delivery in direct proportion to the voltage signals from the O_2S.
   Technician B says that on the General Motors integrator systems, a number larger than 128 represents an increase of pulse width over base.
   Who is correct?
   A. A only
   B. B only
   C. Both A and B
   D. Neither A nor B

9. The throttle position sensor's (TPS) role in the EFI system is to:
   A. Determine idle mode.
   B. Determine off-idle mode.
   C. Determine wide-open throttle mode.
   D. All of the above.
   E. None of the above.

10. *Technician A* says a rich air-fuel mixture will cause increased levels of hydrocarbons and carbon monoxide.
    Technician B says a rich air-fuel mixture will increase Nitrous Oxide (NO_x) emissions.
    Who is correct?
    A. A only
    B. B only
    C. Both A and B
    D. Neither A nor B

Emission Control Systems

Upon completion and review of this chapter, you should be able to:

❏ Explain the emission laws resulting from the Clean Air Act and their impact on the automotive industry.

❏ Describe the combustion process and resulting emissions.

❏ Explain how hydrocarbon exhaust emissions are produced and what affects their levels.

❏ Explain how carbon monoxide is produced and what affects its levels.

❏ Explain the production of NO_x and what affects its levels.

❏ Describe the non-pollutant exhaust gases and how they are used to determine combustion efficiency.

❏ Detail the importance of the air-fuel ratio in controlling exhaust emissions.

❏ Explain typical types of pre-combustion controls that are used including engine design, computerized engine controls, ignition control, evaporative controls, exhaust gas recirculation, and crankcase ventilation.

❏ Explain how evaporative emissions are contained.

❏ Describe how exhaust gas recirculation (ERG) is used to control NO_x emissions and the operation of different types of EGR valves used.

❏ Explain the function of the PCV system.

❏ Describe the components used for post-combustion control including air injection and catalytic converters.

Introduction

Harmful automotive emissions originate from three general sources. The first is **evaporative emissions** from the fuel system. The second source is **crankcase emissions** resulting from blowby around the piston rings and engine oil vapors. The third source of emissions is **exhaust emissions**, which is a product of combustion. Exhaust emissions are produced when the engine is running as a by-product of combustion or incomplete combustion. The exhaust by-products we are concerned with controlling are hydrocarbon (HC), carbon monoxide (CO), and oxides of nitrogen (NO_x).

Emission Laws and the Automotive Industry

In the beginning of the automotive era, emissions were not of much concern. The old engines would chug and belch out smoke, but to most people the pollution produced was more of a nuisance with these "new fangled contraptions." However, as the number of automobiles being driven increased, it was soon realized that pollution had to be controlled. According to the EPA, pollutant emission estimates for the period of 1940 to 1990 showed that the U.S. total air pollution included 18.7 million metric tons of volatile organic compounds (mainly hydrocarbons), 60.1 million metric tons of carbon monoxide, and 19.6 million metric tons of oxides of nitrogen. Automobiles are responsible for 17.8 percent of the total hydrocarbons emissions, 30.9 percent of the total carbon monoxide emissions, and 11.1 percent of the total oxides of nitrogen emissions. As a result, legislation was passed and amended several times to limit the pollutant output of the automobile.

Under Title II of the Clean Air Act, manufacturers are required to control motor vehicle emissions of the criteria pollutants: hydrocarbons (HC), carbon monoxide (CO), oxides of nitrogen (NO_x) and particulate matter (PM). The EPA and the California Air Resources Board (CARB)

Evaporative emissions are raw fuel vapors that drift into the atmosphere. This evaporation is a direct result of the property of gasoline to vaporize at low temperatures, not the result of the combustion process.

Crankcase emissions are engine vapors escaping past the piston rings, out of the engine, and into the atmosphere.

Exhaust emissions are those produced as a result of combustion and escape through the vehicle's exhaust system.

have established standards and measurement procedures for exhaust, evaporative, and refueling emissions of these criteria pollutants. Manufacturers are responsible for developing and certifying vehicles with emission systems and emission-related components that are durable for at least the full useful life of the vehicle. Regulations require vehicle emission certification prior to their introduction into commerce and require in-use compliance to assure that standards are met for the useful life of the vehicle.

History of Emission Laws

In 1959, California established the first standards for automotive emissions. Over the years, the federal government has passed legislation in an effort to maintain air quality standards (Table 7-1). The original Clean Air Act was brought into law in 1963. The Clean Air Act Amendments of 1970 formed the EPA and gave the agency broad authority to regulate vehicle pollution. Specific responsibilities were set for government and private industry to reduce emissions. Since then, the agency's pollution and automotive emission standards have become increasingly more stringent. The EPA has passed legislation regulating various automotive systems over the years.

The 1970 legislation required 90 percent reductions in tailpipe emissions for new 1975 and 1976 automobiles. This presented the automakers with major technical and economic challenges. Nevertheless, the EPA successfully forced the adoption of two major control technologies: the catalytic converter in 1975 and the three-way catalyst in 1981.

The Clean Air Act Amendments of 1990 added more clout to many parts of the earlier law. Some features of the new act are:

- Stricter tailpipe emission standards for cars, trucks, and buses
- Expansion of Inspection and Maintenance (I/M) programs with more stringent testing
- Attention to fuel technology (the development of alternate fuels)
- Study of non-road engines (that is, boats, farm equipment, home equipment, construction equipment, lawn mowers, etc.)
- Mandatory alternative transportation programs (car-pooling) in heavily polluted cities

California created the CARB in 1967, prompted by the original Clean Air Act. CARB passed regulation of onboard diagnostics (OBD) for vehicles sold in California in 1988. This regulation required that all vehicles sold in California have onboard diagnostics by 1991. OBD-I, the first

TABLE 7-1   EPA LEGISLATION HISTORY

Year	Legislation
1963	Original Clean Air Act passed into law
1970	Clean Air Act amendments to current policies amended
1970	Environmental Protection Agency formed
1971	Evaporative emission standards enacted
1972	First Inspection and Maintenance program introduced
1973	NO_x exhaust standards enacted
1975	First catalytic converter introduced
1989	Gasoline volatility standards enacted
1990	New Clean Air Act amended to current policies
1995	I/M 240 testing for gasoline vehicles required in non-attainment zones
1996	OBD-II vehicle compliance required

phase, required monitoring of the fuel metering system, the exhaust gas recirculation (EGR) system, and additional emissions-related electrical components. A malfunction indicator lamp (MIL) was required to alert the driver of a malfunction. Along with the MIL, OBD-I required the setting of DTCs identifying the area at fault.

With the passage of the Federal Clean Air Act Amendments in 1990, CARB developed regulation for the second generation of onboard diagnostics (OBD-II). These amendments also prompted the EPA to develop somewhat different onboard diagnostic requirements. All passenger cars, light-duty trucks and medium-duty vehicles and engines sold in the U.S. were required to comply with OBD-II standards by 1996.

In spite of the tremendous progress that has been made in new-car emissions levels, the total U.S. vehicle population continues to be a significant source of air pollution because the number of vehicle-miles traveled on American roads has doubled in the last 20 years to 2 trillion miles per year. The total volume of emissions offsets a great deal of the technological progress made over the same period. As the number of vehicle-miles continues to increase, continuing efforts will have to be made to reduce emissions levels from individual vehicles.

The Clean Air Act Amendments of 1990 address this problem by providing for the following approaches:

- Cleaner new cars (TeVan)
    TLEV—Transitional Low-Emitting Vehicles
    LEV—Low-Emitting Vehicles
    ULEV—Ultra Low-Emitting Vehicles
    ZEV—Zero-Emitting Vehicles
- Cleaner fuels (compressed natural gas; methanol)
- Improved I/M testing to control evaporative, HC and CO exhaust, and NO_x emissions.
- Second-generation onboard diagnostics (OBD-II)

The first two points are an evolution of a process that has been going on for more than two decades. Modern technology allows manufacturers to build cleaner burning engines without having to go through the same learning curve that was experienced in the early years of pollution control. Cleaner vehicle and fuel technology have already helped to reduce tailpipe exhaust emissions and gasoline evaporation. In some parts of the country, oxygenated fuels are used, particularly during the winter months.

During the 1990s, emission standards in the United States have become increasingly stringent (Table 7-2). In 1994, an ambitious emission program began in California. This program specified emission standards for transitional low-emitting vehicles (TLEV). In 1997, the California emission program specified a further reduction in emission levels for low-emitting vehicles (LEV), and in the year 2000, emission levels are specified for ultra low-emission vehicles (ULEV).

In 1998, California emission regulations required 2 percent of the cars sold in the state to be zero emission vehicles (ZEV). The regulations also called for a higher percentage in the years

TABLE 7-2   EMISSION STANDARDS, 1990 TO 2000

	HC	CO	NO$_x$
1990 — U.S.	0.41	3.4	1.0
1994 — U.S.	0.25*	3.4	0.4
1993 — Calif.	0.25	3.4	0.40
1994 — TLEV	0.125	3.4	0.40
1997 — LEV	0.075	3.4	0.20
2000 — ULEV	0.040	1.7	0.20
*non-methane HC			

2003 and 2005. With present technology, only electric cars meet the ZEV standards. These regulations were repealed in 2002.

In addition to the above emission program vehicles, there is the National Low-Emission Vehicle (NLEV) program that introduced California low-emission cars and light-duty trucks into the **Northeast Trading Region (NTR)**. Section 177 of the Clean Air Act allows states with non-attainment areas to adopt standards identical to California standards in lieu of federal standards. Standards must be adopted at least two model years prior to implementation, and the state action cannot result in a different motor vehicle (third car) than those certified in California.

This program began with the 1999 model year vehicles in the NTR, and the rest of the country started in the 2001 model year. NLEV incorporates the California LEV Program vehicle categories and standards: TLEV, LEV, and ULEV.

Present government goals call for reductions of more than 98 percent for hydrocarbons (HC), 96 percent for carbon monoxide (CO), and 95 percent for nitrogen oxides (NO_x), compared to pre-1970 engine emissions. These standards are referred to as Tier 2.

Most of the publicity surrounding the Clean Air Act Amendments of 1990 centers around improved I/M testing of vehicles on the road. Increased I/M testing requirements are important because, despite all the improvements in new car emissions and fuel quality, it will be many years before these cleaner cars dominate the overall population of vehicles in the United States.

In many cities, vehicles contribute 35 to 70 percent of hydrocarbon emissions and at least 90 percent of carbon monoxide emissions. Concentrations of one or both of these pollutants exceed air quality standards in almost every major city in the United States.

Under this law, more cities began I/M programs. With the Clean Air Act Amendments of 1990, 154 areas require either basic or enhanced I/M testing. I/M testing as it relates to the 1990 amendment is discussed in Chapter 9.

Canadian Requirements.   Through the 1997 model year, Canada voluntarily adopted the Tier I standards (in grams/kilometer) including alternative fuel vehicles with a Memorandum of Understanding. The Tier I standards were regulated beginning in the 1998 model year. The NLEV emission requirements were adopted with a Memorandum of Understanding for 2001–2003 model year vehicles. OBD and evaporative requirements also parallel U.S. federal requirements. The government plans to regulate to Tier 2 standards commencing the 2004 model year.

Canadian fuel economy requirements are a voluntary program that parallel U.S. fuel economy protocol and standards (measured in liters per 100 kilometers). A voluntary standard "Ener-Guide" fuel efficiency label was implemented for 2000 model year vehicles.

In British Columbia, Canadian/U.S. federal emission requirements apply through the 2000 model year. Beginning in 2001, light-duty vehicles must be California certified vehicles. Medium- and heavy-duty vehicles continue to meet the Canadian/U.S. federal emission standards.

The province mandates that vehicles in the Vancouver area pass the "AirCare" Emission Inspection and Maintenance requirements. Beginning with the 1992 model year, vehicles including all-wheel-drive vehicles, are dynamometer tested on the transient IM240 test in government operated facilities. OBD is monitored on 1998 and newer vehicles.

The province of Ontario mandates those vehicles in southern Ontario pass the "Drive Clean" Emission Inspection and Maintenance requirements. Vehicles 20 years and newer are dynamometer tested on a steady state ASM2525 test at independent facilities. OBD is monitored on 1998 and newer vehicles.

The Combustion Process and Exhaust Gases

The power of a gasoline engine is derived from the controlled combustion of the air-fuel mixture in the combustion chamber. For a burn to occur, three elements are required: fuel, oxygen, and heat. If any one of these three elements is missing, combustion will not occur.

Perfect combustion (Figure 7-1) results in the formation of three elements: heat, carbon dioxide (CO_2), and water vapor (H_2O). Carbon atoms from the hydrocarbon fuel (HC) combine

Shop Manual
Chapter 7,
page 284

The **Northeast Trading Region (NTR)** includes Massachusetts, New York, Maine, and Vermont.

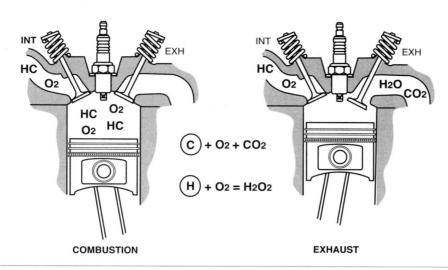

COMBUSTION   EXHAUST

$$\text{C} + O_2 + CO_2$$

$$\text{H} + O_2 = H_2O_2$$

Figure 7-1   Perfect combustion results in CO_2 and H_2O.

with oxygen from the atmosphere (O_2) to form CO_2. Hydrogen atoms from the hydrocarbon fuel combine with oxygen from the atmosphere to form H_2O. In a perfect combustion reaction, there is no excess fuel remaining, and all the available oxygen is used. In other words, there is exactly the right amount of oxygen to burn the existing fuel and exactly the right amount of fuel to consume the existing oxygen. With perfect combustion, the only exhaust gases produced would be harmless carbon dioxide and water vapor.

If it were possible to obtain perfect combustion all of the time, then all of the emission controls that have been developed over the last three decades would not have been necessary. However, perfect combustion rarely occurs in a gasoline engine (Figure 7-2). Although the air-fuel ratio is being controlled more precisely now with the use of computer controls, it is not always perfect. Compounding the problem is the fact that even the highest-quality gasoline has impurities that travel along with the hydrocarbons. In addition, the air that is drawn into the combustion chamber contains more than just oxygen. Air contains about 21 percent oxygen and 78 percent nitrogen. Nitrogen, under certain conditions, can produce harmful oxides of nitrogen (NO_x).

Other factors that impede perfect combustion (even when the proper air-fuel ratio exists) include insufficient heat of combustion and the timing of the spark may not always be ideal.

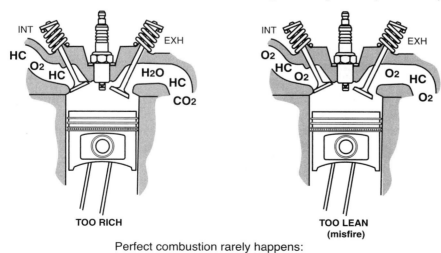

TOO RICH   TOO LEAN
(misfire)

Perfect combustion rarely happens:

- Air-fuel ratio not always perfect
- Air contains elements other than oxygen (78% nitrogen)
- Combustion heat may be insufficient
- Spark timing not always perfect
- Gasoline contains impurities

Figure 7-2   There are many factors that prevent perfect combustion.

These conditions lead to incomplete combustion and the production of carbon monoxide and hydrocarbons in the exhaust.

The Exhaust Gases

There are generally five exhaust gases being observed in today's automobile:

- Carbon monoxide (CO)
- Hydrocarbons (HC)
- Oxides of nitrogen (NO_x)
- Carbon dioxide (CO_2)
- Oxygen (O_2)

Of these five exhaust gases, CO, HC, and NO_x are pollutants. The CO_2 and O_2 are not pollutants, but are informational gases which are observed for diagnostic purposes.

Studies indicate that the automobile accounts for about 50 percent of the hydrocarbons, over 75 percent of the carbon monoxide, and nearly 50 percent of the oxides of nitrogen that pollute our atmosphere (Figure 7-3). Legislation places limits on how much of these pollutants can be produced by vehicles sold in the United States and Canada.

Hydrocarbons. Hydrocarbons consist of hydrogen and carbon atoms in various combinations. Gasoline is a hydrocarbon fuel and is the source of automotive hydrocarbon emissions. Hydrocarbons, in combination with other compounds, can produce ground-level ozone, or **photochemical** smog.

As a tailpipe pollutant, HC emissions are unburned fuel caused by incomplete combustion. Excessive HC emissions can be the result of engine conditions that cause incomplete combustion such as low compression, defective valves or lifters, inaccurate spark timing, improper spark duration, ignition system problems, improper air-fuel ratio, or vacuum leaks. When a cylinder misfires, only a portion, or none, of the fuel is burned. The unburned fuel passes through the exhaust and results in high HC emissions.

High HC emissions are also caused by the **"quenching"** effect inside the combustion chamber. Temperatures are not consistent throughout the combustion chamber. The areas near the cylinder head and engine block are cooler than the other areas since the metal surfaces absorb much of the heat of combustion (Figure 7-4). In these cooler areas, the combustion flame tends to extinguish, or quench. Since the heat is not sufficient to complete the burn of the air-fuel mixture, unburned hydrocarbons pass through the exhaust. Engine designers reduce the effects of quenching through combustion chamber design and piston ring location.

In order to reduce hydrocarbon emissions, there must be a sufficient mass of oxygen during the combustion process. Proper **oxidation** results in the creation of harmless carbon dioxide and water. This oxidation takes place when the air-fuel ratio is correct (Figure 7-5). Notice that hydrocarbon emissions are high as the air-fuel mixture treads toward a rich condition. This is due to the lower ratio of oxygen in the mixture, leaving unburned fuel in the chamber after combustion. At stoichiometric, the hydrocarbon emissions are low but will go even lower at a slightly lean air-fuel mixture. Finally, hydrocarbon emissions increase again as the air-fuel mixture

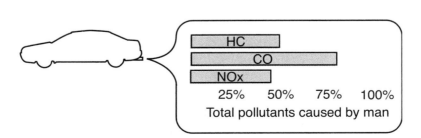

Figure 7-3 Vehicle exhaust gases as source of pollution.

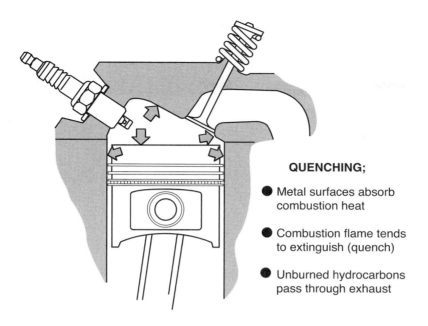

QUENCHING;

● Metal surfaces absorb combustion heat

● Combustion flame tends to extinguish (quench)

● Unburned hydrocarbons pass through exhaust

Figure 7-4 Quenching areas in the combustion chamber cool the air-fuel charge.

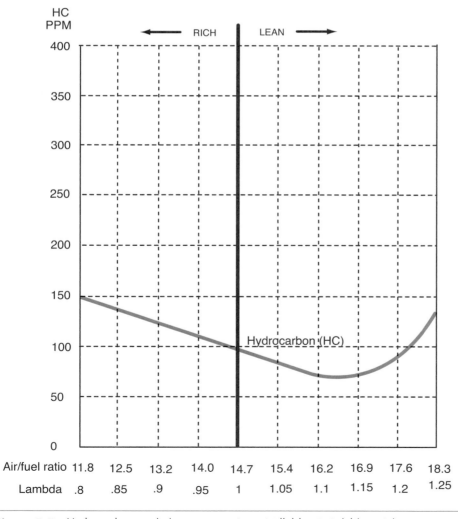

Figure 7-5 Hydrocarbon emissions are most controllable at stoichiometric.

becomes leaner. This is due to cylinder misfire resulting from a lack of enough oxygen to have complete combustion. Since the cyclinder had incomplete combustion, unburned fuel is pushed into the exhaust.

Carbon Monoxide.   Carbon monoxide (CO) exists as a molecule containing one carbon atom and one oxygen atom. It is a colorless, odorless, and very poisonous gas resulting from incomplete combustion of the air-fuel mixture. With perfect combustion, one carbon (C) atom from the fuel will combine with atmospheric oxygen (O_2). This will form harmless carbon dioxide (CO_2). CO_2 is used by plants to manufacture oxygen.

If the air-fuel mixture is too rich, there is a shortage of oxygen in the combustion chamber. Without sufficient oxygen, the burning of the mixture stops prematurely. In this case, most of the oxygen was used up, so the burn stops. This results in one carbon atom combining with one oxygen atom to form harmful carbon monoxide (CO).

Anything that restricts airflow to the combustion chamber, such as a blockage in the intake manifold or cylinder heads, can cause high CO emissions levels. Carbon monoxide is a direct indication of the air-fuel mixture (Figure 7-6). Generally, the lower the CO is reading on an exhaust-gas analyzer, the leaner the air-fuel mixture. This means that high CO levels indicate a rich mixture. Keep in mind that the rich mixture may be caused by leaking injectors and higher-than-normal fuel pressures.

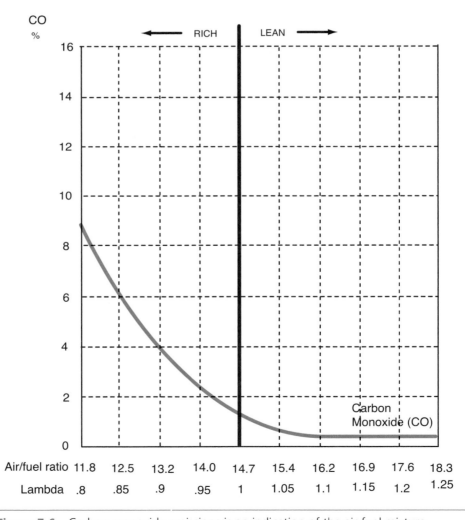

Figure 7-6   Carbon monoxide emissions is an indication of the air-fuel mixture.

Since combustion is required to create CO, cylinder misfire does not increase CO emissions. In fact, the CO emission level may decrease slightly while a misfire is occurring.

Oxides of Nitrogen.   Oxides of nitrogen (NO_x) are gases composed of molecules that contain one nitrogen atom combined with a varying number of oxygen atoms. Combining a nitrogen atom with one oxygen atom results in the formation of nitrous oxide (NO). The combining of one nitrogen atom with two oxygen atoms forms nitrous dioxide (NO_2). The variable number of oxygen atoms results in the formula "NO_x" with "x" being the unknown number of oxygen atoms. Oxides of nitrogen are the result of the combustion process, but they are formed differently than CO and HC.

NO_x is an environmental hazard because ultraviolet radiation from the sun acts upon a combination of NO_x and hydrocarbons in the atmosphere to produce photochemical smog. This smog appears as a brownish haze that irritates the eyes and respiratory system. It is also a contributor to acid rain.

The formation of NO_x is dependent on the combustion temperature. Nitrogen is normally inert and does not tend to readily combine with other elements. However, when the combustion temperature exceeds approximately 2,500°F (1,371°C), nitrogen combines with oxygen to form various oxides, most notably nitrous oxide (NO).

High combustion temperatures can result from heavy engine loads, high compression ratios, excessively lean air-fuel mixtures (Figure 7-7), advanced ignition timing, engine overheating, and vacuum leaks.

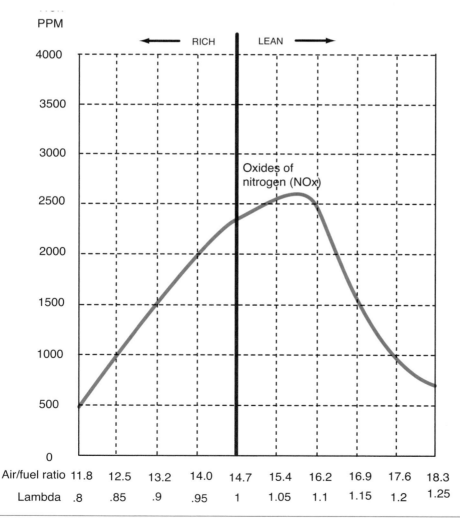

Figure 7-7 NO_x emissions and the air-fuel mixture relationship.

NO_x emissions are controlled by lowering the combustion temperature. This has been achieved through lower compression ratios, camshaft timing adjustments, air-fuel mixture control, and EGR systems.

> **AUTHOR'S NOTE:** Prior to these controls for NO_x, combustion temperatures in excess of 4,000°F (2,204°C) were not uncommon.

Breaking the connective bonds between the nitrogen and oxygen elements can minimize the NO_x that is still produced. This is done by removing the oxygen.

Non-Pollutant Exhaust Gases

The process of removing the oxygen is called a reduction reaction and takes place inside a three-way catalytic converter.

CO_2 is a greenhouse gas and may be one of the causes of global warming.

As mentioned earlier, there are two major exhaust gases that are not considered as pollutants; carbon dioxide and oxygen. Technicians should be aware of these two gases since their presence helps to diagnose certain engine conditions by determining combustion efficiency.

Carbon Dioxide. Carbon dioxide is a by-product of perfect combustion. Carbon dioxide is formed when one atom of carbon bonds with two atoms of oxygen from the air during combustion. An essentially harmless gas, it is present at levels of 14 percent to 15 percent in the exhaust of a properly running engine. Carbon dioxide is also produced when carbon monoxide is oxidized in the catalytic converter.

The amount of carbon dioxide in the exhaust is directly related to the air-fuel ratio (Figure 7-8). As the fuel mixture approaches stoichiometric, the level of CO_2 peaks. It decreases when the mixture becomes richer or leaner. This fact makes CO_2 in the exhaust an excellent reference that can help determine how efficiently the engine is combusting its fuel. The higher the CO_2 reading, the higher the efficiency of combustion.

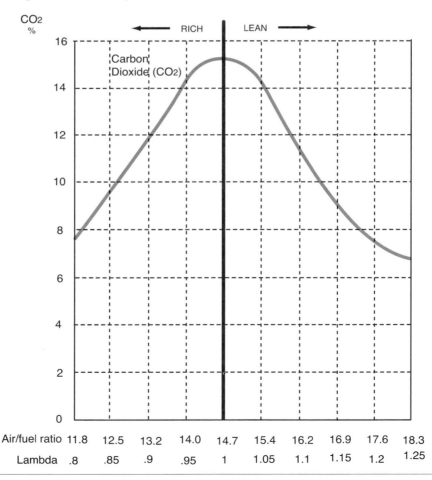

Figure 7-8 CO_2 is the highest at stoichiometric.

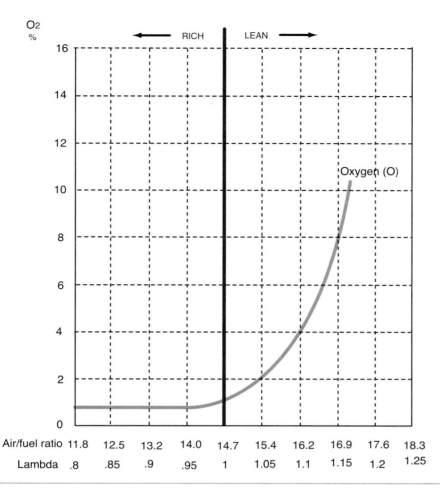

Figure 7-9 O_2 emissions increase with a leaner air-fuel mixture.

Oxygen (O_2).   The air that we breathe is made up of about 21 percent oxygen. Oxygen is mixed with gasoline for combustion and is consumed during the combustion process (Figure 7-9). As long as there is sufficient oxygen, the combustion burn continues. If the combustion chamber is delivered too much fuel for the amount of oxygen, all of the available oxygen is used up. This results in low O_2 content in the exhaust.

As the air-fuel mixture moves toward lean, the amount of oxygen in the exhaust steadily increases. This is due to oxygen being left over since the fuel is burned prior to all of the oxygen being used. Therefore, higher levels of O_2 in the exhaust are a direct indication of leaner air-fuel ratios.

When the air-fuel mixture is either rich or lean, the levels of oxygen and carbon monoxide will be opposite one another (when O_2 is high, CO is low, and vice versa). At the stoichiometric air-fuel ratio, the levels of O_2 and CO in the exhaust are approximately equal. At stoichiometric, O_2 content is about 1 percent to 2 percent.

The Importance of the Air-Fuel Ratio

As just discussed, the air-fuel ratio is the common element in the formation of each of the three main pollutant gases (HC, CO, and NO_x). Each of these emissions can be caused by an improper ratio of air to fuel in the combustion process. Therefore, maintaining the correct air-fuel ratio has become one of the main goals of emissions control system design. As seen in Chapter 6, the control of the air-fuel ratio has become a main function of the PCM.

As related to exhaust emissions, an air-fuel ratio that reduces the level of one pollutant can increase the level of another. Consider the following explanation.

Depending on other circumstances, a lean mixture can be responsible for excess levels of hydrocarbons and oxides of nitrogen. An excessively lean mixture can cause a lean burn misfire resulting when a given volume of fuel mixture does not contain enough gasoline to ignite. What gasoline there is passes through the exhaust system unburned, resulting in high HC levels.

When the fuel mixture is slightly lean, the excess oxygen produces a very hot flame. This results in an increase in combustion temperatures and causes the nitrogen in the intake air to combine with oxygen to form harmful oxides of nitrogen (NO_x).

On the other hand, a lean mixture results in low-to-zero levels of carbon monoxide since all of the fuel is burned. This is because there is sufficient oxygen to burn all of the fuel.

A rich mixture can lead to increased levels of hydrocarbons and carbon monoxide, but NO_x emissions will generally be low. High CO levels are a good indicator of a rich fuel mixture, especially if high HC levels are also present. In a rich mixture, there is insufficient oxygen to combine with the carbon atoms from the gasoline to form harmless carbon dioxide. Instead, each carbon atom combines with one oxygen atom to form poisonous carbon monoxide. NO_x levels will usually be low in a rich mixture situation because the shortage of oxygen results in lower combustion temperatures.

However, hydrocarbon emissions can be caused both by an excessively lean or an excessively rich mixture. In the case of a lean mixture, there is not enough gasoline to ignite, and it passes unburned through the exhaust system. In a very rich mixture, there may not be enough oxygen present to support combustion. The end result is the same unburned hydrocarbons go through the exhaust system.

The relationship between the air-fuel ratio and the emission gases is illustrated in Figure 7-10. As the air-fuel mixture moves away from stoichiometric (either too lean or too rich), a num-

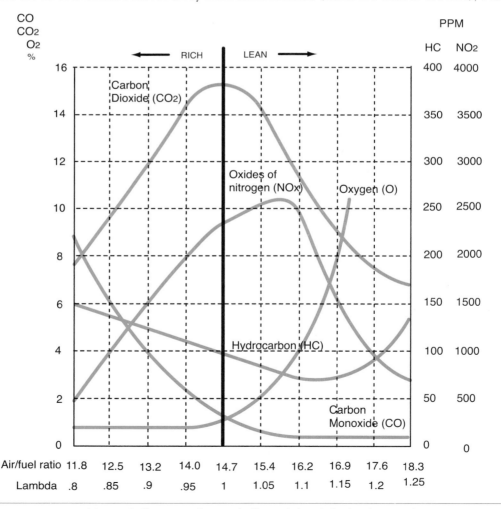

Figure 7-10   This graph illustrates the total effect of the air-fuel ratio on exhaust gases.

TABLE 7-3   AIR-FUEL MIXTURE AND EMISSIONS RELATIONSHIPS

Air-Fuel Mixture	Effect on Vehicle
Too Lean	Increased NO_x emissions Poor engine power Misfiring at cruising speeds Burned valves Burned pistons Scored cylinders Spark knock or ping
Slightly Lean	Low exhaust emissions High gas mileage Reduced engine power Slight tendency to knock or ping
Stoichiometric	Best all-around performance and emissions levels
Slightly Rich	Increased CO emissions Increased HC emissions Maximum engine power Higher fuel consumption Less tendency to knock or ping
Too Rich	Increased CO emissions Increased HC emissions Poor fuel mileage Misfiring Oil contamination Black exhaust

ber of conditions could occur, including emissions test failures and engine problems. Table 7-3 lists some of these relationships.

Pre- and Post-Combustion Controls

All regulated automotive emissions result from the properties and combustion of fuel. Three sources have been identified and are regulated by federal and state governments: evaporative emissions, crankcase emissions, and exhaust emissions.

All three sources produce HC. Crankcase and exhaust emissions also include CO and NO_x. Various designs have evolved to control emissions. All of the systems described in this chapter can be grouped into two general types. The first group consists of those systems which prevent the formation of harmful emissions. This is the preferred approach to pollution control because it is more effective to prevent the formation of pollutants than it is to get rid of them after they have been produced. The methods and devices in this first group are known as **pre-combustion control** systems. The systems in the second group have the job of reducing the level of pollutants after they have been formed. They are known as **post-combustion control** systems.

Table 7-4 identifies the major emissions control systems and classifies them as pre-combustion or post-combustion. The table also shows what pollutants they control.

Pre-combustion control systems are those that prevent the formation of harmful emissions prior to the combustion process.

Post-combustion control systems are those that reduce emissions levels after combustion has taken place.

TABLE 7-4   CLASSIFICATION OF EMISSION CONTROL TYPES

System	Classification	Pollutant Controlled
Engine design/operation	Pre-combustion	HC, CO, NO_x
Computerized engine control	Pre-combustion	HC, CO, NO_x
Spark control	Pre-combustion	HC, CO, NO_x
Exhaust gas recirculation	Pre-combustion	NO_x
Evaporative controls	Pre-combustion	HC
Air injection	Post-combustion	HC, CO
Catalytic converters	Post-combustion	HC, CO, NO_x

Pre-Combustion Control

Pre-combustion control includes fuel formulation, engine design, engine operation, fuel control, ignition control exhaust gas recirculation, and evaporative controls. Fuel formulation was discussed in Chapter 2. The other pre-combustion controls are discussed next.

Engine Design and Operation.   Since it is better to prevent emissions than it is to treat them after they are produced, engine design is the best possible approach to automotive pollution control. A well-designed engine reduces the need for add-on devices and post-treatment methods.

In the 1970s, engineers attempted to meet federal and state emissions regulations with a host of devices and techniques that unfortunately resulted in reduced performance along with the emissions levels. In the 1980s, the trend shifted toward starting with a clean sheet of paper and designing new engines with the express intent of having low emissions levels and good engine performance. Advances in the areas of manifold design, combustion chamber design, fuel injection, and computerization have enabled new engines to provide both performance and lower emissions. For example, CO can be controlled by fuel system design utilizing the delivery of less fuel. This reduces the chance for partial combustion. In addition, a major consideration for CO control is equal fuel distribution. As discussed in Chapter 6, this is accomplished by intake manifold design and the use of fuel injection. Leaner air-fuel mixtures assure enough air for complete oxidation to CO_2 during normal driving. The heated air inlet causes better vaporization of fuel in a cold engine. Advanced ignition timing allows more time for complete burning. However, this can raise HC, and it also raises combustion temperatures, which leads to the formation of NO_x.

NO_x forms at high combustion temperatures. These occur with lean mixtures and advanced timing (which also produces high power output and good fuel economy). Cooler combustion is provided by fuel control and moderate ignition timing advance. It is also achieved through the use of exhaust-gas recirculation to dilute the fresh air-fuel mixture with exhaust gases.

One way of controlling emissions through engine design is by use of a **heated air inlet**. Also called a thermostatic air system, the main component in most of these systems is the thermostatic air cleaner (Figure 7-11). Traditionally, a cold engine required a rich mixture because the fuel did not vaporize well when it was mixed with cold air. A thermostatic air cleaner takes warm air from around the exhaust manifold and directs it through a hot air duct to the air cleaner. This achieves the following results:

- Heated inlet air provides better fuel vaporization.
- Better vaporization means leaner mixtures can be used.
- Leaner mixtures result in reduced levels of HC and CO.

> **Heated air inlet** systems direct warm air into the air cleaner assembly during cold engine operation to help vaporize the fuel.

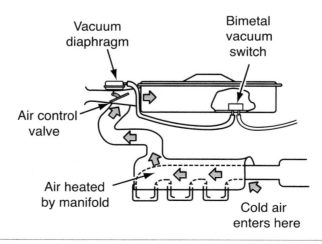

Figure 7-11   Typical air-heated intake system.

AUTHOR'S NOTE: Earlier six- and eight-cylinder vehicles used a heated crossover in the manifold to assist in fuel atomization during cold engine operation.

A bimetal sensor reacts to the air temperature. The sensor is calibrated to provide a specific output vacuum to the air door motor based on air temperature. This allows the amount of heated air to be regulated (Figure 7-12).

Computerized Engine Control. The PCM is considered a precombustion device since its function is to maintain optimum fuel and ignition and to control the operation of other emission control systems. As regulations tightened, more and more pollution control devices were added to the vehicle. Precomputerized systems responded to mechanical feedback such as vacuum signals, and there was a great deal of conflict as to which devices and which conditions should take precedence. The powertrain control module (PCM) brought a degree of precision and control to the emissions control system. The PCM constantly reads electronic signals from various sensors throughout the powertrain and responds by sending control signals to various devices as needed.

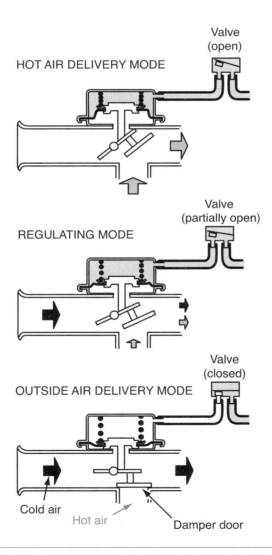

Figure 7-12   The bimetal sensor operates the mode door to direct heated air to the throttle body.

Technicians should be careful not to assume that the PCM is at fault in all repair situations. Items not monitored by the computer that can affect driveability and emissions include:

- Engine mechanical condition
- Ignition system condition
- Exhaust system condition
- Basic timing
- Routing of hoses and tubing

As you learned in Chapter 6, the oxygen (O_2) sensor is located in the exhaust manifold and is used to detect the amount of oxygen in the exhaust stream. This information tells the computer whether the fuel mixture is too rich or too lean. The computer can then have the fuel system adjust the fuel mixture as necessary.

With most computerized engines, the heated intake system is no longer needed. A thermistor in the intake air system provides the PCM with the temperature of the air entering the combustion chamber. The PCM can then alter the air-fuel mixture and/or the ignition timing to provide good performance and lower emissions.

With an ability to analyze several inputs and make corrective decisions every second, the engine computer has made possible the design of smaller, high-performance, low-polluting automotive engines.

Ignition Control. As you learned in Chapter 4, the ignition control system has a tremendous effect on overall emissions output. Ignition timing determines the precise moment at which the electrical impulse is sent to the spark plug to ignite the air-fuel mixture. By controlling this function as precisely as possible before combustion, harmful emissions are prevented so they do not have to be treated after combustion. Improper ignition timing can affect the levels of all three exhaust emissions. Excessive advance can result in partially burned fuel and cause CO to increase. The unburned fuel causes HC to also increase. Finally, high NO_x is created due to excessive combustion temperatures. On the other hand, if the ignition timing is retarded excessively, CO will also increase as a result of partially burned fuel, and high HC results from the unburned fuel.

To maintain optimum ignition control, the engine computer responds to various sensors and is programmed with instructions that tell it how to change the ignition timing for different situations.

Evaporative Emissions. Evaporative emissions result from the direct release of fuel vapors into the atmosphere. This can occur during refueling, while the vehicle is parked, and during engine operation. Gasoline is a volatile substance and vaporizes easily. Prior to the use of evaporative controls, the fuel system was vented directly to the atmosphere. This was necessary to prevent unsafe pressure levels inside the system. This type of atmospheric venting leads to hydrocarbon emissions and the production of photochemical smog. Atmospheric venting of the fuel vapors is no longer allowed by legislation. Today's automobiles have an evaporative emission control system to perform the following functions:

- Contain liquid during a vehicle rollover
- Trap fuel vapors
- Store fuel vapors
- Deliver vapors to the engine for burning

Figure 7-13 illustrates a typical evaporative emission control system. Most systems include the following components (Figure 7-14):

- Fuel tanks designed to limit the amount of fuel that can be put into the tank. This allows for expansion of the fuel as it warms.
- A pressure/vacuum relief fuel caps.
- A purge solenoid.
- An evaporative canister.

Shop Manual
Chapter 7,
page 293

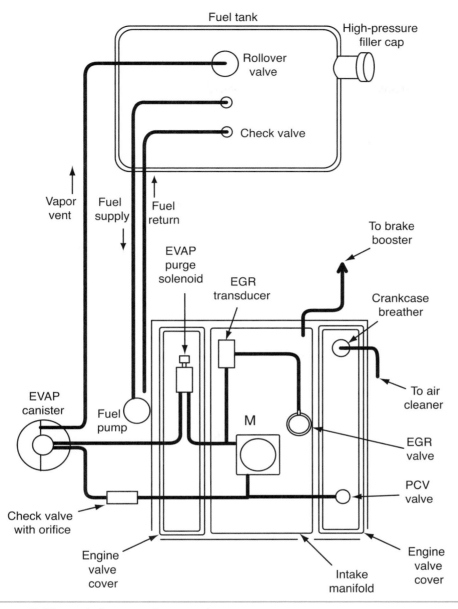

Figure 7-13 Typical evaporative control system.

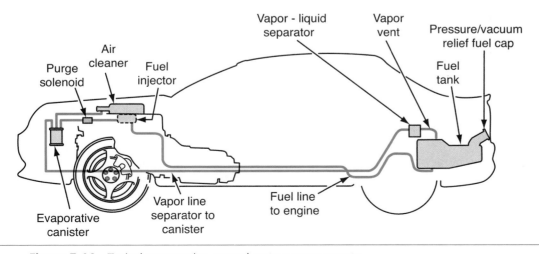

Figure 7-14 Typical evaporative control system components.

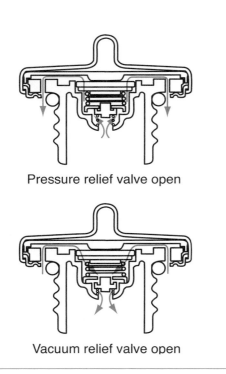

Pressure relief valve open

Vacuum relief valve open

Figure 7-15 Sealed fuel tank cap operation.

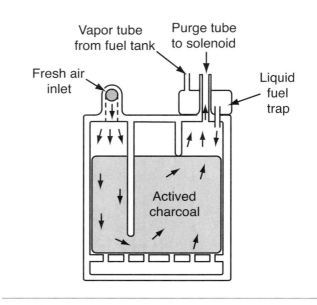

Figure 7-16 Typical evaporative canister.

All of the evaporative controls trap fuel vapors and reduce the tendency for fuel to evaporate to the atmosphere. The fuel tank is sealed with relief valve filler caps to raise the fuel boiling point. The pressure/vacuum filler cap releases to the atmosphere at 44 inches of water column and draws air in at a system vacuum of 4 inches water column (Figure 7-15). Hoses route internal vapors for storage in an **evaporative canister** (Figure 7-16). This canister is a charcoal-filled container. The charcoal absorbs the fuel vapors until a control valve "purges" the vapors from the charcoal. The vapors are then burned in the engine instead of evaporating to the atmosphere. Early vehicles would usually locate the canister in the engine compartment. In recent years, the canister is located near the fuel tank.

The canister contains a liquid fuel trap that holds any liquid fuel to prevent it from saturating the canister. As the fuel tank level lowers or the temperature of the fuel decreases, a vacuum is created that draws any trapped fuel back into the tank.

As fuel is drawn from the tank to be used by the engine, a slight negative pressure (vacuum) is created in the tank. To prevent the tank from collapsing, it needs to be vented to allow air to enter it and replace the fuel. Evaporative controls allow this ventilation to take place within a closed circuit so that the hydrocarbons stay in the fuel system. The sealed fuel system is vented into the charcoal-filled canister that retains the vapors until they can be drawn into the intake manifold and burned.

An evaporative line runs from the fuel tank to the canister. A purge valve is normally used to control the flow of the vapors from the canister into the intake system. On carbureted vehicles, a line also runs from the carburetor fuel bowl to the canister to contain fuel vapors from the float bowl.

Canister purge varies widely between vehicle makes, models, and years. Some systems use a constant purge system that uses a fixed orifice in a line to a **manifold vacuum** port. Others connect the purge line to **ported vacuum** so purge will only occur at open throttle. Other systems use bi-level purging (Figure 7-17) that uses different sized orifices to control purge into a manifold port or to a manifold vacuum port. Still other systems use a thermal delay valve to prevent the canister from purging until the engine reaches a specified temperature.

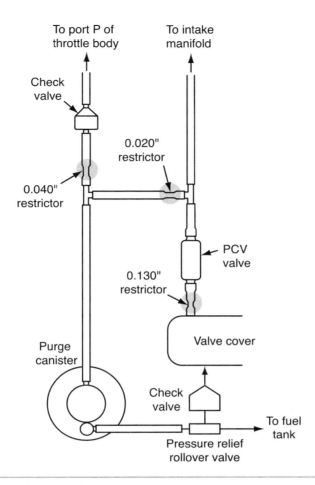

Figure 7-17 Bi-level evaporative control purge system.

A vapor control valve may be incorporated into the system (Figure 7-18). The valve is normally closed to block canister purge. When vacuum is applied to the valve, it opens to allow vapors to exit the canister and enter the intake manifold.

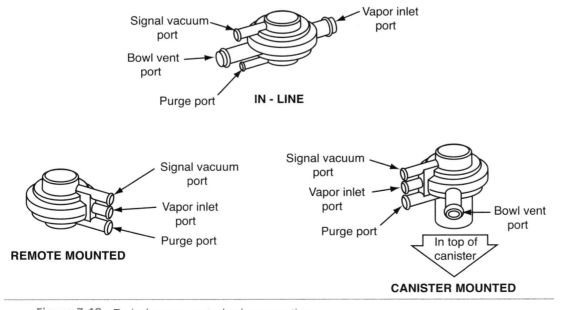

Figure 7-18 Typical purge control valve mountings.

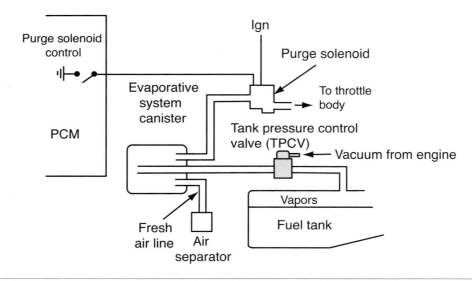

Figure 7-19 Computer-controlled EVAP system.

Today, most canister purge is controlled by the PCM. There are two basic systems that may be used. The first uses a duty-cycled purge solenoid. The electrically controlled canister purge solenoid replaces the conventional vacuum-controlled valves (Figure 7-19). Usually, the solenoid is provided battery voltage to one terminal when the ignition switch is in the RUN position, and ground is provided by the PCM. The purge solenoid is duty cycled to more precisely control vapor flow. When the solenoid is deenergized, the canister is allowed to purge. When the solenoid is energized, the passage from the canister to the intake manifold is blocked. This arrangement allows for purge to occur in the event of a failure with the circuit.

The second method of electronic control uses a linear purge solenoid. This system type may use a high-side driver to pulse width the voltage to the solenoid. As the pulse width is varied, the current is varied. By varying the current to the solenoid, the amount of purge is controlled around the lifted pintel.

Exhaust Gas Recirculation. The **exhaust gas recirculation (EGR)** system consists of the EGR valve (Figure 7-20), a control mechanism, and related hoses and passages. Knowledge of this system is particularly important because EGR directly affects engine performance.

Shop Manual
Chapter 7,
page 299

The **exhaust gas recirculation (EGR)** systems mixes the inert gases of the exhaust into the air-fuel charge to reduce combustion temperatures.

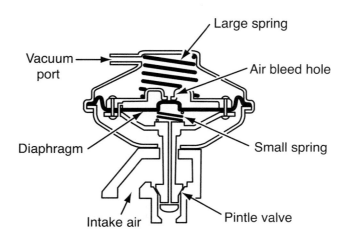

Figure 7-20 Typical EGR valve design.

TABLE 7-5 **EFFECTS THAT IMPROPER EGR HAS ON PERFORMANCE**

Don't Want EGR @	Because
Idle	Causes roughness
WOT	Reduces maximum power
Cold Start } Light Throttle }	Engine not hot enough to produce much NO_x; causes hard starting, roughness, hesitation, stalls

Since NO_x forms in high-temperature combustion, the solution is to prevent NO_x from forming by keeping combustion temperatures low enough to prevent the bonding of nitrogen and oxygen atoms. However, you have already learned that lean mixtures reduce HC and CO but increase NO_x. In addition, advanced ignition timing, which improves fuel economy, also increases NO_x. The EGR valve is used to prevent excessive combustion temperatures and reduce the formation of NO_x by reducing the volume in the combustion chamber. When operating, the EGR system mixes a metered amount of exhaust gases with the combustion mixture in the intake manifold. With the chamber's reduced volume, there is less room for the air-fuel mixture and combustion chamber temperature is reduced. A side benefit of EGR is a slight increase in fuel economy due to less fuel being burned.

Burned exhaust gases are basically inert and will not react chemically or continue to burn. A percentage of exhaust gases mixed with the air-fuel mixture "dilutes" the charge strength. The result is less volume of air-fuel mixture. In turn, this reduces combustion temperature and pressure, resulting in less NO_x.

Since the decrease in the volume of the air-fuel mixture by EGR interferes with combustion, the EGR system must be designed to add just enough EGR to control NO_x but not interfere with engine performance. If EGR operation is allowed to occur when it should not, driveability problems may result (Table 7-5). For best driveability, the EGR valve opens in proportion to throttle opening (but not at wide-open throttle). Also, EGR is not allowed during cold engine operation.

Most EGR valves are vacuum operated, poppet-type valves (Figure 7-21). The valve fastens to the intake manifold delivery port. The mounting block has two passages: one for exhaust gas into the valve, and one for metered exhaust gas flow out of the valve. The stem operates the poppet valve which seats in the mounting block to close off EGR flow.

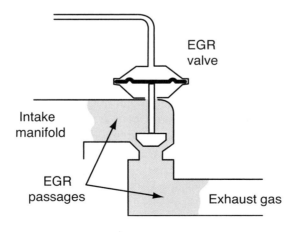

Figure 7-21 Most EGR valves use vacuum controls to lift the valve.

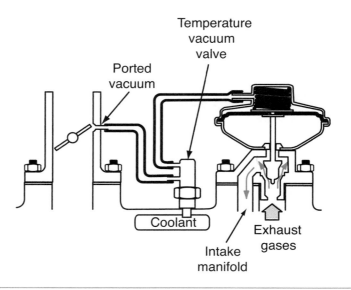

Ported
vacuum

Coolant

Intake
manifold

Exhaust
gases

Figure 7-22   The TVS prevents EGR operation when the engine is cold by blocking vacuum to the valve.

The upper part of the valve contains a vacuum diaphragm, that is attached to the stem. The diaphragm and stem are spring-loaded to close the poppet valve against the seat. The valve can move from closed to open or to metering positions in between depending on the amount of vacuum applied to the diaphragm. In precomputerized systems, a temperature vacuum switch (TVS) was located between the vacuum source and the valve to prevent operation of the EGR valve when the engine was cold (Figure 7-22).

Remembering that NO_x is formed under high heat conditions, increases in engine load requires more EGR flow. A good indication of engine load is exhaust **back pressure**. As the engine is running, each time a cylinder fires and the exhaust valve is open there is a high-pressure pulse in the exhaust system. Between the high-pressure pulses there are low-pressure pulses. At low engine speeds, the pressure in the exhaust system will toggle between high and low. The faster the engine is running, the more firings of the cylinders cause the high-pressure pulses to start overlapping and have less low-pressure pulses. Exhaust back pressure can then be used to regulate EGR flow. There are two types of back pressure EGR valves: **positive back pressure EGR** valve, and **negative back pressure EGR** valve.

The positive back pressure valve uses a bleed port and control valve positioned in the middle of the diaphragm (Figure 7-23). The bleed valve is held open by spring pressure. With the bleed valve open, vacuum is vented to the atmosphere under the diaphragm. The stem has a hollow passage that allows exhaust gases to flow up to the bleed valve. With the engine running, exhaust pressure is applied to the bleed valve. Once the exhaust pressure is great enough (high engine loads), it will overcome the spring and close the bleed valve. With the bleed valve closed, vacuum applied to the top of the diaphragm and atmospheric pressure on the bottom of the diaphragm moves it upward. Since the tapered stem is attached to the diaphragm, the stem is lifted and opens the passage to allow EGR flow into the intake manifold. Once the stem opens to allow flow, back pressure is reduced slightly. As a result of reduced back pressure, the valve will start to close again. With the valve closed, back pressure increases and opens the valve. This oscillation of the valve with changing exhaust back pressure maintains a balanced flow of EGR. EGR will not flow at idle since there is insufficient back pressure.

The negative back pressure EGR valve operates similar to positive, but uses decreasing back pressures. The bleed valve is normally closed and the hollow stem directs exhaust pressure

Shop Manual
Chapter 7,
page 299

Back pressure is the resistance of an exhaust system to the flow of gases.

The **positive back pressure EGR** valve uses exhaust system back pressure to sense engine load.

The **negative back pressure EGR** opens the bleed valve when exhaust back pressure decreases to stop EGR flow.

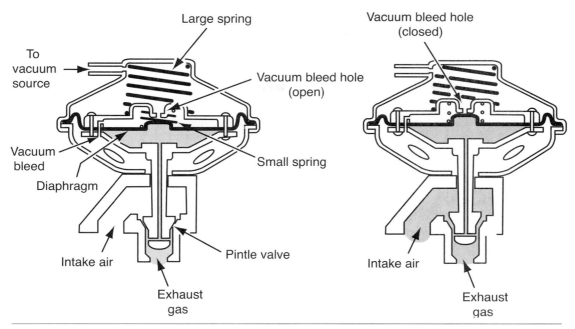

Figure 7-23 Positive back pressure EGR valve operation.

to the valve (Figure 7-24). At low engine speed, the negative pressure pulses from cylinder firing are more predominate. The negative pulses will hold the bleed valve open and prevent EGR flow. At higher speeds, negative pulses decrease and close the bleed valve. With the bleed valve closed, the stem lifts and allows EGR flow.

Computer-controlled EGR systems can use an EGR vacuum regulator (EVR), electronic EGR transducer, or linear solenoids. Some systems will also use a pressure feedback electronic (PFE) sensor.

Shop Manual
Chapter 7,
page 301

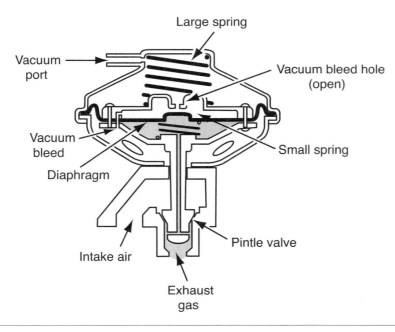

Figure 7-24 Negative back pressure EGR valve.

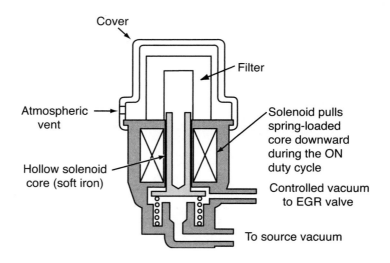

Figure 7-25   Electronic vacuum regulator operation.

An EGR EVR system uses the PCM to control the normally closed EVR solenoid (Figure 7-25). The EVR supplies vacuum to the EGR valve. If the EVR solenoid is not energized, the solenoid plunger tip is seated in the vacuum passage and shuts off vacuum to the EGR valve. In this position, any vacuum in the EGR valve and hose is vented through the EVR solenoid to prevent vacuum from being locked in the system and holding open the EGR valve.

When the PCM inputs indicate the EGR valve should be open, the PCM provides a ground for the EVR solenoid winding. This action moves the solenoid plunger and opens the vacuum passage through the solenoid to the EGR valve. In some systems, the PCM pulses the EVR on and off to supply the precise vacuum and EGR valve opening required by the engine.

The PCM uses inputs such as engine temperature, throttle position, and vehicle speed to operate the EGR valve. The PCM will not energize the EVR solenoid if the engine coolant temperature is below a preset value. The EGR valve is opened by the PCM when the vehicle is operating at normal temperature in the cruising speed range. When the vehicle is operating at low speed or near wide-open throttle, the PCM does not open the EGR valve. If the EGR valve is open while the engine is idling or operating at low rpm, engine operation is erratic.

The electronic EGR transducer system contains an electrically controlled solenoid and a back pressure transducer. The EGR transducer is mounted in line above the EGR valve (Figure 7-26). The

Shop Manual
Chapter 7,
page 303

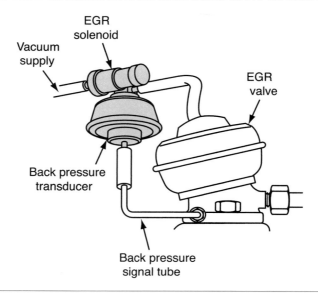

Figure 7-26   Electrically-controlled back pressure transducer.

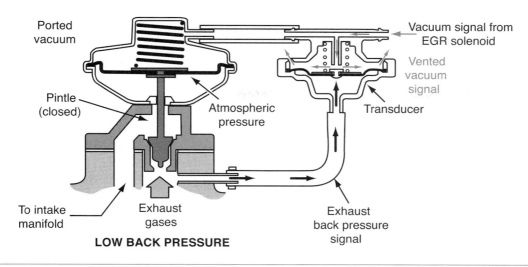

Figure 7-27 With low exhaust back pressure at the transducer, the vacuum supplied through the solenoid is bled off.

transducer controls the operation of the EGR valve when the PCM allows EGR operation. The EGR transducer consists of a diaphragm and three ports; one port receives an exhaust back pressure signal from a vent in the EGR valve housing, one port receives a vacuum signal from the intake manifold, and one port routes controlled vacuum out to the EGR valve diaphragm.

The EGR solenoid is a normally open solenoid that allows the free passage of vacuum to the EGR valve. When the solenoid is energized by the PCM, it restricts vacuum to the EGR transducer.

The PCM controls when EGR can occur by turning the EGR solenoid on and off. However, the actual amount of EGR valve movement depends on the amount of back pressure. If there is low exhaust back pressure reaching the transducer, the transducer diaphragm is not lifted and the vacuum supplied through the solenoid is bled off through the vent (Figure 7-27). As back pressure increases, the diaphragm is pushed up and the vacuum bleed closes. With the bleed closed, vacuum is allowed to reach the EGR diaphragm (Figure 7-28). When the vacuum reaches the EGR valve, the diaphragm is pulled up, causing the EGR valve pintel to move off its seat and allow exhaust gases to enter the intake manifold and finally the combustion chambers. The more back

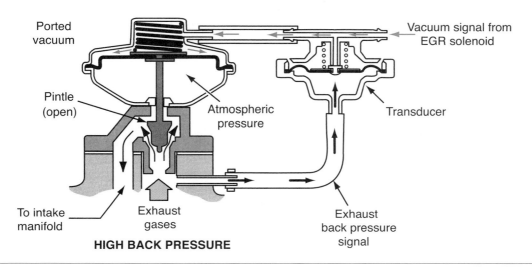

Figure 7-28 EGR valve opens when enough back pressure closes the transducer vent.

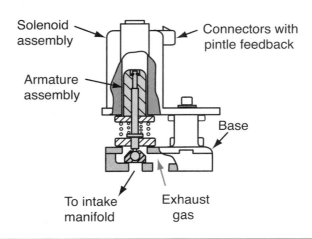

Solenoid assembly

Connectors with pintle feedback

Armature assembly

Base

To intake manifold

Exhaust gas

Figure 7-29   Linear EGR uses a solenoid to control the EGR flow.

pressure, the more EGR flow. The PCM energizes the EGR solenoid during cold engine operation, or wide-open throttle. Under both of these conditions, EGR could reduce vehicle performance, but under either of these two conditions, EGR is not also needed due to the richer fuel mixtures.

The Linear Solenoid EGR valve consists of three major components:

Shop Manual
Chapter 7,
page 304

- Pintle, valve seat, and housing which contains and regulates gas flow
- Armature, return spring, and solenoid coil to provide the operating force to regulate exhaust gas flow by changing the pintle position
- EGR Pintle Position Sensor to return information concerning the position of the pintle to the PCM

The exhaust gas recirculation flow is determined by the PCM. For a given set of conditions, the PCM knows the ideal exhaust gas recirculation flow to optimize NO_x and fuel economy as a function of the pintle position. When the solenoid is energized, the valve is lifted to allow exhaust gases to flow into the intake manifold (Figure 7-29). Pintle position is obtained from a linear potentiometer that is integral to the valve. The PCM adjusts the duty cycle of 128 Hz power supplied to the solenoid coil to obtain the correct position.

Shop Manual
Chapter 7,
page 303

Some manufacturers use a "digital" EGR valve. This type of valve regulates EGR flow according to computer control. The digital EGR valve has three metering orifices that are opened and closed by solenoids (Figure 7-30). By opening various combinations of these three solenoids, different flow rates can be achieved to match EGR to the engine's requirements. The solenoids are normally closed and open only when the computer completes the ground to each.

Some EGR systems have a PFE sensor. These systems have an orifice located in the exhaust passage below the EGR valve that attaches a small pipe to the PFE sensor (Figure 7-31).

The PFE sensor operates in the same fashion as a MAP sensor and changes the exhaust pressure signal to a voltage signal. The voltage signal is sent to the PCM. The exhaust pressure in the orifice chamber is proportional to the EGR valve flow. The PFE signal informs the PCM regarding the amount of EGR flow, and the PCM compares this signal to the EGR flow requested by the input signals. If there is some difference between the actual EGR flow indicated by the PFE signal and the requested EGR flow, the PCM makes the necessary correction to the EVR output signal.

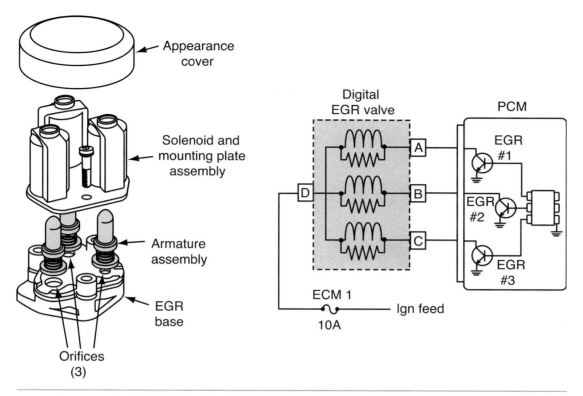

Figure 7-30   Digital EGR valve with three solenoids.

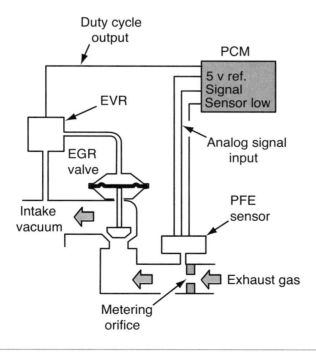

Figure 7-31   PFE sensor measures back pressure.

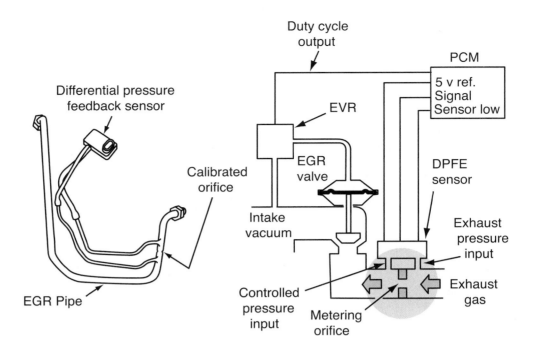

Figure 7-32 Differential PFE with dual pressure pipes connected above and below the metering orifice.

In some EGR systems, two pipes supply exhaust pressure from above and below the orifice under the EGR valve to the differential PFE (DPFE) sensor (Figure 7-32).

AUTHOR'S NOTE: Despite this discussion on EGR systems, some manufacturers are able to control NO_x without the use of an EGR valve. They will use camshaft profile and valve overlap to provide EGR. The camshaft profile will hold the exhaust valve open a little longer as the intake valve is opening. The downward piston movement of the intake stroke will then "suck" exhaust gases around the two valves and into the combustion chamber.

Shop Manual
Chapter 7,
page 308

The leakage of fuel and combustion gases into the engine crankcase is called **blowby**.

Crankcase Ventilation. **Blowby** occurs toward the end of the engine's power stroke. Some unburned fuel and other products of combustion, such as water vapor, leak past the engine's piston rings into the crankcase. Blowby gases must be removed from the crankcase before they condense and react with the oil to form sludge. Sludge, if allowed to circulate with engine oil, corrodes and accelerates the wear of pistons, piston rings, valves, bearings, and other internal working parts of the engine. Blowby gases must also be removed from the crankcase to prevent seal and gasket failures. Since blowby gases pass the piston rings as a result of the pressure formed during combustion, they will pressurize the crankcase. The gases exert pressure on the oil pan gasket and crankshaft seals. If the pressure is not relieved, oil is eventually forced out of these seals.

In addition, blowby also carries some unburned fuel into the crankcase. If not removed, the unburned fuel dilutes the crankcase oil. When oil is diluted with gasoline, it does not lubricate the engine properly, causing excessive wear.

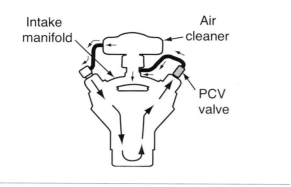

Figure 7-33   PCV valve.

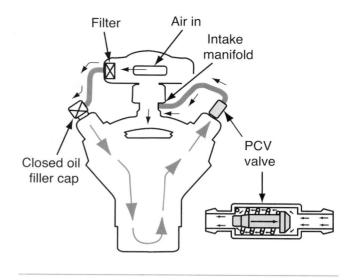

Figure 7-34   PCV system connections used to keep crankcase emissions from escaping.

Combustion gases that enter the crankcase are removed by a **positive crankcase ventilation (PCV)** system that uses engine vacuum and a PCV valve (Figure 7-33) to draw fresh air through the crankcase. The PCV valve is usually mounted in a rubber grommet in one of the valve covers or into the intake manifold. A hose connects the PCV valve to the valve cover and intake manifold (Figure 7-34). A fresh air intake hose is connected from the air cleaner to the valve cover. On V-type engines, the PCV valve is usually connected to one valve cover and the fresh air intake is connected to the other. Some manufacturers will connect the fresh air intake to the oil fill spout. Regardless of the configuration, the fresh air intake is filtered.

When the engine is running, intake manifold vacuum is supplied to the PCV valve. This vacuum moves air through the fresh air intake hose and into the crankcase where it mixes with blowby gases. The mixture of blowby gases and air flows up through cylinder head openings to the rocker arm cover and PCV valve. Intake manifold vacuum moves the blowby gas mixture through the PCV valve into the intake manifold (Figure 7-35). The blowby gases are then moved through the intake valves into the combustion chambers where they are burned.

Because the vacuum supply for the PCV system is from the engine's intake manifold, the airflow through this system must be controlled so it varies in proportion to the regular air-fuel ratio being delivered. If this was not controlled, the additional air that is drawn into the system would cause the air-fuel mixture to become too lean for efficient engine operation. Therefore, the PCV valve is placed in the flow just before the intake manifold to regulate the flow according to vacuum.

On many engines, the PCV system delivers blowby gases to one location in the intake manifold. This type of system may not deliver these gases equally to all the cylinders. This action may

The **positive crankcase ventilation (PCV)** system picks up crankcase gases and sends them to the intake manifold through a control valve.

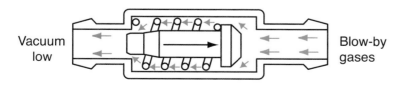

Figure 7-35   Fresh air is drawn into the crankcase as blowby gases flow into the intake manifold.

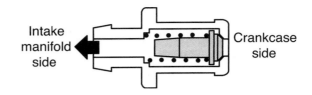

Figure 7-36 PCV valve position with engine off or during a backfire occurrence.

result in an air-fuel ratio variation between the cylinders, which results in rougher idle operation. Some engines, such as the Ford 4.6 L V8 and Dodge 5.7 L Hemi, have passages from the PCV valve through the intake manifold to supply blowby gases equally to each cylinder, resulting in smoother idle operation.

The PCV valve contains a tapered valve surrounded by a spring. When the engine is not running, the spring seats the valve against the valve housing (Figure 7-36). During idle or deceleration, the high intake manifold vacuum moves the tapered valve against the spring tension. With the valve moved this far, the taper of the valve creates a small opening (Figure 7-37). Since the engine is not under heavy load during idle or deceleration operation, blowby gases are minimal and the small PCV valve opening is adequate to move the blowby gases out of the crankcase.

Compared to idle vacuum, intake manifold vacuum is lower during part-throttle operation. Under this condition, the spring pressure overcomes the vacuum against the tapered valve and moves the valve downward. This creates a larger opening between this valve and the PCV valve housing (Figure 7-38). Since engine load is higher at part-throttle operation than at idle operation, blowby gases are increased. The larger opening between the tapered valve and the PCV valve housing allows all the blowby gases to be drawn into the intake manifold.

Idle or deceleration

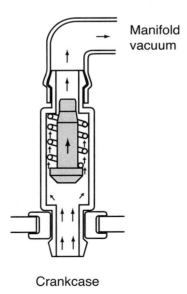

Figure 7-37 PCV valve position at idle with limited flow.

Heavy load

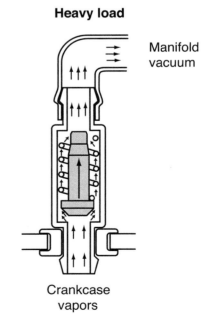

Figure 7-38 PCV valve with engine under load allowing maximum flow.

When the engine is operating under heavy load conditions with a wide throttle opening, the decrease in intake manifold vacuum allows the spring to move the tapered valve further downward in the PCV valve. This action provides a larger opening between the tapered valve and the PCV valve housing. Since higher engine load results in more blowby gases, the larger PCV valve opening is necessary to allow these gases to flow through the valve into the intake manifold.

The earliest PCV systems were an "open" style system. Air entered the crankcase through a breather cap on the valve cover or oil filler tube, but some blowby gases could still escape through the vent. In 1968, "closed" PCV systems became standard. The air inlet vent in a closed system is positioned inside the air cleaner housing. If blowby gases build up faster than the PCV valve can extract them, the blowby gases create a pressure in the crankcase. Some of these gases are forced through the fresh air intake hose and filter and into the air cleaner. This same action occurs if the PCV valve is restricted or plugged.

If the PCV valve sticks in the wide-open position, excessive airflow through the valve causes rough idle operation. If a backfire occurs in the intake manifold, the tapered valve is seated in the PCV valve as if the engine is not running. This action prevents the backfire from entering the engine where it could cause an explosion.

AUTHOR'S NOTE: The PCV system benefits the vehicle's driveability by eliminating harmful crankcase gases, reducing air pollution, and promoting fuel economy. An inoperative PCV system could shorten the life of the engine by allowing harmful blowby gases to remain in the engine, causing corrosion and accelerating wear. Also, the excessive pressures that are built up can cause seals and gaskets to fail.

Some engines do not use a PCV valve. Two examples include the Ford Escort and General Motors Quad 4. These engines use a calibrated orifice in a breather box to siphon blowby vapors back into the intake manifold. The basic system works the same as if it had a valve, except that the system is regulated only by the vacuum on the orifice. The size of the orifice limits the amount of blowby flow into the intake. The engine's air-fuel system is calibrated for this calibrated air leak. Since the action of the PCV allows unmetered air into the intake, the air-fuel system must be set for this amount of extra air.

BIT OF HISTORY

The first emission control device required by law was one that prevented crankcase vapors from escaping to the atmosphere. Prior to the introduction of PCV systems, crankcase blowby gases were dumped into the atmosphere through a "road draft tube." Fresh air entered the crankcase through an open breather cap on the oil filler tube. The fresh air then circulated through the engine, and exited through the road draft tube along with moisture and blowby gases from the crankcase. It was not a very efficient method of venting the crankcase, nor did it help the growing air pollution problem. In fact, prior to PCV, nearly 20 percent of the pollutants vehicles dumped into the atmosphere came from open crankcases.

Post-Combustion Controls

Shop Manual
Chapter 7,
page 311

The **air-injection** system injects air into the exhaust system to heat the catalysts faster and to supply additional oxygen to the center section of some converters.

Air injection systems are also called Air Injection Reaction, or AIR.

Post-combustion controls are used to clean the emissions that are produced during the combustion process. The two post-combustion controls commonly found in vehicles today are air injection and catalytic converters.

Air Injection. One of the earliest of the post-combustion devices for controlling HC and CO emissions was **air injection**. First appearing in 1968, most air injection systems use an engine-driven air pump to force atmospheric air into the exhaust stream (Figure 7-39). The additional oxygen that is introduced causes unburned fuel to continue burning in the exhaust manifold. This will oxidize the HC emissions as well as heat the catalytic converter so it will be more efficient.

Secondary air systems are also used to supply oxygen to the second stage of a dual-bed catalytic converter. The oxygen supplied by the air pump enhances the reaction by which HC and CO is converted to harmless carbon dioxide and water. There are four basic types of secondary air systems that differ by the method in which they supply air to the exhaust system:

- *Air pump systems* (Figure 7-40) employ a belt-driven air pump to inject air into the exhaust system.
- *Aspirated systems* (Figure 7-41) utilize negative pressure from the exhaust system to "pull" air into the system.
- *Pulse air feeder systems* (Figure 7-42) use engine crankcase pressures acting on a diaphragm to draw air from the air cleaner.

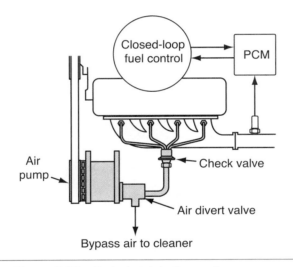

Figure 7-39 Typical air injection system components.

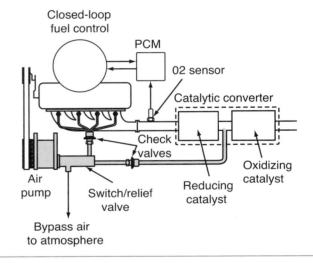

Figure 7-40 Air pump system.

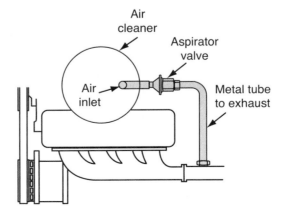

Figure 7-41 Aspirated air injection system.

238

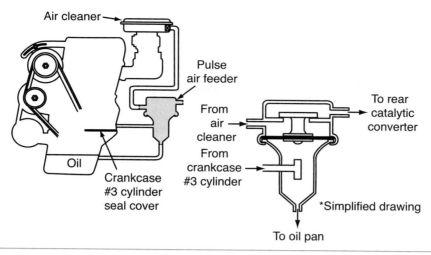

Figure 7-42 Pulse air feeder system.

- *Electronic control air pump* systems are is similar to regular air pump systems, except they use PCM control of an electric air pump to inject air into the exhaust system.

AUTHOR'S NOTE: As an add-on system, air injection does an especially effective job of reducing pollutants without costing much in the way of performance. Its only burden on the engine is a small amount of drag from the air pump. Even though air injection is one of the older emissions control technologies, its effectiveness is such that it is still in use in many vehicles today.

The following are typical components of an air pump system:

- A belt-driven vane pump
- A vacuum-operated diverter valve
- A pressure relief valve
- A one-way check valve
- The plumbing and nozzles necessary to distribute and inject the air

When the engine is started, a belt drives the pump pulley. As the pulley rotates, it drives the rotor. The rotor turns on an axis that is different from that of the pump bore. The vanes slide in and out of slots in the rotor (Figure 7-43). This causes a pumping action that moves air

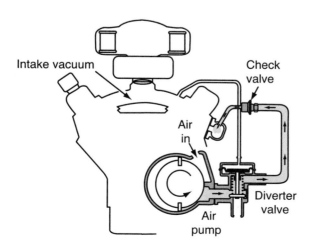

Figure 7-43 Air pump operation.

through the pump. Usually, the air intake is through a centrifugal filter mounted behind the drive pulley. Some applications use a separate intake filter.

Most systems use a relief valve that allows excess pump pressure to escape. The relief valve may be located internal to the pump or may be incorporated into the diverter valve. The pressurized air exits into a large-diameter hose that routes it to this valve.

During all modes of engine operation except deceleration, air from the pump flows into the hose to the check valve. The check valve is a one-way device that allows air to enter the air injection manifold but keeps exhaust from backing up into the pump if a belt should break or if the pump stops working. From the check valve, the air flows into the air injection manifold, which directs it into each exhaust port.

The **diverter valve** vents pump output to the atmosphere during deceleration so the combination of a rich mixture and extra oxygen does not cause backfiring (Figure 7-44). The diverter valve receives a manifold vacuum signal to operate the valve during deceleration. The valve has a vacuum chamber, diaphragm, and spring arrangement. This valve moves the stopper from one of its seats to the other thereby controlling the switching operation. During closed-throttle deceleration, engine vacuum is high enough for the diaphragm in the valve's chamber to overcome the force of the spring. This opens the passage to dump the airflow to a small muffler, which is usually mounted on the pump.

Another anti-backfire device is the gulp valve (Figure 7-45), The gulp valve has a diaphragm chamber, a spring, and a normally closed valve inside. During deceleration, the vacuum signal pulls the diaphragm against spring pressure. This action opens the valve. It also allows some of the air pump's output to flow into the intake manifold to dilute the rich mixture present in this mode.

In air pump systems used with three-way converters (discussed next), an air switching/relief valve directs the air pump output upstream (into the exhaust manifold) while the vehicle is cold and downstream to the catalyst after the vehicle has warmed up. This switching activity is controlled by a coolant vacuum switch, cold open (CVSCO), an air switching solenoid,

Shop Manual
Chapter 7,
page 312

The **diverter valve** directs the air pump's output away from the exhaust system during deceleration.

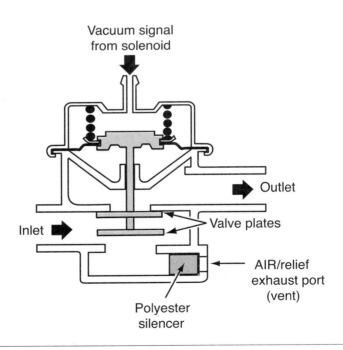

Figure 7-44 Typical diverter valve design.

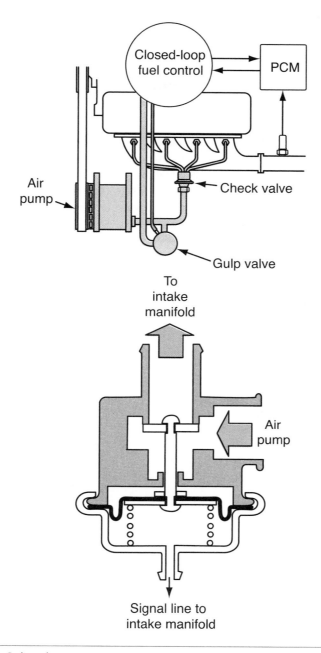

Figure 7-45 Gulp valve system.

or the PCM. Manifold vacuum is applied to the air switching/relief valve at start-up and when the vehicle is cold. Very little NO_x is produced during this time so the efficiency of the reduction process is not important. The vacuum supply is discontinued when the vehicle is warm, causing the air pump output to be directed downstream to the middle bed of the catalytic converter. This boosts the performance of the oxidation catalyst.

Some manufacturers use the PCM to switch air-injection flow from the exhaust ports to the catalytic converter. The PCM makes decisions on where to send the air pump output on the basis of input from various engine sensors, notably the coolant-temperature sensor. It works in conjunction with a pair of solenoid valves: the bypass solenoid, which directs air pump flow to the

Aspirated air
systems use the
exhaust pulses to
bring air into the
exhaust system. The
word *aspirate* refers
to withdrawing a
material with a
negative pressure
apparatus.

atmosphere when energized, and the diverter solenoid, which switches airflow to either the exhaust ports or the catalytic converter.

Some air-injection systems do not use an air pump. These systems use the vacuum that is present in the exhaust manifold momentarily after each cylinder's exhaust stroke (Figure 7-46). When the exhaust valve is open and the piston is moving upward on its exhaust stroke, the volume of exhaust gases rush into the exhaust manifold. This flow of gases results in a positive pressure. When the exhaust valve closes, the inertia of the column of spent gases as it speeds through the exhaust system creates a negative pressure (vacuum). This phenomenon is the basis of **aspirated** air-injection systems. A one-way valve allows fresh air from the air cleaner to enter the exhaust stream whenever vacuum opens it (Figure 7-47). The valve stops the hot gases from escaping by closing when it encounters positive pressure (Figure 7-48). Depending on application, there may be only one aspirator valve or one for each cylinder (Figure 7-49). When the air is injected into the exhaust ports by the pulsed secondary air-injection system, most of the unburned hydrocarbons coming out of the exhaust ports are ignited and burned in the exhaust manifold. Since the duration of low-pressure pulses in the exhaust ports decrease with engine speed, this system is more effective in reducing HC emissions at lower engine speeds.

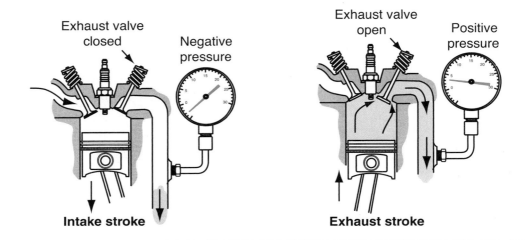

Figure 7-46   Vacuum created in the exhaust manifold as a result of exhaust pulses.

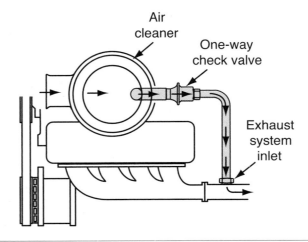

Figure 7-47   The aspirated air system has very few components and is simple in its operation.

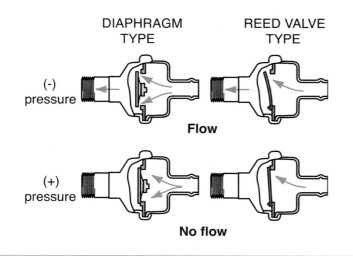

DIAPHRAGM TYPE   REED VALVE TYPE

(-) pressure

Flow

(+) pressure

No flow

Figure 7-48   Pulse air check valve operation.

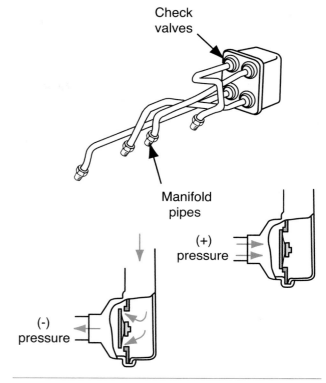

Check valves

Manifold pipes

(+) pressure

(-) pressure

Figure 7-49   Aspirated air-injection system with one valve per cylinder.

Some of today's vehicles are equipped with an electric air pump system that is PCM controlled (Figure 7-50) that injects air into the exhaust to reduce the emissions during engine warm-up. The air injected into the exhaust will cause the catalytic converters to heat up more quickly.

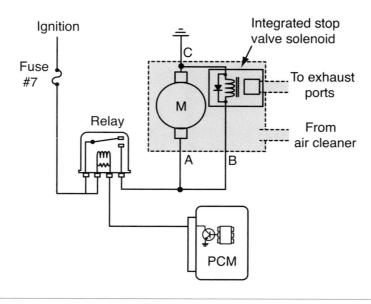

Ignition

Fuse #7

Relay

Integrated stop valve solenoid

C

M

To exhaust ports

From air cleaner

A   B

PCM

Figure 7-50   An electric air pump circuit.

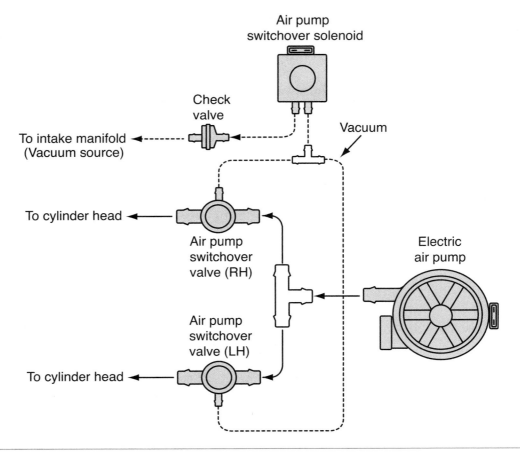

Figure 7-51 Electric air injection system components.

This will improve the emission levels during a cold start. The following is an example of the Mercedes system and is offered to illustrate typical operation.

The system consists of the following components (Figure 7-51):

- An electric air pump
- Two air-pump switchover valves
- An air-pump switchover solenoid
- An air pump relay
- A vacuum check valve
- PCM

The air pump draws in air past a filter and pumps it to the two **air pump switchover valves**. The air pump switchover solenoid is supplied with vacuum from the intake manifold through a check valve. When the air pump switchover solenoid is activated, it allows engine vacuum to be applied to the air pump switchover valves. The air from the pump is forced through the valves into the cylinder head openings to the exhaust. The injected air reacts with the hot exhaust gases in the outlet port. An oxidation of CO and HC takes place and results in an additional increase in the exhaust temperature.

The **air pump switchover valves** prevent exhaust gases from flowing back into the air pump.

Air is allowed to enter the exhaust when the PCM actuates the air pump after engine start-up for up to 2¹/₂ minutes. The following conditions must also be met in order for the system to become active:

- Coolant temperature >50°F (10°C) but <140°F (60°C)
- Engine speed <3,000 rpm
- Throttle valve not wide open

After an actuation, the air injection system will remain deactivated until the coolant temperature drops from >60°C (140°F) to <40°C (104°F).

Catalytic Converter. The most critical element of the emission control system is the **catalytic converter**. Potentially harmful amounts of the three pollutant gases still enter the exhaust system. The catalytic converter is the "last chance" to render these gases harmless before they enter the environment. Automotive engines emit CO and un-reacted hydrocarbons that have to be oxidized, as well as NO_x that has to be reduced.

The catalytic converter is a unique emissions controlling device. It has no moving parts. It is not dependent on engine vacuum, mechanical devices, nor any other special control technology. It is designed for the sole purpose of decreasing exhaust emission levels.

The catalyst uses a **monolithic** construction (commonly referred to as the "biscuit") that is thinly coated by certain "precious metals," such as platinum, palladium, and/or rhodium (Figure 7-52). The effectiveness of the catalytic converter is largely controlled by the performance of cerium oxide, which provides oxygen needed to oxidize CO and HC.

The role of ceria is very important since it enhances the reaction between NO and CO to produce CO_2 and N_2. In addition, ceria improves the thermal resistance of the precious metals. High temperatures can lead to a drastic loss of catalytic performance caused by sintering. This phenomenon, due to the mobility of surface species under high temperatures, is responsible for the expansion of metal particles and consequently the decrease of the contact surface between active elements and exhaust pollutants. Another role of ceria is its ability to store oxygen under lean operating conditions and release it under rich conditions. This action provides improved oxidation reaction.

Shop Manual
Chapter 7,
page 315

As the name implies, the catalytic converter is simply a catalyst that converts emissions into harmless gases. A catalyst triggers a chemical reaction that changes the exhaust gases that pass through it without itself being changed by the reaction.

Monolithic is a type of construction in which a ceramic honeycomb is coated with a thin layer of the catalyst metals.

The monolithic construction in the converter is commonly referred to as the "biscuit."

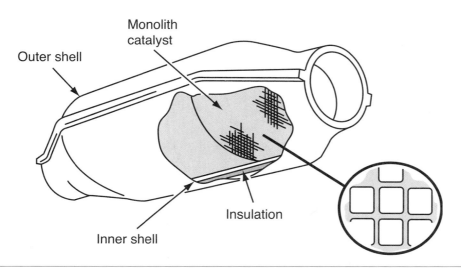

Figure 7-52 Monolith oxidation converter construction.

Internally, one very important design requirement of a catalytic converter is to expose the exhaust gases to the greatest possible amount of catalytic surface area without causing an excessive exhaust restriction. Thus, the biscuit has a honeycomb shape. To protect the fragile biscuit from road shock and to insulate from rapid temperature changes, it is cradled in a stainless steel mat.

There are two main types of catalytic converter that differ in the chemical reactions that take place and in the number of pollutants that they minimize. The two types are oxidation catalysts (or two-way catalysts) and three-way catalysts.

Oxidation catalysts are designed specifically to change HC and CO emissions into CO_2 and water vapor. By supplying a sufficient amount of oxygen to HC and CO in the presence of heat, a chemical reaction results (Figure 7-53). This reaction breaks the HC and CO bonds and results in the formation of harmless carbon dioxide (CO_2) and water (H_2O). In some engine systems, the needed oxygen is supplied by an air pump or aspirator from the secondary air system.

There are two types of two-way converters: monolith and pelletized. Both perform the same function; the difference is in design only. The monolith design uses a honeycomb of small ceramic passageways coated with a thin layer of platinum and palladium as catalysts. Pelletized converters use a bed of aluminum oxide pellets coated with platinum and palladium. Both designs direct exhaust gases over the honeycomb or the pellets where they contact the catalysts. The temperatures of the exhaust gases increase (approximately 500°F or 260°C) and they continue to oxidize. This converts the HC and CO into H_2O and CO_2 before they enter the muffler.

Two special types of oxidation catalyst are known as the mini-oxidation and the close coupled converter (Figure 7-54). These have essentially the same function and are named as such

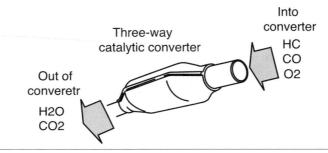

Figure 7-53   The reaction that occurs in an oxidation converter.

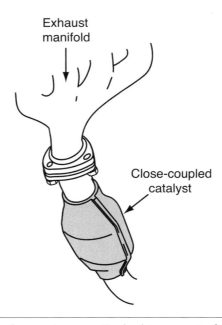

Figure 7-54   Close-coupled converters are attached to, or part of, the exhaust manifold.

246

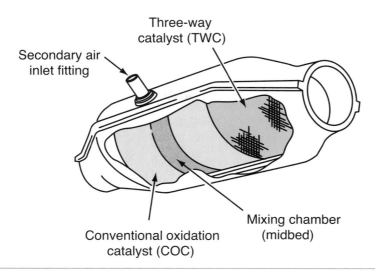

Figure 7-55 Typical three-way converter.

from their small size and the fact that they are installed close to the engine. The mini-oxidation and close coupled converters are *fast-light* off oxidation catalysts that help reduce start-up emissions and pretreat the exhaust before it reaches the main converter.

As government regulations became more stringent, automobile manufacturers needed a more effective means to reduce NO_x emissions. To solve this problem, oxidation and reduction (three-way) catalysts (Figure 7-55) were introduced in the late 1970s. **Three-way converters** use rhodium, which is an effective catalyst for NO_x pollutants (Figure 7-56). The chemical reaction

The purpose of the **three-way converter** is to help clean up the NO_x, HC, and CO emissions in the engine exhaust (hence the term three-way).

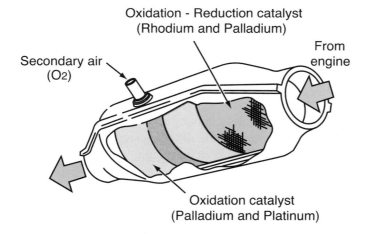

From engine		Between chamber	Exhaust
CO	0.1 - 1.0%	0 - 0.5%	0
HC	50 - 200 ppm	0 - 50	0
CO2	13.6 - 14.3%	14.7 -15.5%	9.6 - 12%
O2	0.3 - 0.7%	0 - 0.5%	2.5 - 5.5%
NOx	1200 ppm	120	120
			Water vapor
			Nitrogen

Figure 7-56 Oxidation and reduction reaction of the three-way converter.

that occurs is quite different from the oxidizing process used to control HC and CO. In this process, the NO_x is "reduced" to harmless nitrogen and oxygen. Instead of oxygen being added, it is taken away. Thus, this chemical reaction is just the opposite of the oxidation reaction described above. Excessive amounts of oxygen in the exhaust stream could actually inhibit the reduction process. To decrease all three pollutants, the three-way catalyst relies heavily on proper fuel-air control. It is the most common type of catalyst used today.

There are two types of three-way converters: without air and with air. A three-way converter without air looks like a two-way converter inside, but the substrate is coated with rhodium and palladium.

Because of varying loads on the vehicle, additional air may sometimes be required to provide enough oxygen for the catalyst to work. To effectively reduce all three pollutants, a three-way converter with air may be used. This converter has two chambers that are separated by an inlet tube from the air-injection system or aspirator. The process of the air-injected three-way converter to reduce NO_x, HC, and CO emissions in the engine exhaust occurs in two stages. The first stage uses the front chamber of the converter. This section is coated with rhodium and palladium to break down NO_x into free nitrogen (N_2) and free oxygen (O_2). The oxygen will be used to further oxidize CO into harmless carbon dioxide (CO_2). In the second stage, additional air is injected into the exhaust gases. The oxygen-enriched emissions pass over a second chamber of the catalyst that is coated with palladium and platinum and is located downstream of the air inlet. This chamber encourages further oxidation of HC and CO into water vapor and carbon dioxide. As a result, the amount of all three pollutants is decreased.

Effective catalytic control of all three pollutants is possible only if the exhaust gas contains a very small amount of oxygen. It is necessary, therefore, to maintain precise control of the air-fuel mixture entering the engine and to keep it very close to the stoichiometric level. Like the engine, the three-way catalytic converter is sensitive to air-fuel ratio, providing maximum overall (HC, CO, and NO_x) conversion efficiency when a stoichiometric air-fuel ratio is being supplied to the engine. A richer-than-stoichiometric ratio will reduce HC and CO converter efficiency *more* than it increases NO_x converter efficiency. A leaner-than-stoichiometric ratio will reduce NO_x converter efficiency *more* than it increases HC and CO converter efficiency. Accordingly, the O_2 feedback system will command a stoichiometric air-fuel ratio whenever engine and driving conditions permit.

The curve of efficiency, as a function of air-fuel ratio, indicates that when the air-fuel ratio is lean, the control of HC and CO is very good but control of NO_x is poor. On the other hand, when the air-fuel ratio is rich, the control of NO_x is very good but control of HC and CO is poor. At the chemically correct mixture, a narrow window exists where the control of all three pollutants is quite good. Maintaining the exhaust contents at this precise value at which the three-way catalyst is most effective is the purpose of **O_2 feedback systems**.

Catalyst types vary, not only by their internal makeup, but by their design configurations as well (Figure 7-57). Catalysts that are mounted near the engine are referred to as "close coupled." Other catalysts, mounted further from the engine, are called "under-floor" catalysts. Some systems have only one catalyst, some have two or three. As discussed, some catalysts also require air injection. To help the catalyst become hotter faster, many are now mounted in the exhaust manifold.

Exhaust temperatures are extremely important to the function of catalytic converters. The desired chemical reactions most readily occur at high temperatures. In fact, the catalyst starts to operate when the exhaust temperatures reach about 500°F (260°C). Catalytic converters experience reductions in conversion efficiency through the normal course of operation. The three primary causes of these reductions are:

- *High temperatures* will oxidize rhodium (Rh), irreversibly affecting the conversion of HC, CO, and NO_x.
- *Contamination* or poisoning of catalyst materials is caused by low catalytic converter temperature. Contaminants collect on the catalyst surface restricting exhaust gases to

O_2 feedback systems use an oxygen sensor to determine if the proper air-fuel mixture was delivered to the combustion chamber.

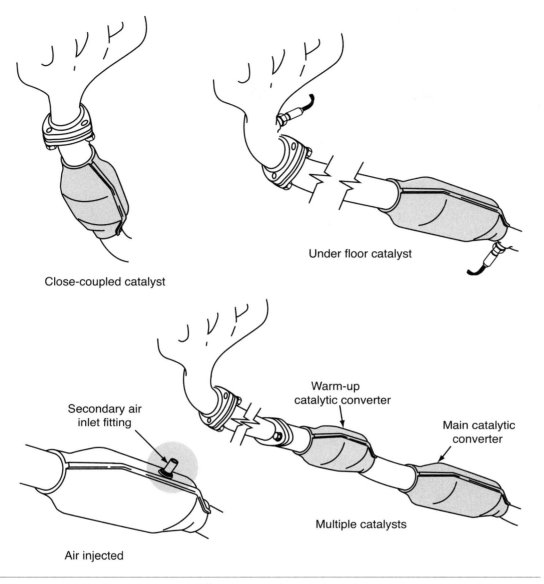

Close-coupled catalyst

Under floor catalyst

Secondary air
inlet fitting

Warm-up
catalytic converter

Main catalytic
converter

Air injected

Multiple catalysts

Figure 7-57 There are several different catalyst converter exhaust system configurations.

and from active catalyst sites where the metals are located. Hydrocarbons are primarily affected because they are larger molecules than CO and NO_x molecules. Table 7-6 shows the common source of contaminants and their effect on emissions.

TABLE 7-6   TABLE OF COMMON CONTAMINANTS AND THEIR EFFECT

COMMON CONTAMINANTS		
Contaminant	**Major Source**	**Gas Most Affected**
Lead (Pb)	Fuel	HC
Phosphorous (P)	Oil	HC, CO
Silicone (Si)	Fuel	HC, CO, NO_x
Sulfur (S)	Fuel	HC, CO, NO_x (esp. Rich)

- *Inadequate air-fuel control* can result in lean mixtures and high temperatures. A lean air-fuel ratio promotes irreversible reactions of Rh with other metal oxides when the converter bed exceeds temperatures of 932°F (500°C). Also, base metals are irreversibly converted to a different, less reactive form.

Summary

❏ The Clean Air Act and its amendments regulate the amount of pollutants that are created by the automobile.

❏ Unburned HC, CO, and NO$_x$ are three types of emissions controlled in gasoline engines.

❏ HC emissions are unburned gasoline released by the engine because of incomplete or lack of combustion.

❏ CO emissions are a byproduct of combustion.

❏ NO$_X$ are formed when combustion temperatures reach more than 2,500°F (1,371°C).

❏ Pre-combustion control systems prevent emissions from being created in the engine either during or before the combustion cycle.

❏ Post-combustion control systems clean up exhaust gases after the fuel has been burned.

❏ The evaporative control system traps fuel vapors that would normally escape from the fuel system into the air.

❏ The PCV system removes blowby gases from the crankcase and recirculates them to the engine intake.

❏ The PCV system benefits the vehicle's driveability by eliminating harmful crankcase gases, reducing air pollution, and promoting fuel economy.

❏ A port EGR valve is opened when vacuum is supplied to the chamber above the diaphragm.

❏ A positive back pressure EGR valve has a normally open bleed valve in the center of the diaphragm. This bleed valve is closed by exhaust pressure at a specific throttle opening.

❏ A negative back pressure EGR valve contains a normally closed bleed valve in the center of the diaphragm. This bleed valve is opened by negative pressure pulses in the exhaust at low engine speed.

❏ A digital EGR valve has up to three electric solenoids operated by the PCM.

❏ A linear EGR valve contains an electric solenoid that is operated by the PCM with a pulse width modulation (PWM) signal.

❏ Many secondary air-injection systems pump air into the exhaust ports during engine warm-up and deliver air to the catalytic converters with the engine at normal operating temperature.

❏ The aspirated secondary air-injection system uses the negative pressure pulses in the exhaust system at low speeds to move air into the exhaust ports.

Review Questions

Short Answer Essays

1. What emissions are manufacturers required to control under Title II of the Clean Air Act?
2. Perfect combustion would result in the formation of what elements?
3. How are hydrocarbon exhaust emissions produced?
4. How are carbon monoxide emissions produced?
5. How are NO_x emissions produced?
6. Describe the basic function and operation of the evaporative emission controls system.
7. Describe how ERG is used to control NO_x emissions.
8. Explain the different PCV valve positions as they relate to vacuum.
9. Explain the purpose of the air-injection system.
10. Explain the function of the three-way catalysts.

Fill-in-the-Blanks

1. Of five exhaust gases, _____ and _____ are not pollutants but are informational gases that are observed for diagnostic purposes.
2. As a tailpipe pollutant, _____ emissions are caused by partially burned fuel.
3. _____ is a direct indication of the air-fuel mixture.
4. The formation of _____ is dependent on the combustion temperature.
5. A _____ air-fuel mixture can lead to increased levels of HC and CO, but NO_x emissions will generally be low.
6. _____ emissions can be caused both by an excessively lean or an excessively rich mixture.
7. When operating, the _____ system mixes a metered amount of exhaust gases with the combustion mixture in the intake manifold.
8. A _____ back pressure EGR valve has a normally open bleed valve in the center of the diaphragm. This bleed valve is closed by exhaust pressure at a specific throttle opening.
9. Combustion gases that enter the crankcase are removed by a(n) _____.
10. _____ are designed specifically to change HC and CO emissions into CO_2 and water vapor.

Multiple Choice

1. Under Title II of the Clean Air Act, manufacturers are required to control which of the following pollutants?
   A. Hydrocarbons
   B. Carbon monoxide
   C. Oxides of nitrogen
   D. Particulate matter
   E. All of the above

2. *Technician A* says that hydrocarbon emissions are high as the air-fuel mixture treads toward a rich condition.
   Technician B says hydrocarbons are a result of the combustion process.
   Who is correct?
   A. A only   C. Both A and B
   B. B only   D. Neither A nor B

3. *Technician A* says high levels of O_2 indicate a rich air-fuel mixture.
   Technician B says the formation of NO_x is dependent of the combustion temperature.
   Who is correct?
   A. A only   C. Both A and B
   B. B only   D. Neither A nor B

4. *Technician A* says depending on other circumstances, a lean mixture can be responsible for excess levels of HC and NO_x.
   Technician B says high CO levels are a good indicator of a rich fuel mixture.
   Who is correct?
   A. A only   C. Both A and B
   B. B only   D. Neither A nor B

5. *Technician A* says hydrocarbon emissions can be caused both by an excessively lean or an excessively rich mixture.
   Technician B says high NO_x emissions result from rich air-fuel mixtures.
   Who is correct?
   A. A only   C. Both A and B
   B. B only   D. Neither A nor B

6. Of the following items, which would be considered pre-combustion control?
   A. Oxidation catalytic converter
   B. Air injection
   C. Three-way catalytic converter
   D. EGR

7. Which of the following components are responsible for sending vapors in the evaporative canister into the intake manifold?
   A. Purge solenoid
   B. PCV valve
   C. Switchover valve
   D. EGR transducer

8. Which pollutant is the EGR valve used to control?
   A. Hydrocarbons
   B. NO_x
   C. Carbon monoxide
   D. Carbon Dioxide

9. *Technician A* says the positive back pressure EGR valve contains a normally closed bleed valve in the center of the diaphragm.
   Technician B says the negative back pressure EGR valve contains a normally open bleed valve in the center of the diaphragm.
   Who is correct?
   A. A only   C. Both A and B
   B. B only   D. Neither A nor B

10. *Technician A* says on air-injected exhaust systems, air is injected into the exhaust manifold to heat the catalysts faster.
    Technician B says air is sent to the catalytic converter to oxidize HC and CO.
    Who is correct?
    A. A only   C. Both A and B
    B. B only   D. Neither A nor B

On-Board Diagnostics Second Generation (OBD-II)

Upon completion and review of this chapter, you should be able to:

❏ Define the purpose of On-Board Diagnostics Second Generation (OBD-II).

❏ Define the terms used in reference to OBD-II.

❏ Explain the requirements to illuminate the MIL in an OBD-II system.

❏ Describe the (DLC) used in an OBD-II system.

❏ Describe how the MIL is extinguished after a fault is detected.

❏ Explain the procedure for self-erasure of DTCs.

❏ Define a trip in an OBD-II system.

❏ Explain the difference between one- and two-trip DTCs.

❏ Describe how the downstream HO_2S sensors monitor the catalytic converter efficiency.

❏ Describe the purpose and function of the Comprehensive Components Monitor.

❏ Explain the purpose and function of the HO_2S monitor.

❏ Explain how engine misfiring is monitored in an OBD-II system.

❏ Describe how the EGR flow is monitored in an OBD-II system.

❏ Describe the purpose and function of the EVAP monitor.

❏ Explain how OBD-II monitors are run in common systems.

Introduction

As was discussed in Chapter 7, the federal government has passed legislation over the years in an effort to maintain air quality standards (Figure 8-1). In 1988, California required all automotive manufacturers to provide a system capable of identifying faults in the computer-controlled systems and to notify the driver by means of a light that a fault exists. This system was capable of monitoring the functionality of a component within the fuel metering system, EGR system, and additional emission-related components. In addition to illumination of the CHECK ENGINE or SERVICE ENGINE SOON lights, a DTC was stored in the computer's memory. The trouble code would identify the specific area of the fault.

The Clean Air Act amendments also included requirements for a second generation of OBD, referred to as **On-Board Diagnostics Second Generation (OBD-II)**. The CARB found that by the time a computer or emission system component failure occurs and the MIL is illuminated, the vehicle emissions have been excessive for some time. The CARB developed requirements to monitor the performance of emission systems as well as indicating component failure. In 1990, these requirements were accepted by the EPA and the federal government required manufacturers to be compliant with OBD-II by model year 1996. The new requirement mandated that the monitoring results must be available to service personnel without special test equipment marketed by the vehicle manufacturer and only available to franchised dealers.

Although the requirement did not mandate OBD-II compliance until the 1996 model year, vehicle manufacturers began installing OBD-II systems on some 1994 models. For example, in 1994 Ford installed this system on 3.8L Mustangs and 4.6L Thunderbirds and Cougars. In that same year, Chrysler installed OBD-II on the Neon 2.0-L. Since OBD-II systems were not mandated until 1996, some of early OBD-II systems are partial systems that do not have all the monitors required on a complete OBD-II system.

On-Board Diagnostics Second Generation (OBD-II) is a legal regulation that requires vehicle manufacturers to have on-board diagnostic routines to determine if the engine is polluting.

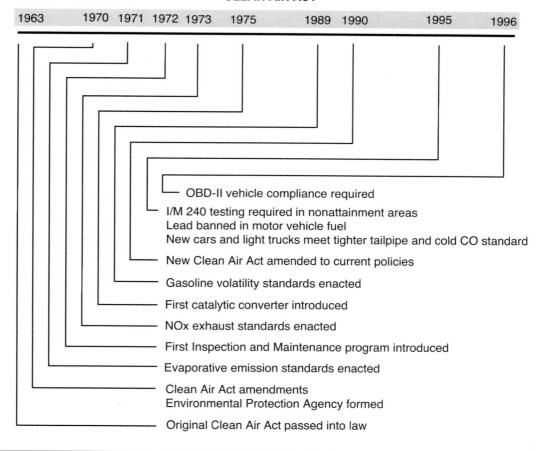

Figure 8-1   Clean air regulations since 1963.

Although OBD-II regulations dictate the function of the system, manufacturers are allowed to use their own methods of accomplishing the task. This chapter will detail the purpose of OBD-II, what it means to the technician, and common OBD-II systems used.

OBD-I versus OBD-II

Shop Manual
Chapter 8,
page 333

Hardware differences between OBD-I and OBD-II are not great. Basically, OBD-II uses the same type of components as OBD-I. The only additional components needed are the post catalyst HO_2S (downstream HO_2S) and an EVAP leak detection pump or vacuum control valve on some vehicles. In addition, some manufacturers had to install faster PCMs with additional storage capacity, drivers, and terminal connections.

From a manufacturing standpoint, the major changes required with OBD-II include the software used by the PCM, standard J1962 DLC, and J1850 communication protocol. Software changes were the most extensive since the PCM has a larger job to do. What this means to the technician, though, is that there is now more information stored in the PCM about a fault. Once the technician understands the purpose, function, and operation of OBD-II, this additional information should make diagnosing a fault easier.

The difference between OBD-I and OBD-II MIL illumination is that on OBD-II vehicles the MIL is illuminated only for emission-related failures, whereas OBD-I allowed the MIL to be illuminated for non-emission-related failures such as cruise control. In addition, OBD-I allowed the MIL to be extinguished as soon as the fault was gone. Under OBD-II, the MIL must remain on until the monitor passes three consecutive tests with no additional failures.

Computer systems with OBD-I have the ability to detect component and system failure. Computer systems with OBD-II are capable of monitoring the ability of systems and components to maintain low emission levels. OBD-II requires monitoring emission-related components for both functionality and **rationality**. This means the OBD-II systems must illuminate the MIL if the vehicle emissions exceed one-and-one-half times the allowable standard for that model year based on a **federal test procedure (FTP)**. When a component or strategy fails or deteriorates to the point that allows emissions to exceed this level, the MIL is illuminated to inform the driver regarding the problem and a DTC is stored in the PCM.

Testing for **rationality** is a measure of whether a particular sensor signal makes sense.

The **federal test procedure (FTP)** is a series of test procedures, including drive cycles, that are used to measure a vehicle's emissions.

Federal Test Procedure Standards

BIT OF HISTORY

In 1961, the first automotive emissions control technology in the nation, PCV, was mandated by the California Motor Vehicle State Bureau of Air Sanitation to control hydrocarbon crank case emissions.

As mentioned, OBD-II is the result of the Clean Air Act and its amendments. In recent years, additional mandated and voluntary guidelines for vehicle emissions have been put into place. These include the requirements for Tier 1, national low-emission vehicles (NLEV), and Tier 2. The federal test procedure standard details the allowable emission levels of the vehicle over its useful life. The allowable emissions are based on Tier 1 and Tier 2 requirements.

Tier 1 regulations were published as a final rule on June 5, 1991, and were fully implemented in 1997. Tier 2 standards were adopted on December 21, 1999, to be phased in beginning in 2004.

Tier 1 Standards

Tier 1 light-duty standards apply to all new light-duty vehicles (LDV), such as passenger cars, light-duty trucks, sport utility vehicles (SUV), minivans, and pick-up trucks. The LDV category includes all vehicles of less than 8,500 pounds gross vehicle weight rating (GVWR). LDVs are further divided into the following sub-categories:

- Passenger cars
- Light light-duty trucks (LLDT), below 6,000 pounds GVWR
- Heavy light-duty trucks (HLDT), above 6,000 pounds GVWR

TABLE 8-1   *TIER 1 STANDARDS*

Category	50,000 miles/5 years						100,000 miles/10 years[1]				
	THC	NMHC	CO	NO_x diesel	*NO_x gasoline*	PM	THC	NMHC	CO	NO_x diesel	*NO_x gasoline*
Passenger cars	0.41	0.25	3.4	1.0	*0.4*	0.08	—	0.31	4.2	1.25	*0.6*
LLDT, LVW <3,750 lbs	—	0.25	3.4	1.0	*0.4*	0.08	0.80	0.31	4.2	1.25	*0.6*
LLDT, LVW >3,750 lbs	—	0.32	4.4	—	*0.7*	0.08	0.80	0.40	5.5	0.97	*0.97*
HLDT, ALVW <5,750 lbs	0.32	—	4.4	—	*0.7*	—	0.80	0.46	6.4	0.98	*0.98*
HLDT, ALVW >5,750 lbs	0.39	—	5.0	—	*1.1*	—	0.80	0.56	7.3	1.53	*1.53*

1—Useful life 120,000 miles/11 years for all HLDT standards and for THC standards for LDT

Abbreviations

LVW—loaded vehicle weight (curb weight + 300 lbs)
ALVW—adjusted LVW (the numercial average of the curb weight and the GVWR)
LLDT—light light-duty truck (below 6,000 lbs GVWR)
HLDT—heavy light-duty truck (above 6,000 lbs GVWR)

The SFTP includes additional test cycles to measure emissions during aggressive highway driving, and also to measure urban driving emissions while the vehicle's air-conditioning system is operating.

The NLEV is a voluntary program that came into effect through an agreement by the northeastern states and the auto manufacturers.

Tier 1 standards were phased in beginning in 1994, and were totally implemented by 1997. They apply to a full vehicle useful life of 100,000 miles. The regulation also defines an intermediate standard to be met over a 50,000-mile period (Table 8-1).

Car and light truck emissions are measured over the Federal Test Procedure (FTP) test and expressed in grams of pollutants per mile (g/mile). In addition to the FTP test, a Supplemental Federal Test Procedure (SFTP) has been phased in between 2000 and 2004.

National LEV Program

On December 16, 1997, the EPA finalized the regulations for the NLEV program. The NLEV program provides more stringent emission standards for the transitional period before Tier 2 standards are introduced.

Starting in the northeastern states in model year 1999 and nationally in model year 2001, new cars and LLDTs have to meet tailpipe standards that are more stringent than the EPA can legally mandate prior to model year 2004. However, after the NLEV program was agreed upon, these standards are enforceable in the same manner as any other federal new motor vehicle program.

The NLEV program uses the standards as required by the California Low Emission Vehicle program. The program is phased in through schedules that require vehicle manufacturers to certify a percentage of their vehicle fleets to increasingly cleaner standards.

Tier 2 Standards

Tier 2 standards bring significant emission reductions relative to Tier 1 regulation. In addition to more stringent numerical emission limits, the regulation introduces a number of important changes that make the standard more stringent for larger vehicles. Under the Tier 2 standard, the same emission standards apply to all vehicle weight categories. This means cars, minivans, light-duty trucks, and SUVs have the same emission limit. Since light-duty emission standards are expressed in grams of pollutants per mile, large engines (such as those used in light trucks or SUVs) will have to utilize more advanced emission control technologies than smaller engines in order to meet the standard.

In Tier 2, the applicability of light-duty emission standards has been extended to cover some of the heavier vehicle categories. Tier 1 standards applied to vehicles up to 8,500 pounds GWVR. The Tier 2 standard applies to all vehicles that were covered by Tier 1 and, additionally, to "medium-duty passenger vehicles" (MDPV). The MDPV is a new class of vehicles that are rated between 8,500 and 10,000 GVWR and are used for personal transportation. This category primarily includes larger

SUVs and passenger vans. Engines in commercial vehicles above 8,500 pounds GVWR, such as cargo vans or light trucks, will continue to certify to heavy-duty engine emission standards.

The same emission limits apply to all engines regardless of the fuel they use. That is, vehicles fueled by gasoline, diesel, or alternative fuels must all meet the same standards.

The Tier 2 tailpipe standards are structured into eight certification levels of different stringency, called "certification bins," and an average fleet standard for NO_x emissions. Vehicle manufacturers will have a choice to certify particular vehicles to any of the eight bins. At the same time, the average NO_x emissions of the entire vehicle fleet sold by each manufacturer will have to meet the average NO_x standard of 0.07 g/mi. Additional temporary certification bins (bin 9, 10, and an MDPV bin) of more relaxed emission limits will be available during the transitional period. These bins will expire after the 2008 model year.

The Tier 2 standards are being phased in between 2004 and 2009. For new passenger cars and light LDTs, Tier 2 standards were phased in beginning in 2004, with the standards to be fully phased in by 2007. For heavy LDTs and MDPVs, the Tier 2 standards will be phased in beginning in 2008, with full compliance in 2009. During the phase-in period from 2004 to 2007, all passenger cars and light LDTs not certified to the primary Tier 2 standards will have to meet an interim average standard of 0.30 g/mi NO_x, equivalent to the current NLEV standards for LDVs. During the period 2004 to 2008, heavy LDTs and MDPVs not certified to the final Tier 2 standards will phase in to an interim program with an average standard of 0.20 g/mi NO_x, with those not covered by the phase-in meeting a per-vehicle standard (that is, an emissions "cap") of 0.60 g/mi NO_x (for HLDTs) and 0.90 g/mi NO_x (for MDPVs).

The emission standards for all pollutants (certification bins) are shown in Table 8-2. The vehicle "full useful life" period has been extended to 120,000 miles. The EPA bins cover California LEV II emission categories to make certification to the federal and California standards easier for vehicle manufacturers.

TABLE 8-2 **TIER 2 STANDARDS**

Bin #	50,000 miles					120,000 miles				
	NMOG	CO	NO_x	PM	HCHO	NMOG	CO	NO_x*	PM	HCHO
Temporary Bins										
MDPV[c]						0.280	7.3	0.9	0.12	0.032
10[a,b,d,f]	0.125 (0.160)	3.4 (4.4)	0.4	—	0.015 (0.018)	0.156 (0.230)	4.2 (6.4)	0.6	0.08	0.018 (0.027)
9[a,b,e]	0.075 (0.140)	3.4	0.2	—	0.015	0.090 (0.180)	4.2	0.3	0.06	0.018
Permanent Bins										
8[b]	0.100 (0.125)	3.4	0.14	—	0.015	0.125 (0.156)	4.2	0.20	0.02	0.018
7	0.075	3.4	0.11	—	0.015	0.090	4.2	0.15	0.02	0.018
6	0.075	3.4	0.08	—	0.015	0.090	4.2	0.10	0.01	0.018
5	0.075	3.4	0.05	—	0.015	0.090	4.2	0.07	0.01	0.018
4	—	—	—	—	—	0.070	2.1	0.04	0.01	0.011
3	—	—	—	—	—	0.055	2.1	0.03	0.01	0.011
2	—	—	—	—	—	0.010	2.1	0.02	0.01	0.004
1	—	—	—	—	—	0.000	0.0	0.00	0.00	0.000

*—average manufacturer fleet NO_x standard is 0.07 g/mi
a—Bin deleted at end of 2006 model year (2008 for HLDTs)
b—The higher temporary NMOG, CO and HCHO values apply only to HLDTs and expired after 2008
c—An additional temporary bin restricted to MDPVs, expires after model year 2008
d—Optional temporary NMOG standard of 0.195 g/mi (50,000) and 0.280 g/mi (120,000) applies for qualifying LDT4s and MDPVs only
e—Optional temporary NMOG standard of 0.100 g/mi (50,000) and 0.130 g/mi (120,000) applies for qualifying LDT2s only
f—50,000 mile standard optional for diesels certified to bin 10

AUTHOR'S NOTE: As you can see, pinpointing exactly what the FTP standard is depends on the year of manufacturer, weight classification of the vehicle, and the emission classification. All that is important to consider at this time is that OBD-II defines excessive emissions as exceeding one-and-one-half times the FTP standard.

OBD-II Component Requirements

OBD-II requires standardized components and systems be used by the automotive manufacturers. These include DLCs, data circuits, diagnostic tests, and DTCs. In addition, in 1991 the SAE published standards for electrical/electronic system diagnostic terms, definitions, abbreviations, and acronyms. The resulting publication, J1930, applies to the following:

- Diagnostic, service, and repair manuals
- Bulletins and updates
- Training manuals
- Repair databases
- Under-hood emission labels
- Emission certification applications

Test Connector

OBD-II implemented the use of "generic" scan tools to access the emission-related items and DTCs. This would allow access to this information without the use of a manufacturer-specific scan tool. The standardized DLC, developed by the SAE, allows for these tools to communicate with the PCM. Standards require the DLC to be mounted in the passenger compartment out of sight of vehicle passengers (Figure 8-2). The DLC must be a 16-terminal connector with some terminals defined by the SAE (Figure 8-3). This connector is referred to as the J1962 connector.

AUTHOR'S NOTE: Just because a vehicle has a 16-terminal DLC does not necessarily mean the vehicle is equipped with OBD-II.

Cavity	General Assignment
1	Ignition Control
2	BUS (+) SCP
3	Discretionary
4	Chassis Ground
5	Signal Ground (SIG RTN)
6	Discretionary
7	K Line of ISO 9141
8	Discretionary
9	Discretionary
10	BUS (–) SCP
11	Discretionary
12	Discretionary
13	FEPS (Flash EEPROM)
14	Discretionary
15	L Line of ISO 9141
16	Battery Power

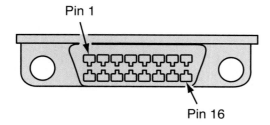

Figure 8-2 J1962 16 terminal DLC.

Figure 8-3 DLC terminal identification.

A new requirement for the automotive manufacturers started being implemented in the 2004 model year. It is the use of the controller area network (CAN) bus for diagnostics. The requirement (J2284) is that this bus system be connected to pins 6 and 14. These pins were previously discretionary. Thus, some manufacturers that were using these two pins for their own purpose have had to start migrating to a new DLC configuration.

Serial Data Circuits

As part of the legislation adopted by the United States Federal Government, the Clean Air Act of 1990 required all PCMs to be able to communicate with any generic scan tool using a common protocol for vehicle communications. These requirements were to be implemented no later than the 1996 model year.

However, the manufacturer can use any communications method it prefers when using its own scan tool. For this reason, there may be more than one data link connected to the DLC on OBD-II systems.

OBD-II Monitors

The OBD-II monitors discussed previously provide a general description of each monitor. In this section, the monitor testing requirements and methods are discussed to provide an overall understanding of each monitor operation. Later in this chapter, detailed monitor information will be provided for each system discussed.

OBD-II monitors include:

1. Catalyst efficiency monitor
2. Engine misfire monitor
3. Fuel system monitor
4. Heated exhaust gas oxygen sensor (HO₂S) monitor
5. Evaporative system monitor
6. EGR monitor
7. Secondary air-injection monitor
8. Comprehensive component monitor
9. Chlorofluorocarbons (CFC) monitor

Catalyst Efficiency Monitor

When a lean air-fuel ratio is present for an extended length of time, the oxygen content in the catalytic converter will reach maximum. When a rich air-fuel ratio is present, the oxygen is depleted. When this occurs, the catalyst fails to convert the harmful emission gases. The **catalyst monitor** detects the catalyst's ability to store and give off oxygen through the use of a **post-catalyst oxygen sensor** (Figure 8-4). A good catalyst should have 95 percent hydrocarbon conver-

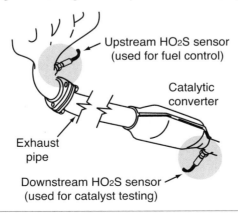

Upstream HO₂S sensor (used for fuel control)

Catalytic converter

Exhaust pipe

Downstream HO₂S sensor (used for catalyst testing)

Figure 8-4   The efficiency of the catalytic converter is monitored by the use of upstream and downstream oxygen sensors.

Shop Manual
Chapter 8,
page 341

The **catalyst monitor** tests for the ability of the converter to store oxygen efficiently.

Depletion of oxygen in the converter is called catalyst punch through.

The **post-catalyst oxygen sensor** is used to measure the O_2 content in the exhaust after the catalyst to determine the converter's storage capacity of O_2.

The post-catalyst O_2 sensor is also called the downstream O_2 or catalyst monitor sensor.

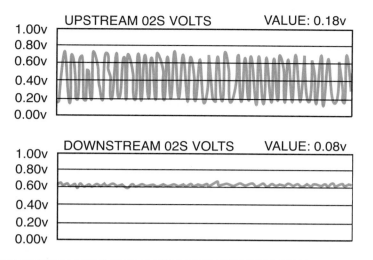

Figure 8-5   Normal upstream and downstream oxygen sensor activity with a good catalytic converter.

sion efficiency. A catalytic converter stores oxygen during lean engine operation and gives up this stored oxygen during rich operation to oxidize excessive hydrocarbons. Catalytic converter efficiency is measured by monitoring the oxygen storage capacity of the converter during closed-loop operation.

The downstream O_2 sensor will show a relatively flat output voltage as compared to the upstream O_2 sensor (Figure 8-5). A high oxygen storage capacity indicates a good catalyst, while low oxygen storage capacity indicates a catalyst that is deteriorating. Voltage spikes that match the upstream O_2 sensor output (Figure 8-6) will indicate a deteriorated catalyst. Peaks on the downstream O_2 sensor indicate the catalyst has lost its ability to store oxygen.

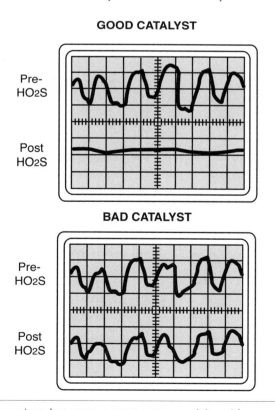

Figure 8-6   Comparison between oxygen sensor activity with a good and bad catalyst.

Misfire Monitor

If a cylinder has a misfire, unburned gas and excess oxygen are exhausted from the cylinder, and these excessive HC emissions enter the catalytic converter. Two adverse conditions happen at this time. First, when the catalytic converter changes these excessive HC emissions to carbon dioxide and water, the catalytic converter is overheated. The honeycomb in the converter may melt together into a solid mass. The excess HC will continue to burn in the catalytic converter, increasing its temperature. If this action occurs, the converter is no longer efficient in reducing emissions. Second, the HO_2S senses the excess oxygen in the exhaust stream and sends a lean condition input to the PCM. The PCM will then increase the injector pulse width, adding more raw fuel to the exhaust stream.

The **misfire monitor** is based on the principle that crankshaft rotation fluctuates as each cylinder is fired. In the event of a misfire, the crankshaft rotation will slow down (Figure 8-7).

The **misfire monitor** detects any external torque disturbances of the crankshaft and indicates whether or not the engine is polluting above FTP or if the converter is being damaged.

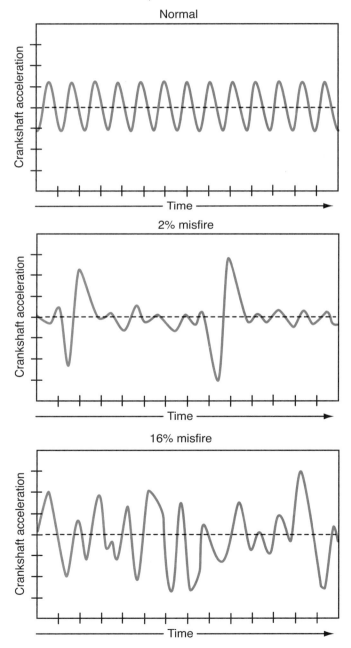

Figure 8-7   The crankshaft position sensor will indicate to the PCM when a misfire occurs.

Using the crankshaft position sensor, the PCM can determine if a cylinder is misfiring by monitoring for **rotational velocity fluctuations**.

There are three types of misfires that must be monitored. A Type A misfire is a percentage of misfire occurring in 200 crankshaft revolutions. This type of misfire can result in catalyst damage and will cause the MIL to flash. A vehicle should not be driven if the MIL is flashing. A Type B is a percentage of misfire in 1,000 to 3,200 crankshaft revolutions. A Type B misfire results in emission in excess of one-and-one-half times the FTP standard. With this type of misfire, the MIL is on steady. A Type C misfire is identical to a Type B except it indicates failure of California I/M test.

> **AUTHOR'S NOTE:** The manufacturers predetermine the percentage of misfire for each test. The percentage of misfire for malfunction criteria varies based on engine rpm and load. Failure percentages vary from engine to engine.

In addition, the system is capable of detecting which cylinder is misfiring and setting individual DTCs for each cylinder. All systems must have a DTC for random or multiple cylinder misfires.

Crank Sensor Learning. PCM software is used to learn and correct for mechanical inaccuracies in the crankshaft position tone wheel tooth spacing. The sum of all the angles between crankshaft teeth will equal 360 degrees. This constant means that a correction factor can be calculated for each misfire sample interval that makes all the angles between individual teeth equal. To prevent any fueling or combustion differences from affecting the correction factors, learning is done during fuel cut-off decelerations. During this time, the crankshaft is still rotating but the engine does not produce any power pulses.

The misfire monitor is not active until the crank sensor corrections have been learned. In the event of battery disconnection or loss of KAM, the correction factors are lost and must be relearned.

Fuel System Monitor

The **fuel system monitor** detects malfunctions that can result in fuel system lean or rich conditions, resulting in excessive emissions (over one-and-one-half times the FTP standard). The OBD-II regulations require that the fuel delivery system be monitored continuously for the ability to provide compliance with emission standards. The regulations require monitoring of the long-term fuel trim limits.

As the engine is running, the PCM monitors HO_2S activity to determine the quality of combustion burn. If the HO_2S indicates a lean condition, the PCM will initiate a fuel trim program to add an additional amount of injector pulse width to richen the mixture until the HO_2S indicates normal activity. If the HO_2S indicated a rich condition, the fuel trim program would reduce the injector pulse width to correct the reading. The monitor is watching changes in the fuel trim to determine if the engine is running too rich or too lean. Most manufacturers use short-term and long-term fuel trim values. These values are displayed on the scan tool as percentage of correction. Different cells are used based on engine rpm and MAP sensor inputs. If the long-term and short-term trim values reach and stay at their maximum limits for a period of time, a malfunction is indicated.

HO₂S Monitor

OBD-II regulations require the monitoring of the HO_2S. The PCM runs at least two tests on the HO_2S; one on the sensor operation, and the other on the heater operation. For both monitors, upstream and downstream oxygen sensors are tested.

As the crankshaft is rotating, every power impulse should increase its speed. If a misfire occurs, the crankshaft slows down. These **rotational velocity fluctuations** are used to determine misfire.

The **fuel system monitor** determines if the running condition of the engine is too rich or too lean to be controlled and brought back to stoichiometric.

Shop Manual
Chapter 8,
page 335

EVAP Monitor

The EVAP system collects fuel vapor from the fuel tanks and then controls the flow of those vapors into the engine to be burned. Since a leak in this system as small as 0.020-inch diameter can result in a yield of 1.35 grams of HC per mile (over thirty times the allowable limits), OBD-II requires the system be monitored for proper operation. OBD-II regulations have been modified to require systems to detect leaks 0.020-inch (0.5 mm) diameter or larger. Vacuum tubing and connections located between the intake manifold and the purge valve are excluded from these regulations. There are two types of EVAP systems that can be used on OBD-II vehicles. Each is monitored differently.

One method of monitoring this system requires the system to be pressurized or put under vacuum to detect a leak. This system is called **enhanced EVAP**. The manufacturer is left to their discretion as to how the system will be tested, but it must detect leaks 0.020-inch (0.5 mm) in diameter or larger and be able to verify airflow through the system.

Non-enhanced EVAP systems are used if the vehicle is not equipped with a leak detection system. Changes in short-term fuel trim shifts, movement in target IAC at idle, or idle speed change are used to monitor the system.

Since the EPA is more in favor of leak detection over stricter EVAP, manufacturers wishing to install stricter EVAP may also be required to have leak detection installed on a sampling of their vehicles to prove system integrity. For example, Chrysler NS body minivans built in November of 1995 (for the 1996 model year) have both leak detection pumps and stricter EVAP.

The EVAP system monitor will not run unless specific fill requirements in the tank are met. As a general rule, the monitor will not run if the tank is filled more than 80 percent or is less than 20 percent filled.

EGR Monitor

OBD-II requires the EGR system be monitored for low and high flow rates that can result in excessive emission levels. The EGR system is considered failed if an EGR system component fails or a change in the EGR flow rate results in the vehicle exceeding one-and-one-half times the FTP standard. The DTC must set within two vehicle trips once the malfunction has been detected.

AIR System Monitor

If the vehicle is equipped with an AIR system, OBD-II requires monitoring for the presence of airflow in the exhaust stream and for the functional monitoring of the pump and any switches or solenoids.

Comprehensive Components Monitor

While the input signal of **comprehensive components** is constantly being monitored for electrical opens and shorts, they are also tested for rationality. One strategy for monitoring inputs involves checking certain inputs for electrical defects and out-of-range values by checking the input signals at the A/D converter. The comprehensive component monitor also checks inputs by performing rationality checks. During a rationality check, the monitor uses other sensor readings and calculations to determine if a sensor reading is proper for the present conditions. The following are examples of sensors that would be monitored by this system:

- Engine Coolant Temperature Sensor
- Manifold Absolute Pressure Sensor
- Throttle Position Sensor
- Mass Air Flow Sensor
- Idle control
- Injector control

Shop Manual
Chapter 8,
page 345

Enhanced EVAP systems use either positive or negative pressures to detect a 0.020-inch (0.5 mm) leak.

Non-enhanced EVAP systems must be certified by the manufacturer to the EPA to have an expected useful life of 100,000 miles.

Non-enhanced EVAP systems are also referred to as stricter EVAP.

Shop Manual
Chapter 8,
page 340

Comprehensive components are any components that may affect emissions. These are generally components that were tested under OBD-I.

In addition, comprehensive components that are outputs are monitored for functionality along with electrical testing. When the PCM provides a voltage to an output component, it can verify that the command was carried out by monitoring specific input sensors for expected changes.

OBD-II Terms

There will be some terms used to describe a test, a driving method, or monitor criteria. The following identifies some of these terms.

Shop Manual
Chapter 8,
page 338

A **drive cycle** is a specific driving method to verify a symptom or the repair for the symptom. A drive cycle is also a specific driving method to begin and complete a specific OBD-II monitor.

Drive Cycle

A minimum **drive cycle** is from engine start-up and vehicle operation until after the PCM enters closed loop. To complete a drive cycle, all five trip monitors must be completed followed by the catalyst monitor. The catalyst monitor must be completed after the other five monitors are completed in a trip. A steady throttle opening between 40 to 60 mph (64 to 96 kph) for 80 seconds is required to complete the catalyst efficiency monitor.

The following conditions must occur to complete all OBD-II monitors:

- The misfire, comprehensive component, and fuel monitors are checked continuously from engine warm up and can complete at any time.

- The misfire monitor, on applications with fuel deceleration shutoff, requires a deceleration at closed throttle for 10 seconds following the acceleration to 55 mph at one-quarter to one-half throttle. Decelerations following acceleration must be performed at least twice consecutively to satisfy this misfire requirement.

- A transmission component functional verification in the comprehensive component monitor requires at least six complete stops in the normal city portion of the drive cycle.

- The EGR and secondary air injection monitors require a series of idles and accelerations.

- The HO_2S monitor requires a steady speed drive for approximately 1 minute at 30 to 40 mph.

- The secondary air injection monitor requires almost 12 minutes of vehicle operating time from initial startup.

- The catalyst efficiency monitor requires a steady speed drive for 5 minutes at 40 to 60 mph, followed by a normal city drive between 25 and 40 mph for 10 minutes.

- The evaporative emissions monitor requires at least 3 minutes of the steady throttle part of the drive cycle (10 minutes) between 45 to 60 mph to test the evaporative system.

- The OBD-II readiness function is available on all scan tools to indicate the status of the once-per-trip monitors.

Trip

The term **trip** is a difficult concept to define since the requirements for a trip varies depending on the test being run. The minimum requirement for a trip is that an ignition switch-off period must precede an OBD-II drive cycle. Basically, in order to have a trip, there must be an ignition off period and then the engine must be started and the vehicle driven.

Vehicle tests vary in length and may be either a **once-per-trip-monitor**, or a **continuous monitor**. For example, the misfire, comprehensive, and fuel system monitors are checked continually once enabling conditions are met. During a trip, these monitors can complete at any time. A series of engine idle and accelerations is required for the EGR monitor to complete. A steady speed for 20 seconds between 20 to 45 mph (32 to 64 kph) after engine warm up is required to complete the HO$_2$S monitor (Figure 8-8). The catalyst efficiency monitor is not required to define an OBD-II trip.

What constitutes a particular OBD-II test is important since it is possible that the vehicle may need to fail more than once in order to illuminate the MIL and record a DTC. Tests can be either **one-trip faults** or **two-trip faults**.

There are times when diagnostic tests are stopped. The test can be **pended** if an existing condition will cause the test to fail. For example, if the MIL is illuminated due to an oxygen sensor fault, the catalyst monitor will not run since it requires signals from the oxygen sensor. Running the test now would cause inaccurate results, so the test is not run pending the resolution of the related problem.

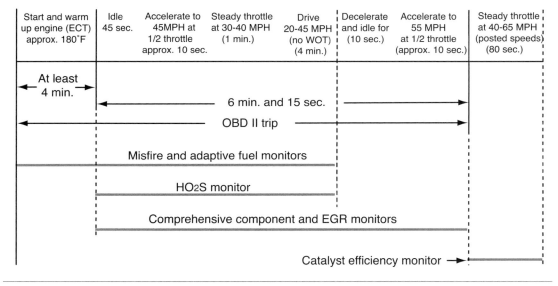

Figure 8-8   Drive cycle chart.

Shop Manual
Chapter 8,
page 334

A **trip** is a set of vehicle operating conditions that must be met for a specific monitor to run. All trips require a key cycle.

Once-per-trip monitor is one that runs once for every driving trip defined by an engine start-stop cycle.

A **continuous monitor** is one that runs all the time to determine the readiness of OBD to set DTCs.

One-trip faults are tests that illuminate the MIL after one failure is recorded. These are also called "A monitors."

Two-trip faults require two failures in a row before the MIL is illuminated. This allows the system to double check itself to prevent the MIL from illuminating when a fault does not really exist. These are also called "B monitors."

A monitor is **pended** if the diagnostic routine has found fault in another monitor that will cause this monitor to fail. The monitor is stopped pending repair of the first fault.

A monitor **conflict** occurs when two monitors have met the enabling conditions to run. One of the monitors will stop running since the testing of the other monitor may cause fault failures. A conflict can also occur in some monitors when the A/C compressor is turned on.

A **suspended** monitor means it has recorded a failure in the first test and now the second test is indicating a failure. The fault code maturing may be suspended until other components are run again to check their function since they may be what is causing this monitor to fail.

Shop Manual
Chapter 8, page 332

At times there may be other tests of running or existing faults that **conflict** with the operation of a test. In this event, the system will not run the test. In addition, a two-trip fault may not be allowed to mature by having the test result **suspended** until the results from another monitor are received.

Warm-Up Cycle

A **warm-up cycle** is defined by vehicle operation after an engine shut down period. During the vehicle operation, the engine coolant temperature must rise at least 40° F, and the coolant temperature must reach at least 160° F. Most DTCs are erased after forty warm-up cycles if the problem does not reoccur after the MIL light is turned off for that problem.

Diagnostic Trouble Codes (DTC)

Once a test has been run, the OBD-II software must determine whether the system passed or failed. If the system failed, it must then determine if it failed enough times to illuminate the MIL. If not, the PCM will store a maturing code. When this test is run again on the next trip, the results will once again be pass or fail. If the test fails, the code matures, a DTC is set, and the MIL is illuminated. Once the MIL is turned on, the MIL is turned off if the system passes the test three consecutive times. After forty warm-up cycles with no other failures, the DTC is erased.

SAE published J2012 to describe industry-wide standards for a uniform DTC format. This format allows a generic scan tool to access any OBD-II system. The OBD-II codes contain four

EXAMPLE: P0137 LOW VOLTAGE BANK 1 SENSOR 2

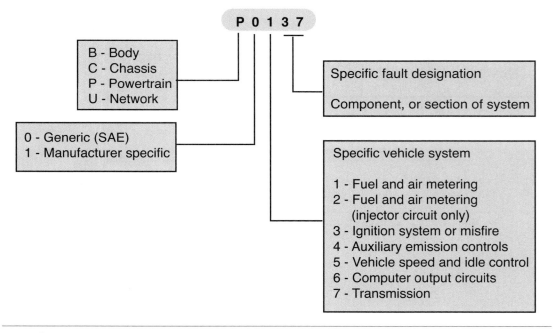

Figure 8-9 DTC structure as dictated by J2012.

characters, one letter and three numbers (Figure 8-9). The letter identifies the function of the device that has generated the fault code:

- P = Power train
- C = Chassis
- B = Body
- U = Network or DLC

The first digit of the code indicates whether the DTC is generic or manufacturer-specific. A "0" here indicates the code is generic, while a "1" indicates it is manufacturer specific. The second digit indicates the vehicle system that generated the fault code:

- 1 = Fuel and air metering
- 2 = Injector circuit malfunctions
- 3 = Ignition system or misfire
- 4 = Auxiliary emission controls
- 5 = Vehicle speed control and idle control systems
- 6 = Computer output circuit
- 7 = Transmission
- 8 = Transmission

The last two digits indicate the specific fault designation.

Freeze Frame

All OBD-II-monitored systems provide **freeze frame** data on the vehicle's operating conditions when a maturing code is set. This data is retrievable by the use of a scan tool. The data is for the conditions at the time the system *first* failed, not necessarily at the time the MIL was illuminated. In the event of multiple monitor failures, the first to occur is stored in the freeze frame. Since only one freeze frame is stored, the monitor with the highest priority will overwrite a lower priority freeze frame. Information contained within the freeze frame includes:

- DTC
- Engine rpm
- Engine load
- Fuel trim
- ECT
- MAP
- Operation mode (open or closed loop)
- Vehicle speed
- Freeze frame priority

Freeze frame priorities are based on the severity of the malfunction. For example, when a fault in the system is detected, the maturing code or DTC is prioritized for freeze frame data. Maturing codes or DTCs with a higher priority overwrite lower priority codes in freeze frame. The freeze frame priority DTCs are as follows:

- Priority 0—Non-emission-related trouble codes
- Priority 1—One-trip failure of a two-trip fault for non-fuel system and non-misfire
- Priority 2—One-trip failure of a two-trip fault for fuel system or misfire
- Priority 3—Two-trip failure of a two-trip fault for non-fuel system and non-misfire
- Priority 4—Two-trip failure of a two-trip fault for fuel system or misfire

Shop Manual
Chapter 8,
page 335

The **warm-up cycle** counts are used to erase the DTC from the PCM's memory. The number of warm-up cycles does not increase until the MIL is turned off by three good trips.

A maturing code is also referred to as a one-trip failure or pending code.

Shop Manual
Chapter 8,
page 335

Freeze frame is a snapshot of the vehicle's operating conditions when a fault is first detected.

Although the priority levels are the same, some manufacturers use numbers while others use letters to distinguish the priority.

Shop Manual
Chapter 8,
page 333

The **similar conditions window (SCW)** is a frame of data that is stored recording the operating conditions at the time the fault originated.

Shop Manual
Chapter 8,
page 335

The **CARB readiness indicator** identifies whether or not all of the once-per-trip monitors have been run.

Similar Conditions Window

If a DTC is stored due to misfire or fuel system-related problems, the PCM requires the engine be returned to a **similar conditions window (SCW)** to retest for a pass of the monitor. After a misfire or fuel system fault is detected and the MIL is illuminated, the PCM must perform the tests once the similar condition window is entered. The monitor must pass three consecutive times within the similar conditions window to turn off the MIL. Similar conditions are defined as within 375 rpm and within 20 percent of the MAP reading at the time of the fault. In addition, the engine must be in the same fuel control mode and close to the same temperature.

> **AUTHOR'S NOTE:** It is important to note that the vehicle does not need to be driven in the similar conditions window in order to mature a code. However, the vehicle must be in the similar conditions window and pass to increment the Good Trip counter for turning off the MIL.

CARB Readiness Indicator

The **CARB readiness indicator** is used when an emissions I/M test is performed. EPA regulations mandate that certain cities perform these tests before a license registration is issued. One part of this regulation states in order to pass the I/M test, the vehicle must have had the entire CARB readiness indicator monitors run at least once. If any monitor fails, the MIL is illuminated and the vehicle cannot be registered until it is repaired. The readiness indicator only flags once-per-trip monitors. Some cities will use the CARB readiness indicator in lieu of sampling emissions from the tail pipe.

> **AUTHOR'S NOTE:** The CARB readiness indicator will indicate that the monitor ran, not necessarily that it passed. The emission inspector will check to see if the monitors ran, and it will also check to see if the MIL is on.

Exponentially Weighted Moving Averaging

Exponentially Weighted Moving Averaging (EWMA) is a well-documented, statistical data processing technique that is used to reduce the variability on an incoming stream of data.

EWMA may be thought of as a deterioration or decay factor.

Some manufacturers use **exponentially weighted moving averaging (EWMA)** for some of their monitors such as catalyst and EGR. The EWMA uses both instantaneous and history data to determine trends. This prevents transit conditions from causing the MIL to be illuminated. Use of EWMA does not affect the mean of the data. However, it does affect the distribution of the data. Use of EWMA serves to "filter out" data points that exhibit excessive and unusual variability and could otherwise erroneously light the MIL. The simplified mathematical equation for EWMA implemented in software is as follows:

New Average = [New data point × "filter constant"] + [(one–"filter constant") × Old Average]

This equation produces an exponential response to a step change in the input data. The "filter constant" determines the time constant of the response. A large filter constant of 0.90 (for example) means that 90 percent of the new data point is averaged in with 10 percent of the old average. This produces a very fast response to a step change.

Conversely, a small filter constant of 0.10, for example means that only 10 percent of the new data point is averaged in with 90 percent of the old average. This produces a slower response to a step change.

When EWMA is applied to a monitor, the new data point is the result from the latest monitor evaluation. A new average is calculated each time the monitor is evaluated and stored. This normally occurs each driving cycle.

OBD-II Scan Tool

Shop Manual
Chapter 8,
page 334

The SAE document J1978 describes the minimum requirements for an OBD-II scan tool. This document covers the required capabilities of and conformance criteria for OBD-II scan tools. Tool manufacturers are allowed to add additional capabilities if they desire. Basic OBD-II requirements are:

- Automatic determination of the communication interface used
- Automatic determination and display of inspection and maintenance readiness information
- Display the emission-related DTC, current data, freeze frame data, and oxygen sensor data
- Ability to perform Expanded Diagnostic Protocol function described in J2205

OBD-II System Samples

The following provides an overview of some of the OBD-II systems designed by different manufacturers to help you become familiar with the different methods used to accomplish the requirements of OBD-II. Although the requirements for monitors are in place, the manufacturer is left to their own discretion on how they will perform the monitor.

DaimlerChrysler OBD-II

The Chrysler group of DaimlerChrysler Motors Corporation uses three different PCMs: Single Board Engine Controller (SBEC) for passenger cars, Jeep/Truck engine controller (JTEC) used on Jeep and Dodge truck vehicles, and Next Generation Controller (NGC) starting in the 2002 model year and will replace SBEC and JTEC by the 2005 model year. All three OBD-II systems are similar. The SBEC and JTEC systems will be presented first, then the differences in the NGC controller will be discussed.

The Chrysler group of DaimlerChrysler Motors Corporation uses a **task manager** software program within the PCM to organize and prioritize the diagnostic procedures and the protocol for recording and displaying their results. During the course of the vehicle operation, many diagnostic steps must be performed. Most of these steps must be performed under separate operating conditions to be accurate. The task manager is responsible for determining when these diagnostic tests will run. The following is a list of the responsibilities of the task manager:

The **task manager** is a software program that directs the functions of the OBD-II system.

- Trip indicator
- Test sequence
- Readiness indicator
- DTC identification, maturation, and erasure
- Freeze frame data storage and erasure
- Freeze frame priority
- MIL illumination
- Test status

In summary, the task manager determines if the conditions are appropriate for a test to be run, knows the definition of a trip for each test, and records the results of each test when it is completed.

Basically, to increment a **global good trip**, all monitors that run once per trip must have run and passed. This usually means the HO$_2$S and Catalyst Efficiency monitors must run and pass.

A **fuel system good trip** is used to turn the MIL off after a fuel system DTC has been set. It requires three good trips

A **misfire good trip** is used to turn the MIL off after a misfire DTC has been set. It requires three good trips.

An **alternate good trip** is used in place of global good trips for comprehensive components and major monitors.

Trips are criteria used to turn the MIL off. The scan tool displays the number of good trips under a trip counter. Chrysler has defined four good trip counters:

- **Global Good Trip**
- **Fuel system good trip**
- **Misfire good trip**
- **Alternate good trip**

The **fuel system good trip** counter increments a good trip if the following conditions are met and the test passes:

- Engine in closed loop
- Operating in similar conditions window
- Short-term trim multiplied by long-term trim is less than threshold
- Less than threshold for predetermined time

The **misfire good trip** counter is incremented if the vehicle is operating in the similar conditions window and there are 1,000 engine revolutions with no misfire.

Since a global good trip could not occur if there is a fault with a comprehensive component (because the test is pended) the MIL could never be extinguished. To prevent this from happening, an **alternate good trip** is counted if there are 2 minutes of engine run time without any other faults occurring. The task manager counts an alternate good trip for a major monitor when the monitor runs and passes. Only the major monitor that failed needs to pass to count an alternate good trip.

If a fault is detected, the maturing code or DTC is prioritized. Maturing codes or DTCs with a higher priority overwrite lower priority codes. This is true concerning freeze frame data only. Chrysler systems are capable of storing several different DTCs; however, pre-2000 model year OBD-II systems only stored one freeze frame. The MIL is illuminated for only priority 3 and priority 4 DTCs. A freeze frame is stored when a fault is *first* detected. If the fault matures so as to turn the MIL the data on in freeze frame does not change, except for the DTC priority.

Comprehensive Components Monitor.   While the input signal of comprehensive components is constantly being monitored for electrical opens and shorts, they are also tested for rationality. In addition, comprehensive components that are outputs are monitored for functionality along with electrical testing. When the PCM provides a voltage to an output component, it can verify that the command was carried out by monitoring specific input sensors for expected changes. On Chrysler-built vehicles, these are one-trip faults that carry a priority 3 DTC, thus the MIL is illuminated after the first failure. The only exception is on JTEC vehicles where the EVAP purge solenoid is a two-trip fault. On Mitsubishi-built vehicles, the comprehensive components are all two-trip faults. The first time a failure is detected, a priority 1 freeze frame is stored. At this time, the MIL is off and no DTC is stored. If the component fails on the next test, the freeze frame is updated to a priority 3, the MIL is illuminated, and a DTC is stored.

Fuel System Monitor.   The fuel system monitor is a two-trip monitor. The fuel monitor must fail twice in succession before the MIL is illuminated. Once the MIL is on, the fuel monitor must pass three times within a similar conditions window.

The PCM monitors the short-term and long-term adaptive memories on a constant basis. If a lean condition exists, then both adaptive memory values will increase until normal HO$_2$S inputs are received again. If during this lean condition the combined short-term and long-term fuel correction values exceed a predetermined threshold (which varies according to engine), the PCM sets a fuel system failure for that trip and a freeze frame is entered.

If a rich condition exists, the PCM combines short term and long term. If the result is less than a predetermined value, the PCM checks the **purge free cells**. At least two purge free cells are used; one corresponds to an adaptive memory cell at idle in drive, the others to a cell at off-

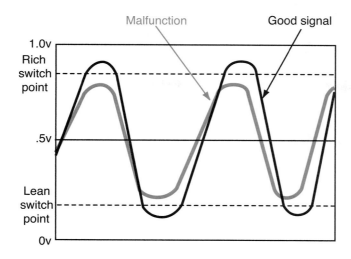

Figure 8-10   Oxygen sensor switch points.

idle or cruise. If all purge free cells are less than a certain percentage, the PCM sets a fuel system rich fault for that trip and a freeze frame is entered. If the purge free values are above the threshold, the fault code maturing is suspended until the EVAP system is tested.

Heated Oxygen Sensor Monitor.   For the PCM to detect a shift in the air-fuel ratio, the voltage input from the HO_2S must change beyond its threshold. These thresholds are the lean and rich **switch points** (Figure 8-10). As the voltage crosses the switch points, the PCM alters the air-fuel mixture accordingly.

The HO_2S is tested for **big slope** counts and **half cycle counts**. The big slope is determined by the formula:

$$Slope = \Delta \ voltage / \Delta \ time$$

The PCM will check the voltage level of the HO_2S every 11 to 12 milliseconds (Figure 8-11). If the slope is great enough, the counter is incremented by one. If the count reaches a predetermined value within a specified length of time, the monitor passes.

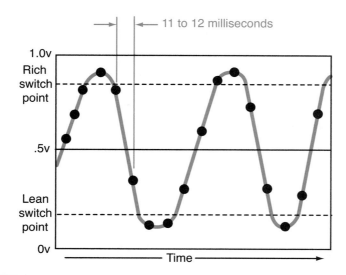

Figure 8-11   Big slope is the change in voltage over time.

Purge free cells are values placed in adaptive memory cells when the EVAP purge solenoid is off.

The **switch points** are the upper and lower ranges of the O_2 sensor voltage.

The rate of change that an oxygen sensor experiences is called the **big slope**. It is determined by voltage change over time.

The **half cycle counts** indicate the frequency of the O_2 sensor.

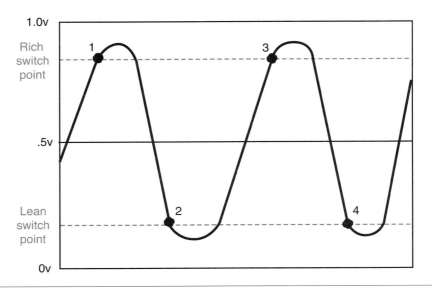

Figure 8-12 The half cycle counter counts the number of times the oxygen sensor voltage crosses the switch points.

The PCM also monitors the number of times the HO_2S signal goes beyond the switch points (Figure 8-12). Each time the voltage surpasses the threshold, a counter is incremented by one. This is referred to as the half cycle counter. If it counts a sufficient number of half cycles within a defined timeframe, the monitor passes.

AUTHOR'S NOTE: Both the big slope and half cycle counters do not have to pass for the monitor to pass. If either of the two tests passes, the monitor passes. The HO_2S Monitor is a two-trip monitor that is tested once per trip.

Oxygen Sensor Heater Monitor. In order for the HO_2S to operate properly, it is heated to approximately 572° to 662°F (300° to 350°C). OBD-II requirements mandate the monitoring of the heater circuit of the oxygen sensor. Most of Chrysler's oxygen sensors contain heater elements that are PTC devices. The PTCs in the sensor are fed direct battery voltage through the automatic shutdown (ASD) relay. As the PTCs receive voltage, their internal structure is changed and the temperature increases.

Chrysler does not check the heater element directly. Instead, the oxygen sensor is heated while its bias voltage is monitored. Basically, the resistance in the oxygen sensor output circuit is tested to determine heater operation. The internal circuit of the oxygen sensor itself is an NTC device. At room temperature, it has about 4.5 M ohms of resistance, but as it warms the resistance drops to about 100 ohms. The PCM sends a 5-volt bias voltage through the oxygen sensor to ground to monitor this circuit. The voltage value received by the PCM during the pulsing of the 5 volts will depend on the voltage drop over the internal resistance of the sensor, which depends on its temperature.

On Chrysler passenger car vehicles that use the SBEC PCM, the O_2 heater monitor is run after the vehicle has been driven and is then shut off. After key off, the O_2 sensor is still warm enough to produce a voltage. The voltage will be low due to high O_2 content in the exhaust pipe. A 5-volt signal is pulsed every 1.6 seconds and is on for 35 ms each time. As the sensor cools, the internal resistance increases and the pulsed bias voltage begins to increase (Figure 8-13). The PCM uses the difference in voltages between high bias voltage and low bias voltage to determine if the sensor was heated sufficiently during the drive cycle. The difference must be less than 1.57 volts.

Next, the PCM checks to see if the HO_2S is cooled down enough to allow the monitor to run. This is done in two steps. First, the PCM monitors the low bias voltage. As the HO_2S cools,

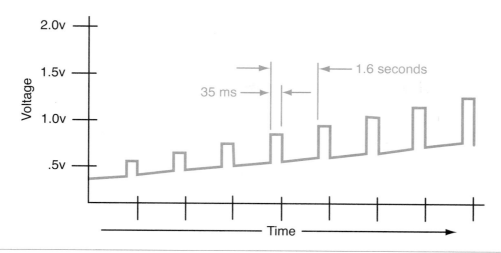

Figure 8-13   Checking to see if the HO$_2$S is cooled enough by measuring the difference between high-bias voltage and low-bias voltage.

the bias voltage will increase since the sensor's internal resistance increases. The low bias level must increase over 0.49 volt. Once low bias voltage is over this threshold, the PCM enters the second step. During this step, the PCM monitors the difference between high bias voltage and low bias voltage as the circuit is pulsed. Sufficient sensor cooldown is indicated when the difference between these voltages is greater than the parameters.

If these two tests pass, the oxygen sensor is cool enough to run the heater monitor. The PCM energizes the ASD relay, turning on the heater element which begins to increase the sensor's temperature. The PCM will continue to bias the 5-volt signal to the sensor. The PCM takes a low bias voltage reading before and after each pulse (Figure 8-14). For the monitor to pass, the voltage difference of the low bias voltage must be greater than 0.16 volts within 32 pulses.

Jeep and Dodge trucks that use the JTEC-type PCM run the heater monitor on initial engine start-up. The engine coolant temperature must be below 140°F in order for the monitor to run. Also, the engine temperature and the outside ambient temperature must be within 20°F of each other for the monitor to run. After the engine is started, 5 volts is applied to the O$_2$ sensor circuit. As the O$_2$ sensor heats, this voltage will drop to zero and the O$_2$ sensor will become active. This should occur within 45 seconds for the monitor to pass.

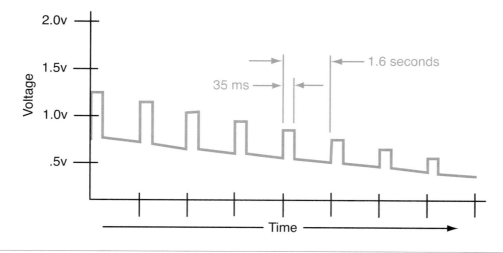

Figure 8-14   The PCM turns on the HO$_2$S heaters and monitors the difference between pulses to see if the low-bias voltage is dropping.

The O_2 heater monitor is a two-trip fault. On the first failure, a priority 1 freeze frame is recorded. If the heater fails on the next test, it will mature the freeze frame to priority 3 and turn the MIL light on.

Catalyst Monitor. The catalyst monitor failure threshold is different from other monitors. Most monitors fail when tailpipe emissions exceed one-and-one-half times the FTP standard. The catalyst monitor, however, is based on the vehicle's emission certification standard (that is, LEV, NLEV, SULEV, etc.). In most cases, the catalyst monitor fails when the emissions level exceeds one-and-three-quarter times the particular standard.

To monitor catalyst efficiency, the PCM expands the rich and lean switch points of the HO_2S. The expanded switch point causes the air-fuel mixture to be richer and leaner than normal. This will overburden the catalytic converter. Once the test is initiated, the PCM counts the number of upstream and downstream HO_2S switches. If at any point during the test the **switch ratio** reaches 90 percent for vehicles equipped with automatic transmissions or 70 percent for vehicles with manual transmissions, a counter is incremented by one. The monitor is enabled to run another test during the same trip. If the test fails a second time, the counter is incremented to two. When the test fails three times within the same trip, a freeze frame is entered and a maturing code is set. The monitor runs once per trip, however there are three tests per trip if the tests fail.

When the test fails the first trip, the MIL is not on since this is a priority 1 fault. If the test fails on the second consecutive trip, the freeze frame priority is matured to 3 and the MIL is illuminated.

In order to increment a good trip after MIL illumination, the downstream oxygen sensor switch rate must be less than 80 percent of the upstream sensor for vehicles equipped with automatic transmissions and 60 percent for vehicles with manual transmissions.

Later-year SBEC and JTEC controllers use an EWMA method for determining catalyst failures. The EWMA catalyst monitor differs from earlier catalyst monitoring technology in that it requires six separate test sequences run before a failure is recorded. The methods in which the tests are run differ between SBEC and JTEC vehicles. On SBEC vehicles, all six tests are conducted within the same trip. On JTEC vehicles, testing is suspended until the next trip if the PCM detects three failures. Once enabling conditions are met during the next trip, the PCM resumes testing. If three additional failures are detected, the PCM stores a DTC and illuminates the MIL. JTEC uses this strategy because the under-floor cats can become wet enough during a rain storm to result in a cooling effect and a false failure.

This monitor uses a **slow moving average** and a **fast moving average** along with instantaneous switching frequency. The slow moving average is determined by adding 90 percent of the "old" slow moving average and 10 percent of the "new" instantaneous switching frequency. The fast moving average is determined by adding 75 percent of the "old" fast moving average and 25 percent of the "new" **instantaneous switching frequency**.

A suspicious counter is incremented when the slow moving average is 20 percent or more different from the instantaneous switching frequency. On SBEC vehicles, if the suspicious counter increments six times and the slow or fast moving average is above a calibrated frequency, a DTC sets and the MIL illuminates. On JTEC vehicles, the third suspicious count sets a one-trip failure. Counting resumes on the next trip. After three more suspicious counts, a DTC is declared and the MIL is illuminated.

When the catalyst monitor is about to fail, the fault code maturing is suspended and the oxygen sensor response monitor is retested during the catalyst monitor to verify that a lazy oxygen sensor is not corrupting the results of the catalyst monitor.

The **switch ratio** is calculated by dividing the number of downstream switches by the number of upstream switches.

Slow moving average is used to detect a naturally aging high-mileage catalyst.

The **fast moving average** is used to detect a major catalyst failure, such as a severe misfire that has melted the catalyst substrate.

The **instantaneous switching frequency** is the rate at which both the upstream and downstream O_2 sensors cross the rich-to-lean switching points.

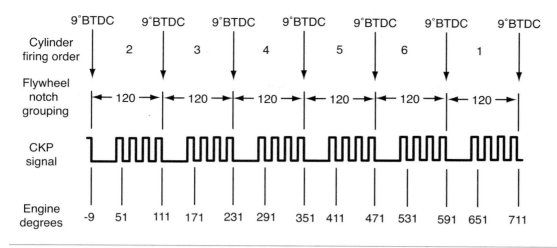

Figure 8-15 The PCM monitors the crankshaft position sensor to determine the position of the pistons and rotational speed.

Misfire Monitor. Chrysler uses a crankshaft speed fluctuation method to monitor for misfire. The crankshaft position sensor can detect slight variations in engine speed that result from cylinder misfire. The monitor must take into account such variables as component wear, sensor fatigue, and machining tolerances. In a perfect design, the crankshaft signal produces a series of 0- to 5-volt pulses to identify the position of two cylinders. In a six-cylinder engine, there are three groups of slots. Each group is 120 degrees apart, and each slot within a group is 20 degrees apart (Figure 8-15). This perfect pattern is programmed into the PCM. Also, the PCM is programmed to know it will never see a perfect design. In order for the misfire monitor to run, it must first learn the **adaptive numerator**. To do this, the PCM uses the first signal set as a point of reference. It then measures where the second set of signals is compared to where engineering data has determined it should be (Figure 8-16). The PCM calculates the adaptive numerator during engine full fuel cutoff. It may take several attempts before the adaptive numerator is learned.

The **adaptive numerator** is the learned value of the crankshaft tone wheel.

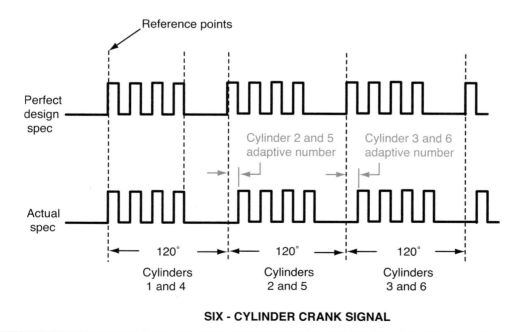

Figure 8-16 The adaptive numerator is the variance between a perfect crankshaft position sensor signal and the actual signal.

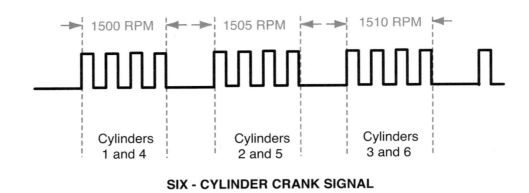

SIX - CYLINDER CRANK SIGNAL

Figure 8-17 The rpm error is the amount of speed fluctuations within a group of crankshaft tone wheel slots.

The PCM also checks the machining tolerances for each group of slots. By monitoring the speed of the crank from the first slot to the last slot in a group, the PCM can calculate engine rpm. The variance between groups of slots is known as the rpm error (Figure 8-17). In order for the PCM to run the misfire monitor, rpm error must be less than approximately 5 percent.

To detect misfire, the PCM monitors the crankshaft speed. If a misfire occurs, the PCM will detect a decrease in crankshaft speed and counts the misfire. Every 1,000-revolution window contains five 200-revolution windows (Figure 8-18). The PCM counts the number of misfires in every 200 revolutions of the crankshaft. The value of the 200-revolution window is carried over for the 1,000-revolution window. If during five 200 counters (1,000 revolutions) the misfire percentage exceeds the threshold, a maturing fault is set and freeze frame is entered. The data in the freeze frame corresponds to the last 200 revolutions of the 1,000-revolution period. Also, a similar conditions window is recorded. A second failure on the next trip will mature the code, set a DTC, and illuminate the MIL. The PCM will attempt to identify which cylinder is misfiring at the time of the fault.

Since the 1,000-revolution counters are two-trip monitors, the MIL is not illuminated on the first trip even if the misfire continues. However, if the misfire exceeds the malfunction percentage in any 200-revolution period, this can result in catalyst damage and the MIL will be flashed. The

200 Revs	400 Revs	600 Revs	800 Revs	1000 Revs	
5	5	15	20	25	1000 Rev counter
5	0	10	5	10	200 Rev counter

Figure 8-18 The 1,000-revolution window contains five 200-revolution windows. The PCM counts the misfires for each window and carries that value over for the 1,000-revolution window.

Type A misfire is a one-trip monitor. While the MIL is flashing, the PCM converts to open loop fuel strategy to prevent additional fuel flow to the cylinders. In order to extinguish the MIL, the test must pass under the similar conditions three consecutive times.

EVAP Monitor.   Emission tests have determined that EVAP system leaks the size of 0.020 inch diameter can result in the release of about 1.35 grams of HC per mile. This is thirty times the allowable exhaust emissions standard. OBD-II standards have dictated that the EVAP system be tested for leaks 0.020 inch in diameter or larger.

SBEC and JTEC Chrysler vehicles can be equipped with either an EVAP leak detection pump (Figure 8-19) or stricter EVAP systems. The **leak detection pump (LDP)** performs two functions. First, it pressurizes the EVAP system for testing. Second, it seals the charcoal canister. The leak detection portion of the EVAP monitor is run immediately after a cold start if ambient temperatures are within predetermined parameters. The PCM uses the leak detection pump solenoid to operate the system. The solenoid does this by connecting the pump to vacuum and to atmospheric pressure. The leak detection pump uses vacuum to operate and force a low pressure

Shop Manual
Chapter 8,
page 345

The **leak detection pump (LDP)** is a vacuum-driven pump that pressurizes the EVAP system to test for leaks.

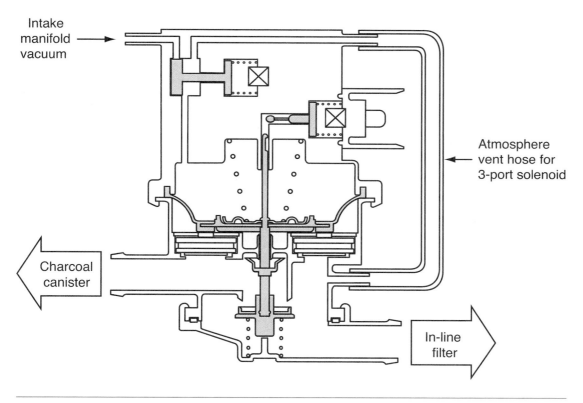

Figure 8-19   The leak detection pump pressurizes the EVAP system to test for leaks.

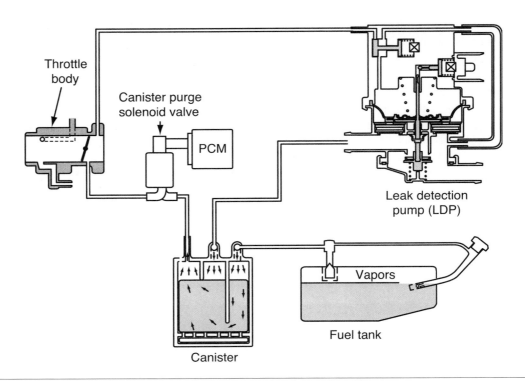

Figure 8-20 Leak detection pump system.

into the EVAP system (Figure 8-20). When the solenoid is energized, vacuum pulls up the pump diaphragm. This pulls atmospheric pressure into the pump. When the solenoid is turned off, the pump is sealed and spring pressure forces the diaphragm down. The air is then pumped into the system. The solenoid is cycled to cause the pump to pressurize the EVAP system to about 7.5 inches of water column pressure. If no leaks are present in the system, pressure will equalize and the pump stops.

The first portion of the leak detection section of the monitor is a test to determine if a line is pinched. This is done by monitoring the pump rate. Pump rate is monitored by the PCM through the use of a reed switch. If the pump rate falls to zero too soon, the PCM determines there is insufficient space to pressurize (Figure 8-21). This indicates the possibility of a pinched

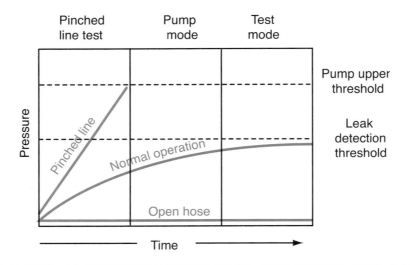

Figure 8-21 The rate of pressure buildup determines if there is a pinched line, a leak, or normal operation.

line. Before the PCM will store a one-trip failure for the pinched line, the leak detection pump monitor and the purge flow test must both be run.

Next, the PCM checks the system for leaks. During this test, the leak detection pump is controlled in two modes: **pump mode** and test mode. During the pump mode, the PCM is capable of determining the amount of time required to pressurize the system based on fuel level inputs. If it takes longer to pressurize the system than determined, the PCM enters a low **pressure check mode**. With the system pressurized, the PCM switches the pump to **test mode** to determine if there are any leaks. If a leak larger than 0.020 inch is in the system, the pump rate falls to a rate proportional to the size of the leak. When the PCM determines the pump rate reveals the presence of a leak, a one-trip failure occurs and freeze frame is entered. If no leaks are detected, the purge solenoid is activated to depressurize the system.

If the vehicle is not equipped with a leak detection pump, it is certified by the stricter EVAP standards. This EVAP monitor uses two stages to test the system. Stage one of the test is a **passive test**. The PCM compares adaptive memory values between purge and purge-free cells. These values are used by the PCM to determine the amount of fuel vapors entering the system. In order to pass this stage of the test, the difference between purge and purge-free cells must be greater than a predetermined value. This would indicate that vapors are entering the combustion chamber, so the EVAP system must be intact. If this stage fails, the monitor progresses to stage two.

During stage two, the PCM de-energizes the purge solenoid then waits until engine rpm, short-term adaptive, and the idle air control have all stabilized. When this is accomplished, the PCM increments up the purge solenoid duty cycle about 6 percent every eight engine revolutions. If during this test any one of the following three conditions occurs prior to the purge cycle reaching 100 percent, the EVAP test passes. The three conditions are:

- rpm rises by a predetermined value
- Short-term adaptive value drops by a predetermined value
- Idle air control closes by a predetermined value

If none of these conditions occur, the test fails and the counter is incremented by one. The PCM will run the test three times during a trip. If the counter increments three failures within the trip, the monitor fails and a freeze frame is entered.

EGR Monitor.   During this monitor, the PCM activates the solenoid allowing vacuum to be applied to the transducer. With negative back pressure, the manifold vacuum is vented; however, with positive pressure, the transducer will modulate the vacuum supplied to the EGR valve. While the monitor is running, the EGR solenoid is turned off and the PCM monitors the short-term adaptive memory values. If the PCM does not detect a change in short-term memory within a given percentage range, the monitor fails and a freeze frame is stored.

Chrysler OBD-II with Next Generation Controller

Beginning in the 2002 model year, Chrysler began to phase in its new Next Generation Controller (NGC). This controller offers faster computer speed than the SBEC or JTEC and uses high-side quad and dual drivers for many output functions. Another unique feature of this system is that the O_2 sensor's return circuit is biased to 2.5 volts. Some of the OBD-II monitors run differently than those of the SBEC and JTEC. These differences are discussed here.

Freeze Frame.   During the 2000 model year, SBEC and JTEC controllers began to support two freeze frames. Frame 1 is the CARB-mandated freeze frame. Rules of priority and how long the failure has been present apply to this freeze frame. With two freeze frames, the failure snapshot will first attempt to occupy freeze frame 1. If freeze frame 1 is occupied with an equal or greater priority, then the snapshot will move to freeze frame 2. There is no data transfer possible between the two freeze frames.

In the **pump mode**, the pump's on and off periods are cycled at a fixed rate to pressurize the system to 7.5 inches of water (Figure 9-22).

The **pressure check mode** means the PCM will loop back and run in the pump mode again.

In **test mode**, the on period only is cycled at a fixed rate. The pump is operated until the system is pressurized and the pump stalls.

The **passive test** monitors the system or components during normal operation.

TABLE 8-3   **RULES OF ORDER FOR MULTIPLE FREEZE FRAMES**

Freeze frame 1	Freeze frame 2	Freeze frame 3	Most recent freeze frame	CARB freeze frame
First failure	Second failure	Third failure	Last failure	Order of occurance does not matter
Priority does not matter	Priority does not matter	Priority does not matter	Priority does not matter	Priority does matter
Could be 1 trip failure or DTC	Could be 1 trip failure or DTC	Could be 1 trip failure or DTC	Could be 1 trip failure or DTC	Could be 1 trip failure or DTC
				Highest priority failure

The NGC controller has five freeze frames (Table 8-3). The first one is still the CARB-mandated freeze frame but has been expanded to include more data than originally provided. This freeze frame is titled "CARB Freeze Frame." Freeze frame 1 is the first failure regardless of priority. Freeze frame 2 is the second failure, freeze frame 3 is the third failure, and the "Most recent freeze frame" is the last failure. An intermittent or chronic condition has the potential to fill all five freeze frames with the same P code, but snapshot condition and priority could vary. The current rules dictate that for a two-trip fault, only the priority and not the data is updated. EPA/CARB rules change often, so consult the Service Information for the latest information.

The priority of the faults has also changed. The fault code priority for the NGC system is:

Priority 1—The first occurrences of a two-trip fault, except fuel system rich/lean or misfire

Priority 3—The second consecutive occurrence of a two-trip fault (matured fault), except fuel system rich/lean and misfire or a one-trip fault of a comprehensive component

Priority 4—The first occurrence of a two-trip fault for fuel system rich/lean or misfire

Priority 6—The second consecutive occurrence of a two-trip fault (matured fault) for fuel system rich/lean or misfire

This was done so the technician can still have the freeze frame for the first occurrence of a fuel system or misfire fault. On the former systems, the priority 2 freeze frame would be erased if a priority 3 fault matured. This means the first occurrence of the priority 2 failure would be lost. On this system, the MIL is illuminated on priority 3 or priority 6 faults, not on priority 2 or priority 4 faults.

Good Trip Counter.   Like previous controllers, a good trip means that the monitor that failed was tested and passed. On NGC systems, in order for the good trip to count up (as seen on the scan tool) 2 minutes of run time is also required. This 2-minute run time is required for all failures, not just the comprehensive component monitor. In addition, the NGC processes a good trip on shut down. This means that the ignition key must be cycled to see a good trip increment on the scan tool.

O_2 Response Monitor.   Like SBEC and JTEC, the O_2 sensor response monitor is a two-trip fault and a once-per-trip monitor. On the NGC system, the oxygen sensor voltage range is 2.5 to 3.5 volts. The O_2 response monitor runs under light engine load at approximately 25 to 50 mph. Big slope is not used on NGC; the method of judging pass/fail is a voltage switch counter that replaces the half cycle and big slope method. The monitor takes 20 seconds to run. However, there is fast pass mode if the specification is met within 10 seconds. This would indicate that the trend is good enough for a pass condition. Two consecutive failures result in an oxygen sensor slow response failure DTC.

Oxygen Sensor Heater Monitor.   The oxygen sensor heater can fail in two ways. An electrical circuit failure results in a heater circuit fault. The electrical circuit failure is a one-trip fault and results in a DTC and illumination of the MIL on the first occurrence.

A monitor failure results in a performance failure. The O_2 sensor heater monitor is a two-trip fault and a once-per-trip-monitor. Unlike previous controllers, the NGC measures the temperature in the heater circuit. This monitor runs on cold start-up (coolant temperature less than 122°F).

Fuel System Lean/Rich Monitor.   The NGC system does not have to update the purge-free cells during a rich failure before maturing a fault code as SBEC and JTEC controllers did. All long-term cells of the NGC are purge free. This means the cells do not update while EVAP purge is occurring. A saturated EVAP system will not set a fuel system rich fault. The NGC controller updates the speed density pulse width equation through purge vapor ratio and purge adaptive instead of long-term/short-term adaptives.

Misfire Monitor.   The NGC misfire monitor is a continuous monitor that is a two-trip fault. Like previous controllers, the adaptive numerator must be learned in order for the misfire monitor to detect any misfires. There are two misfire categories; the 200-revolution counter is looking for catalyst damage from misfire, and the 1,000-revolution counter is looking for FTP emission failure due to misfire.

If misfire is at a rate high enough to cause catalyst damage, the MIL will flash. The percentage of misfire within 200 revolutions needed to log a misfire is dependent on NGC calibration, which is different for each application. If the vehicle is operating under 3,000 rpm, a group of three 200-revolution blocks is necessary to set a one-trip failure. If the vehicle is operating over 3,000 rpm, one 200-revolution block with set a one-trip failure. To set a DTC, a second consecutive trip with misfire is required. The time to set the DTC is dependent on the second trip rpm operating range. For these reasons, it is possible to have a vehicle complaint of a flashing MIL without a set DTC.

The FTP emission misfire is set if a percentage misfire is exceeded in a 1,000-revolution block. A misfire that is present upon start-up will set a one-trip failure after 1,000 revolutions. A misfire that occurs after the vehicle has run 1,000 revolutions without a misfire takes four blocks of 1,000 revolutions to set a one-trip failure. To set a misfire, the DTC requires two consecutive trips with a misfire present. On the second trip, regardless of whether the misfire was present at start-up, it will only take one 1,000-revolution block to set a misfire DTC.

A good trip takes at least 2 minutes of run time with a pass through the similar conditions window with no misfire occurring.

EVAP and Purge Monitors.   The NGC system uses natural vacuum leak detection (NVLD) instead of the leak detection pump. This system relies on the **gas law principle**. Even small leaks will cause the pressure to equalize with the outside ambient pressure.

A vent valve seals the canister vent during engine off conditions. If the evaporative system has a leak of less than the failure threshold, the evaporative system will be pulled into a vacuum due to a cooldown from ambient temperatures. When the vacuum in the system exceeds about 1 inch H_2O (0.25 kPa), a vacuum switch closes. If the PCM detects the switch is closed, the small leak monitor will record a pass condition on the next start. If the switch state does not change, either the system has a leak or the proper temperature change did not occur. This will result in the test being flagged as "inconclusive."

The **gas law principle** states that in a sealed container, a vacuum will be created as the temperature of the gas decreases.

Shop Manual
Chapter 8,
page 349

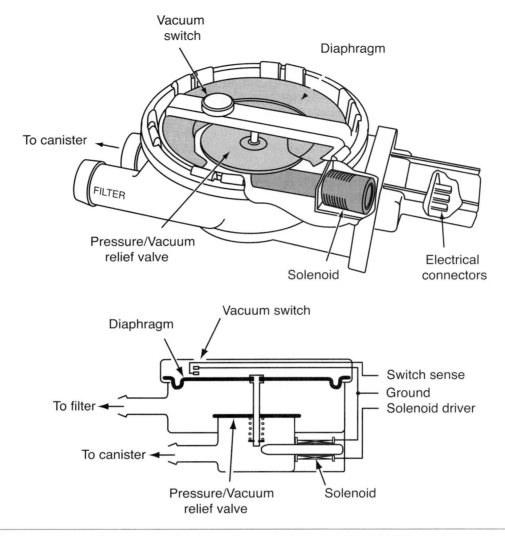

Figure 8-22 Components of the natural vacuum leak detection (NVLD) assembly.

The NVLD assembly (Figure 8-22) is designed with a normally open vacuum switch, a normally closed (de-energized) solenoid, and a pressure/vacuum relief valve. Both the solenoid and a diaphragm actuate the pressure/vacuum relief valve.

AUTHOR'S NOTE: The NVLD is located on the atmospheric vent side of the canister. The NVLD assembly may be mounted on top of the canister outlet or in line between the canister and atmospheric vent filter.

The normally open vacuum switch will close when about 1 inch H_2O (0.25 kPa) vacuum lifts the diaphragm. The normally closed pressure/vacuum relief valve in the NVLD is intended to maintain the seal on the evaporative system during engine off conditions. If vacuum in the evaporative system exceeds 3 inches to 6 inches H_2O (0.75 to 1.5 kPa), the valve will be pulled off the seat, opening the seal. This will protect the system from excessive vacuum as well as allowing sufficient purge flow in the event that the solenoid became inoperative.

The PCM activates the solenoid when the engine is running so the valve will unseat and allow the canister to vent. The PCM will de-energize the solenoid to close the vent during the medium and large leak tests and during the purge flow test. Also, the diaphragm will open the seal in the NVLD if the pressure in the evaporative system is too great (about 0.5 inch H_2O, or 9.012 kPa). This will permit the venting of vapors through the canister during refueling or if there is an increase in temperature.

The NVLD assembly has three wires: switch sense, solenoid driver, and ground. The NGC utilizes a high-side driver to energize and duty cycle the solenoid.

The small leak test is a passive test that is performed while the engine is off. The NGC logs ignition off time as well as ignition on time to determine how often the test is run and must pass. A predetermined amount of off time is required to make the test valid. If the test does not pass in a specified amount of time, a DTC is set immediately. The test is passed when the NVLD switch closes; this indicates vacuum in the fuel tank. If the switch did not close during the engine off period, the NGC has inconclusive results. This means the PCM must perform the **intrusive test** if the enabling conditions are met.

The engine running test is an intrusive test that will only run if the small leak test is inconclusive. It is used to check for 0.040-inch and 0.060-inch diameter leaks. Closing the EVAP vent and drawing a vacuum into the tank using the purge solenoid initiates this test. The purge solenoid is turned on and the amount of purge is slowly increased. With engine vacuum being drawn into the tank, the NVLD switch should close. If the switch fails to close, a general EVAP failure DTC is set. This fault can set if there is a large leak (loose gas cap), if the switch contacts are faulty, or if the circuit to the switch is open.

If the switch closes, the purge monitor passes. A closed switch indicates the system is in tact from the intake manifold to the tank. Next, the NGC controller will turn off the purge solenoid to seal the system. The NGC will then monitor the length of time, in seconds, for how long the switch remains closed. If the switch opens before a specified time period, a failure is recorded.

If the vehicle passes the small leak test, the PCM will perform a purge flow monitor that is similar to previous controllers. During stage one, the NGC controller monitors purge vapor ratio. If the ratio is above the specification, the monitor passes. If it fails to meet the specification, stage two of the test will be run. During stage two, the NGC commands the purge solenoid to flow at a specified rate (in grams per second) to force the purge vapor ratio to update. The vapor ratio is compared to the specification; if it is below the specification, a one-trip failure is recorded.

The **intrusive test** is a special type of active test in that the action taken by the PCM may affect vehicle performance and emissions.

General Motors OBD-II

The PCM performs two functions: control vehicle systems using sensor input and software programming, and to perform system diagnostics. Both of these management systems communicate with each other. The PCM uses the **diagnostic management system** to perform diagnostic testing, record the results of the tests, and request the test fail actions. The different types of diagnostic tests performed by the diagnostic management system include passive, **active**, and intrusive (Figure 8-23).

The **diagnostic management system** is the software that manages the OBD-II system.

Active testing requires the PCM to control the system or components in a specific manner while monitoring the results.

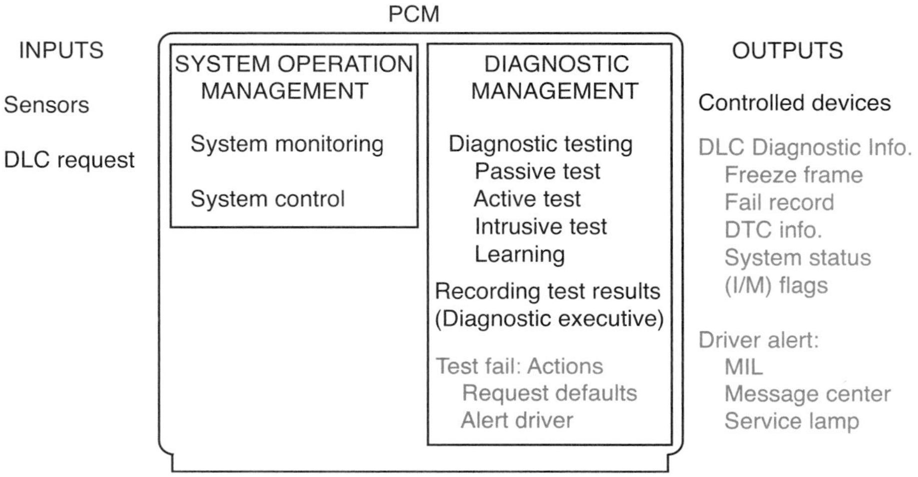

Figure 8-23 The different types of diagnostic tests performed by the diagnostic management system.

The diagnostic tests occur during a trip, and the PCM is capable of "learning" from the results of the tests. The PCM charts the results and creates baseline (the normal) results of the test. This allows the PCM to filter out information that could result in erroneous DTCs. The **diagnostic executive** stores the following information:

- DTC information
- Freeze frame data
- Fail records
- System status
- Warm-up cycles

DTC information indicates the status of the diagnostic testing for the specified DTC. This information includes pass/fail status of the test, along with information concerning when the test failed and if the DTC requires any diagnostic lamps to be illuminated. The diagnostic executive will record freeze frame records on one frame of data about the operating conditions at the time for the *first* failed test of an emission-related system or component. A history DTC is stored, and the MIL is requested on. Freeze frame will not be updated if the test fails a second time. However, fuel trim and misfire DTCs have a higher priority than other DTCs and will overwrite lower priority freeze frames. Fail records store up to five DTCs whether they are emission related or not. The fail records contain the same type of information that would be stored in the freeze frame. The system status program stores data on which diagnostic tests have been run.

Test Fail Action.   If a diagnostic test fails and a DTC is set, the diagnostic management system performs a series of **test fail actions**. These actions are dependent upon the DTC. The actions the Diagnostic Management System may perform are:

- Illumination of the MIL
- Illumination of the service lamp
- Data message sent to the Drive Information Center (DIC)
- Initiate usage of default values stored in program
- Request PCM default operations
- Store a freeze frame
- Store a fail record

Illumination of the MIL is done only if the DTCs impact vehicle emissions. Upon a failure, the diagnostic executive will turn the MIL on as requested by the diagnostic management system. In the event of a misfire malfunction resulting in possible catalytic converter damage, the diagnostic executive will flash the MIL once per second. The MIL will continue to flash until the engine operating conditions are such that catalytic damage is no longer imminent; however, at this time the MIL will be on steady.

If there are three consecutive trips in which the diagnostic test passes (for the test that turned the MIL on originally), the MIL will be turned off. However, for fuel trim or misfire DTCs, the test must run and pass in similar operating conditions that the DTC was originally set under. The window is within 375 rpm and within 10 percent of the engine load and similar engine temperature.

On some General Motors vehicles, a service lamp is also available. This lamp alerts the driver to non-emissions-related malfunctions, such as air conditioning and cruise control faults. This lamp will only illuminate if the DTC is a type C. In addition, some vehicles are equipped with a DIC that alerts the driver to type C failures.

DTC Types.   If there is a failure during a trip, the diagnostic management system sets a DTC that is directly related to the diagnostic test. Some DTCs will not set until there have been two consecutive failures. The following are the four types of DTCs and the characteristics of the code:

- Type A
  —Emissions-related failure
  —MIL illuminated on the first trip with a failure
  —Store a history DTC on the first trip with a failure
  —Stores a freeze frame if empty
  —Stores a fail record
  —Updates the fail record each time the test fails
- Type B
  —Emissions-related failure
  —"Armed" after one failure
  —Cleared after one trip with a pass
  —MIL illumination after two consecutive trips that fail
  —Stores a history DTC on the second trip failure
  —Stores a freeze frame after the second trip failure (if empty)
  —Stores fail record on the first trip failure
  —Updates the fail record each time the test fails
- Type C
  —Non-emission-related fault
  —Requests illumination of the service lamp or DIC (not the MIL) on the first failure
  —Stores a history DTC on the first failure
  —No freeze frame stored
  —Stores fail record at first failure
  —Updates the fail record each time the test fails
- Type D
  —Non-emissions-related fault
  —No illumination of any lamps
  —Stores a history DTC on the first failure
  —No freeze frame stored
  —Stores fail record at first failure
  —Updates the fail record each time the test fails

Comprehensive Component Monitor.   Input and output components are monitored for minimum circuit continuity and out-of-range values. Comprehensive components are also monitored for functionality, such as the TPS showing a wide-open throttle condition while other inputs indicate a closed throttle position.

Heated O$_2$ Sensor Monitor.   To comply with OBD-II requirements of HO$_2$S monitoring, General Motors has designed diagnostic tests to monitor time to activity, response time, and sensor voltage. The results of the **time to activity test** are compared to calibrated parameters within the PCM. The heater circuit indicates a failure when the time to activity increases to a point outside the parameters. This test is only run following a cold start.

The **time to activity test** monitors the heater system by measuring the time the sensor requires to become active.

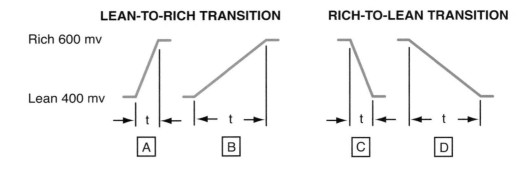

LEAN-TO-RICH TRANSITION   **RICH-TO-LEAN TRANSITION**

Rich 600 mv

Lean 400 mv

A   B   C   D

Figure 8-24   Both chart A and chart C illustrate correct HO$_2$S transitions. Chart B and chart D illustrate HO$_2$S transitions that may cause a test to fail.

The **response time test** monitors the transition times between lean-to-rich and rich-to-lean of the upstream O$_2$ sensors.

The **sensor voltage test** monitors upstream and downstream HO$_2$S for voltages that are out of range.

If the **response time test** indicates too much time elapses between the lean-to-rich and rich-to-lean transitions, the test fails (Figure 8-24).

The **sensor voltage test** looks for sensors that are inactive, shorted to voltage, shorted to ground, or have rich-to-lean shifts of the fuel control system. Operation of the HO$_2$S can also be affected by such factors as quality of the fuel, vehicle operating conditions, and the condition of the catalytic converter.

General Motors uses an HO$_2$S with a high signal wire and a low signal wire that are connected to a comparator in the PCM (Figure 8-25). The low signal wire provides the comparator with reference low. This is different from the HO$_2$S used in OBD-I vehicles in that the return low circuit went directly to ground.

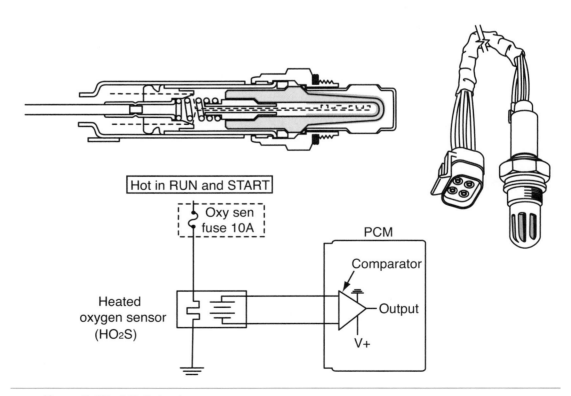

Figure 8-25   HO$_2$S circuit.

Catalyst Monitor. The catalyst monitor determines the oxygen storage capacity of the catalytic converter. To test this capacity, the PCM runs a two-stage diagnostic test. In the first stage, the PCM calculates the oxygen storage capacity of the converter and compares it to calculated values stored in the PCM. This stage is done when the engine and catalyst have reached normal operating temperatures and the vehicle is being driven at a steady cruising speed. By monitoring the upstream and downstream HO_2S the PCM determines if the activity of the two is similar. Similar activity indicates the storage capacity of the catalyst has deteriorated. If the calculated storage capacity is within the fail parameters, the test fails Stage 1. This does not necessarily indicate a failed catalyst.

If the catalyst monitor fails Stage 1, Stage 2 is entered. Stage 2 is additional monitoring of the storage capacity of the catalyst. If the monitor fails Stage 2, a DTC is set. Staged testing of the catalyst allows the PCM to statistically filter the test data to prevent erroneous pass or fails while still passing or failing a marginal catalyst.

Misfire Monitor. To detect engine misfire, the PCM stores a record of the previous 3,200 engine revolutions. The 3,200 revolutions are then divided into sixteen 200-revolution counters. If a misfire occurs during any one of the 200-revolution samples, a DTC is set and the MIL flashes (Figure 8-26). If the misfire is inconsistent, the misfire must fail five of the test samples. If the test fails in a second trip, a DTC is stored and the MIL is turned on. This type of misfire will not result in catalyst damage. Non-catalyst damaging misfires will set a DTC and turn on the MIL under the following conditions:

1. A second consecutive trip contains a failed test.
2. A second non-consecutive trip contains a failed test that occurred under the same operating conditions window as the first test within eighty trips of the first failure.

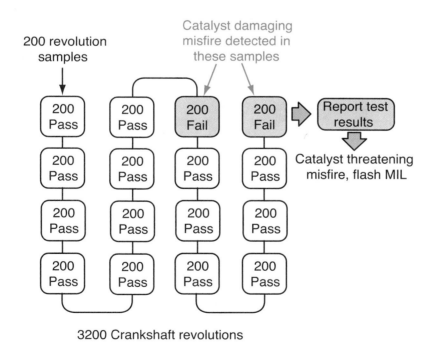

Figure 8-26 The PCM maintains a record of the previous 3,200 revolutions which are divided into sixteen 200-revolution counters. If a misfire occurs during any 200-revolution sample that could damage the catalyst, the MIL flashes.

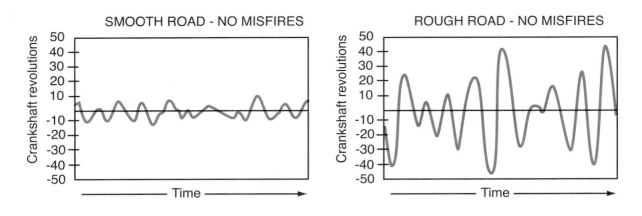

SMOOTH ROAD - NO MISFIRES

ROUGH ROAD - NO MISFIRES

Figure 8-27   Rough road conditions can alter the crankshaft position sensor input to look like a misfire.

Since rough roads can cause false misfire failures, the software in the PCM is enhanced to detect most rough road conditions using the crankshaft sensor (Figure 8-27). General Motors will also use the Anti-lock Brake System (ABS) to detect rough road conditions. Monitoring the wheel speed sensors does this. As the wheels go over the rough surfaces, variations in wheel speed occur. The ABS system will inform the PCM of the rough road conditions to prevent false failures.

In addition, General Motors disables the Torque Converter Clutch (TCC) when a misfire is suspected. This isolates the engine from the drivetrain. On rough roads, the torque applied to the drive wheels and drivetrain can temporarily decrease engine speed. This can be detected by the PCM and falsely indicate a misfire. By isolating the engine from the drivetrain, the PCM will not fail the misfire test because of road conditions. If no misfire activity is detected after the TCC has been disabled, it will be engaged after 3,200 engine revolutions. If a misfire is detected with the TCC disabled, the PCM will continue to evaluate the system and set a DTC if out of limits.

Misfire counters record each cylinder's activity (Figure 8-28). Each cylinder has a history misfire and current misfire counter. History misfire counters store the total number of misfires that

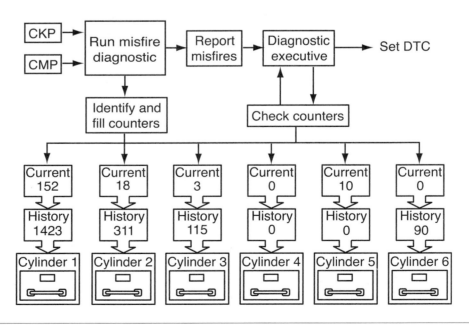

Figure 8-28   Illustration showing misfire counters for each cylinder.

288

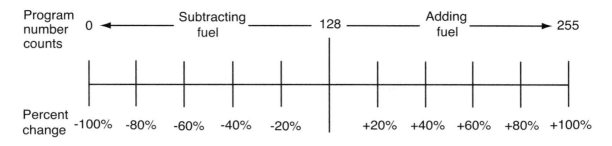

Figure 8-29 Fuel trim conversion to percentage.

occurred within the cylinder since the time the DTC was set. Current misfire counters contain only the number of misfires from the current 200-revolution sample. After 200 revolutions, the number is added to the history misfire counter.

At the time of a cylinder misfire, the detection system counts the misfire. In the Figure 8-28, there are misfires counted in several of the cylinders. However, cylinder number one has the greatest number of history misfires. This indicates that number one is the probable cylinder and the others are the result of erratic crankshaft rotation because of number one misfiring.

Fuel Trim Monitors.   With OBD-II, General Motors had to change their method of displaying fuel trim values (Figure 8-29). The values are now displayed in percentages. The PCM uses fuel trim cells to determine if a fuel trim failure occurs. The cells are determined by different operating conditions of the engine (Figure 8-30). Some of the cells carry a certain amount of **diagnostic weight**. Only the cells that are assigned a diagnostic weight value are used to set DTCs. A cell that has a higher diagnostic weight value will set a DTC easier than one with a lower value. For example, in Figure 8-30, fuel trim cells 4, 5, and 9 are the weighed cells. No fuel trim DTC will set regardless of the fuel trim count unless that count is located in one of the weighted cells.

Diagnostic weight means that some conditions have higher precedence, and events that happen in these conditions are weighted.

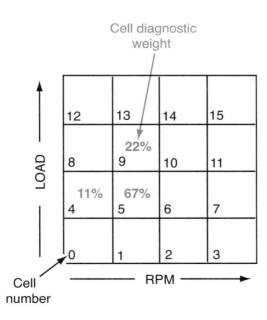

Figure 8-30   Fuel trim cells are determined by the operating conditions of the engine. Some cells carry a higher diagnostic weight than others.

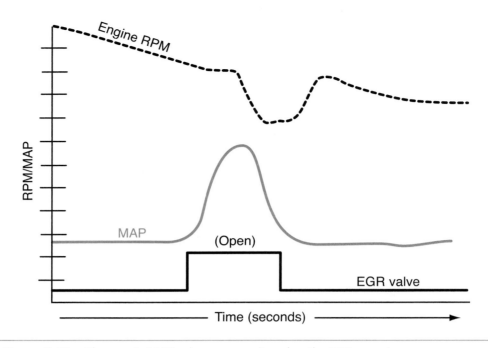

Figure 8-31 Changes in MAP values are monitored as the EGR operates.

The fuel trim diagnostic compares an average of long-term trim values and short-term trim values to rich and lean limits that are the calibrated fail thresholds for the test. If either value is within the fail thresholds, a pass is recorded. If both values are outside the thresholds, then a failure exists.

Exhaust Gas Recirculation Monitor. Changes in the EGR valve will have an effect on MAP and rpm inputs. Opening the valve increases manifold pressure, while closing the valve decreases the pressure (Figure 8-31). The amount of MAP sensor input change is correlated to the amount of EGR flow.

The EGR monitoring system runs an intrusive test by forcing the EGR valve open during a closed throttle deceleration. Closing the valve during a steady cruise condition can also run the test. The PCM records the average MAP changes over several EGR valve actuations. If the resulting average is outside parameter limits, the test fails.

Secondary Air Injection Monitor. General Motors' AIR monitoring system uses both passive and active tests to evaluate the AIR system. The AIR diagnostic system will indicate a test pass if the passive portion of the AIR system observes the correct lean reading from the upstream HO_2S after start-up and prior to closed-loop operation. The HO_2S will read low voltage if the AIR pump is delivering air to the exhaust system. When the AIR system is turned off, the test looks for HO_2S voltage to toggle.

If the AIR monitor passes during the passive test, the test is completed. However, if the passive test fails, the active test is initiated. The active test activates the AIR pump during closed-loop operation and observes the response from the upstream HO_2S. It may also monitor short-term fuel trim values. A lean HO_2S voltage or a fuel trim value that increases indicates the AIR system is operating properly.

Evaporative Emission System Monitor. General Motors vehicles can be equipped with either a non-enhanced or enhanced EVAP system. The non-enhanced system includes the following components:

- EVAP emission pressure control valve
- Vented canister
- Diagnostic switch
- EVAP purge solenoid

The EVAP solenoid is pulse-width modulated and controls the amount of canister purge. When the solenoid is energized, vapors are allowed to flow from the canister into the intake manifold. The **diagnostic switch** is normally closed and will open when the solenoid is energized and there is sufficient intake manifold vacuum. If the solenoid is stuck in the open position, vacuum will be present at the switch and it will remain in the open position. A solenoid that is stuck in the closed position will constantly indicate atmospheric pressure. In addition, a clogged vent could trap vacuum between the switch and the canister, causing the switch to open even with the solenoid de-energized.

The enhanced EVAP system includes the following components (Figure 8-32):

- EVAP canister
- Fuel tank pressure sensor
- Canister purge solenoid
- Canister vent solenoid
- Service port
- Fuel level sensor
- Fuel cap

The purge solenoid operates in the same manner as in the non-enhanced system. The vent solenoid is a normally open solenoid that allows fresh air to the canister during purge and allows

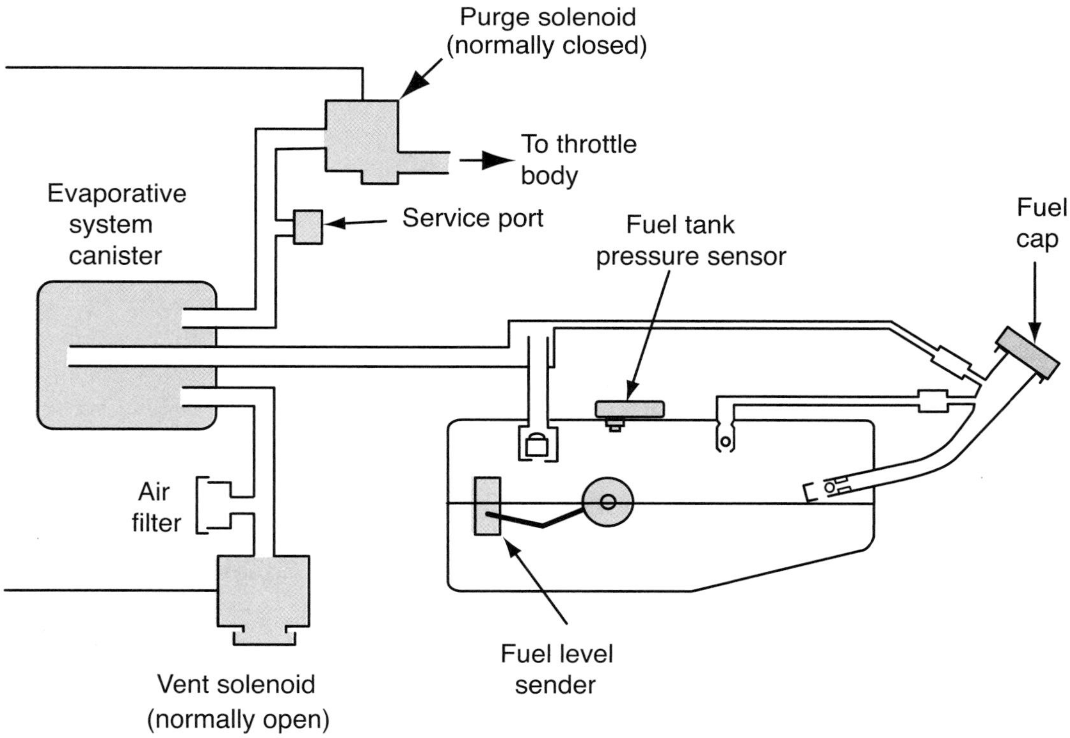

The **diagnostic switch** is used to determine correct operation of the solenoid.

Figure 8-32 Typical enhanced EVAP system.

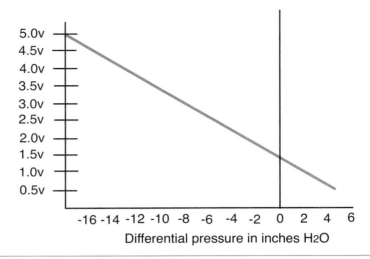

Differential pressure in inches H2O

Figure 8-33 A fuel pressure sensor that is mounted on top of the fuel sending unit measures the difference in pressure between atmospheric and what is in the fuel tank.

The **power up vacuum test** is a passive test used to detect restrictions in the vent path.

The **excess vacuum test** is also a passive test designed to detect restrictions in the vent path.

The **loaded canister test** is a passive test designed to detect leaks in the entire EVAP system.

The **weak vacuum test** is an active test designed to detect large leaks.

The **purge solenoid leak test** is an active test designed to detect a leak through the purge solenoid.

the diagnostics to pull a vacuum on the fuel tank when closed. The fuel tank pressure sensor is a three-wire strain gauge that measures the difference between the air pressure in the tank and the atmospheric pressure (Figure 8-33).

Enhanced EVAP systems must be capable of detecting leaks in the system that are 0.20 inch (0.5 mm) diameter or larger. The PCM uses several tests to perform this task:

- Power up vacuum test
- Excess vacuum test
- Loaded canister test
- Weak vacuum test
- Small leak test
- Purge solenoid leak test

The **power up vacuum test** is run when the vent solenoid is open with the purge solenoid closed. The engine must also be cold and the ignition in the run position but the engine off. Under these conditions, the fuel tank pressure sensor should not indicate pressure or vacuum. This test will not run if the fuel level sensor indicates more than 85 percent full or 15 percent empty.

The **excess vacuum test** will run during normal purge operations with the solenoid valve open. During this test, the pressure sensor in the tank should not indicate excessive vacuum.

The **loaded canister test** monitors the HO_2S voltage levels and fuel trim values during normal EVAP operation. If large amounts of fuel vapors are being purged, the canister is sufficiently loaded and the system is functioning properly.

The **weak vacuum test** will only run if the loaded canister test failed. During this test, the vent solenoid is closed during normal EVAP operation. The fuel tank pressure sensor should indicate the presence of vacuum during this test.

If the weak vacuum test passes, the PCM will initiate the small leak test while vacuum is still present in the fuel tank. The vent and purge solenoids are closed to seal the system. With the system sealed, the tank should hold vacuum. The fuel tank pressure sensor is monitored for a decay rate. If the decay is too rapid, the test fails.

During the **purge solenoid leak test,** the vent and purge solenoids are closed when EVAP purge is not occurring. The fuel tank pressure sensor is monitored; if vacuum is present the purge solenoid is leaking.

Ford OBD-II

Ford began installing the ODB-II system on the 3.8 L Mustang and the 4.6 L Thunderbird and Cougar in 1994 with the introduction of Electronic Engine Control V (EEC V). The EEC V system hardware includes the PCM, sensors, switches, actuators, solenoids, wires, terminals, and the constant control relay module (CCRM). The main hardware difference in the EEC V system is the addition of downstream heated oxygen (HO_2S) sensors mounted behind the catalytic converter. The conventional HO_2S sensor is mounted in the exhaust manifold.

The major difference between EEC V with OBD-II and previous systems is the software containing the monitoring strategies inside the PCM. The OBD Executive is the software program of the PCM that manages the sequencing and execution of all OBD-II diagnostic tests. The components in the EEC V with OBD-II are designed to perform within specifications for 100,000 miles.

Catalyst Efficiency Monitor Steady State. The 20-second steady state catalyst efficiency monitor (used in 1994 through some 1996 vehicles) uses an HO_2S located downstream of the catalytic converter to infer catalyst efficiency based on oxygen storage capacity. A high-efficiency catalyst will result in a lower downstream HO_2S switch rate as compared to the upstream sensors.

The downstream HO_2S sensors have additional protection to prevent the collection of condensation on the ceramic. The internal heater is not turned on until the ECT sensor signal indicates a warmed up engine. This action prevents cracking of the ceramic. Gold plated pins and sockets are used in the HO_2S sensors, and the downstream and upstream sensors have different wiring harness connectors.

When the catalytic converter is storing oxygen properly, the downstream HO_2S sensors provide low-frequency voltage signals. If the catalytic converter is not storing oxygen properly, the voltage signal frequency increases on the downstream HO_2S sensors until the frequency of the downstream HO_2S sensors approaches the frequency of the upstream HO_2S sensors. When the downstream HO_2S sensors voltage signals reach a certain frequency, a DTC is set in the PCM memory. If the fault occurs on three drive cycles, the MIL light is illuminated.

The monitor runs after all enabling criteria have been met. This criteria is based on input from the engine coolant temperature (ECT), intake air temperature (IAT), throttle position sensor (TPS), crankshaft position (CKP), and vehicle speed sensor (VSS), along with a calculated length of engine run time in closed loop. Once the monitor is activated, closed-loop fuel control is temporarily transferred from the upstream HO_2S to the downstream HO_2S. The monitor then analyzes the downstream HO_2S signal switching frequency to determine if the catalytic converter has failed. High catalytic converter efficiency is determined by a slow **test frequency**.

A second **calibrated frequency** is determined based on the engine speed and load. If the test frequency is less than the calibrated frequency, the catalytic converter passes the monitor test. If the test frequency is higher than the calculated frequency, the test fails and a pending DTC is stored. If the test fails on two consecutive drive cycles, the MIL is turned on.

Catalyst Efficiency Monitor Federal Test Procedure. The catalyst efficiency monitor is a once-per-trip monitor. Ford has used different monitoring systems in different model years.

The FTP switch ratio method (used model years 1996 to 2003) of determining catalyst efficiency is activated during part-throttle, closed-loop fuel conditions after the engine is warmed up and inferred catalyst temperature is within limits. When this monitor is activated, the front and rear HO_2S switches are counted to calculate a switch ratio. The switch ratio is determined when the front HO_2S are accumulated in up to nine different air mass regions or cells (although three air mass regions are typical). Rear switches are counted in a single cell for all air mass regions. When the required number of front switches has accumulated in each cell (air mass region), the total number of rear switches is divided by the total number of front switches to compute a switch ratio.

The switch ratio is compared against an emission threshold value. If the switch ratio is greater than the emission threshold, the catalyst has failed. The MIL is turned on after a fault is detected up to six consecutive drive cycles.

This actual measured output frequency of the downstream HO_2S is called the **test frequency.**

The **calibrated frequency** serves as a high limit threshold for the test frequency and is determined by the PCM.

The index ratio method uses the ratio of signal lengths to determine proper activity.

A switch ratio near 0.0 indicates high oxygen storage capacity, hence high HC efficiency. A switch ratio near 1.0 indicates low oxygen storage capacity, hence low HC efficiency. If the actual switch ratio exceeds the threshold switch ratio, the catalyst is considered failed.

Beginning in 2001, Ford began to use the **index ratio** method. Like the switch ratio method, the PCM executive counts front HO$_2$S switches during part-throttle, closed-loop fuel conditions after the engine is warmed up and inferred catalyst temperature is within limits. Front switches are accumulated in up to three different air mass regions or cells. While catalyst monitoring entry conditions are being met, the front and rear HO$_2$S signal lengths are continually being calculated. When the required number of front switches has accumulated in each cell (air mass region), the total signal length of the rear HO$_2$S is divided by the total signal length of front HO$_2$S to compute a catalyst index ratio.

An index ratio near 0.0 indicates high oxygen storage capacity, hence high HC efficiency. An index ratio near 1.0 indicates low oxygen storage capacity, hence low HC efficiency. If the actual index ratio exceeds the threshold index ratio, the catalyst is considered failed.

If the catalyst monitor does not complete during a particular driving cycle, the data that has been accumulated is retained and is used during the next driving cycle. This gives the monitor a better opportunity to complete, even under short or transient driving conditions.

LEV and ULEV vehicles monitor less than 100 percent of the catalyst volume, usually the light-off catalyst in the system. Partial volume monitoring is done on LEV and ULEV vehicles in order to meet the one-and-three-quarter times the FTP standard. The rationale for this practice is that the catalysts nearest the engine will deteriorate first, allowing the catalyst monitor to be more sensitive and properly illuminate the MIL at lower emission standards.

An EWMA algorithm is used to improve the robustness of the FTP catalyst monitor. During normal customer driving, a malfunction will illuminate the MIL, on average, in three to six driving cycles. If KAM is reset (battery disconnected), a malfunction will illuminate the MIL in two driving cycles.

The EWMA catalyst monitor has three parameters that determine the MIL illumination characteristics:

- "Fast" filter constant, used for two driving cycles after DTCs are cleared or KAM is reset
- "Normal" filter constant, used for all subsequent, "normal" customer driving
- Number of driving cycles to use fast filter after KAM clear (normally set to two driving cycles)

Examples for a typical calibration are shown in Table 8-4.

TABLE 8-4   EWMA CATALYST MONITOR CALIBRATION TABLE

Monitor evaluation (new data)	EWMA filter calculation, "normal" filter constant Set to 0.4 Malfunction threshold =.75	Weighted average (new average)	Driving cycle number	Comments
0.15	.15 (0.4) + .15 (1 - 0.4)	0.15		Normal 100K system
1.0	1.0 (0.4) + .15 (1 - 0.4)	0.49	1	Catastrophic failure
1.0	1.0 (0.4) + .49 (1 - 0.4)	0.69	2	
1.0	1.0 (0.4) + .69 (1 - 0.4)	0.82	3	Exceeds threshold
1.0	1.0 (0.4) + .82 (1 - 0.4)	0.89	4	MIL "on"
0.15	.15 (0.4) + .15 (1 - 0.4)	0.15		Normal 100K system
0.8	0.8 (0.4) + .15 (1 - 0.4)	0.41	1	1.5 threshold failure
0.8	0.8 (0.4) + 41 (1 - 0.4)	0.57	2	
0.8	0.8 (0.4) + .57 (1 - 0.4)	0.66	3	
0.8	0.8 (0.4) + .66 (1 - 0.4)	0.72	4	
0.8	0.8 (0.4) + .72 (1 - 0.4)	0.75	5	Exceeds threshold
0.8	0.8 (0.4) + .75 (1 -0.4)	0.77	6	MIL "on"

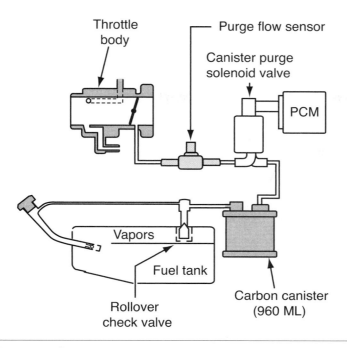

Figure 8-34 Purge flow sensor used within the EVAP system.

Evaporative (EVAP) Emission System Monitor. Some EVAP systems have a purge flow sensor (PFS) connected in the vacuum hose between the canister purge (CANP) solenoid and the intake manifold (Figure 8-34). The PCM monitors the PFS signal once per drive cycle to determine if there is vapor flow through the solenoid to the intake manifold.

Some vehicles that meet enhanced evaporative requirements utilize a vacuum-based evaporative system integrity check that tests for 0.020-inch (0.5 mm) diameter leaks. The EVAP system integrity check uses a fuel tank pressure transducer (FTPT), a canister vent solenoid (CVS), and fuel level input (FLI), along with the vapor management valve (VMV) to find 0.020-inch (0.5 mm) diameter, 0.040-inch (1 mm) diameter, or larger EVAP system leaks (Figure 8-35).

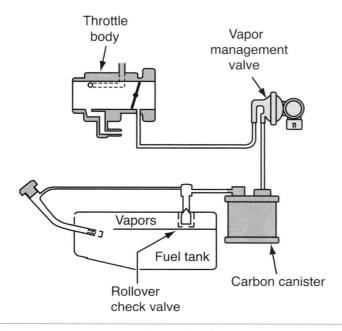

Figure 8-35 Vapor management valve used within an EVAP system.

AUTHORS NOTE: Early vehicles with enhanced evaporative EVAP monitoring only checked for leaks that were 0.040-inch (1 mm) diameter or larger. These were used prior to the more strict requirements of testing for the 0.020-inch (0.5 mm) diameter leaks.

The VMV is a normally closed valve used to control flow of fuel vapors into the engine. The CVS is a normally open solenoid used to control evaporative flow between the carbon canister and atmosphere. The monitor relies upon the CVS to seal the entire EVAP system from atmosphere and the VMV to pull engine vacuum on the fuel tank. The CVS is closed and the VMV is opened a calibrated amount. If the FTP sensor does not sense a target vacuum within a given time, then a leak or flow fault exists. If the target vacuum is obtained, then both solenoids are closed to lock the vacuum for a specified length of time. If the vacuum bleeds up above a fault threshold within that period, the EVAP monitor fails the test. DTCs are set after three unsuccessful attempts to hold vacuum.

The EVAP system leak test is done under two different sets of conditions. The first is a cruise test that is used to detect 0.040-inch (1 mm) diameter leaks and to screen for 0.020-inch (0.5 mm) leaks. If a 0.020-inch (0.5 mm) diameter leak is suspected during the cruise test, an idle test is performed to verify the leak under more restrictive cold start with the engine idling conditions.

The cruise test is run after a six-hour engine off time, and then the vehicle is driven at steady highway speeds. Also, the test requires the ambient air temperatures to be between 40°F and 100°F.

The cruise test is done in four phases:

- *Phase 0*—First, the CVS is closed to seal the entire evaporative system, and then the VMV is opened to pull a 7-inch H_2O vacuum. If this vacuum cannot be reached, a large leak is indicated. In addition, at every engine start-up, a check for refueling events is done. If the fuel level at start-up is at least 20 percent greater than the fuel level when the engine was shut off, a **refueling flag** is set in the PCM executive. The flag will remain set until the EVAP monitor completes Phase 0.

The refueling flag is used to prohibit the 0.020-inch (0.5 mm) idle test until the gross leak check is done during cruise conditions.

If the initial vacuum could not be achieved after a refueling event, a gross leak (fuel cap off) is indicated and the recorded minimum fuel tank pressure during pulldown is stored in KAM. A "Check Fuel Cap" light may also be illuminated.

If the initial vacuum is excessive, a vacuum malfunction is indicated. Plugged or pinched vapor lines between the FTPT and the CVS, or a stuck open VMV could cause this. If there are large leak faults detected during this phase of the test, the EVAP test will not run the other phases of the test.

- *Phase 1*—This portion of the test is referred to as vacuum stabilization. At this time, the VMV is closed and vacuum is allowed to stabilize for a fixed time. If the pressure in the tank immediately rises, the stabilization time is bypassed and Phase 2 of the test is entered.
- *Phase 2*—During this portion of the test, the vacuum is held for a calibrated time. Two test times are calculated based on the fuel level input. The first (shorter) time is used to detect 0.040-inch (1 mm) diameter leaks; the second (longer) time is used to detect 0.020-inch (0.5 mm) diameter leaks.

The initial vacuum is recorded upon entering Phase 2. At the completion of the 0.040-inch (1 mm) diameter test time, the vacuum level is recorded again. The two vacuum levels are checked to determine if the change in vacuum exceeds the 0.040-inch (1 mm) diameter vacuum bleed-up criteria. If the criteria is exceeded on three successive monitoring attempts, a 0.040-inch (1 mm) diameter leak is likely and a final vapor generation check is done to verify the leak (Phases 3 and 4).

If the criteria are not exceeded, the test is allowed to continue until the 0.020-inch (0.5 mm) diameter leak test time has completed. The starting and ending vacuum levels are checked to determine if the change in vacuum exceeds the criteria for a 0.020-inch (1 mm) diameter leak. If the criteria is exceeded on a single monitoring attempt, a 0.020-inch (0.5 mm) diameter leak is likely and a final vapor generation check is done to verify the leak (Phases 3 and 4).

If the vacuum bleed-up criteria is not exceeded, the leak test (either 0.040-inch (1 mm) or 0.020-inch (0.5 mm) diameter is considered a pass.

> ![] **AUTHOR'S NOTE:** For both leak checks, the fuel level and intake air temperature is used to adjust the criteria for the appropriate fuel tank vapor volume and temperature. Steady state conditions must be maintained throughout this bleed-up portion of the test. The monitor will abort if there is an excessive change in load, fuel tank pressure, or fuel level input since these are all indicators of impending or actual fuel slosh. If the monitor aborts, it will attempt to run again (up to twenty or more times) until the maximum time after start is reached.

- *Phase 3*—The CVS is opened for a fixed period of time and releases any vacuum in the system. The VMV will remain closed.
- *Phase 4*—During this time, the sealed system is monitored for a rise in tank pressures due to excessive vapor generation. At the conclusion of Phase 3, the tank pressure reading is recorded. The pressure is monitored for a change from the initial pressure and for absolute pressure.

If the pressure rise is below the threshold limit for the absolute pressure reading and for the change in pressure and a 0.040-inch (1 mm) diameter leak was indicated in Phase 2, a DTC is stored.

If the pressure rise is below the threshold limit for absolute pressure and for the change in pressure and a 0.020-inch (0.5 mm) diameter leak was indicated in Phase 2, a 0.020-inch (0.5 mm) idle check flag is set to run the 0.020-inch (0.5 mm) leak check during idle conditions.

Some 2003 model year PCMs have a "leaking" VMV test. This test is intended to identify a VMV that does not seal properly during Phase 1 but is not fully stuck open. If more than 1-inch H_2O of additional vacuum is developed in Phase 1, the EVAP monitor will bypass Phase 2 and go directly to Phase 3 and open the canister vent solenoid to release the vacuum. It will then proceed to Phase 4, close the canister vent solenoid, and measure the vacuum that develops. If the vacuum exceeds approximately 4-inches H_2O, a DTC will be set.

The long period of time required to detect a 0.020-inch (0.5mm) diameter leak, coupled with the changes in driving conditions, may lead to false leak indications while the vehicle is in motion. To confirm that a leak is really present, the **idle check** repeats Phases 0 and 2 with the vehicle stationary. The idle check is done on cold engine start-up to ensure that the fuel is cool and cannot pick up much heat from the engine, fuel rail, or fuel pump. This test is run only during the first 10 minutes after the engine is started. The test is aborted if vehicle speed exceeds 10 mph. However, to minimize the amount of idle time to complete the test, the initial vacuum pulldown (Phase 0) can start with the vehicle in motion.

Results from the idle check are:

The **idle check** screens out leak indications caused by changes in altitude.

- Loss of vacuum greater than the 0.020-inch (0.5 mm) diameter malfunction criteria is recorded as a failure.
- Vacuum loss below the vacuum malfunction criteria is recorded as a pass, and the pending DTC resulting from the cruise test may be cleared.
- Vacuum loss that is greater than the pass criteria but less that the failure criteria are indeterminate and do not count as a pass or fail.

EVAP bypass logic is programmed into the PCM's executive in the event that the evaporative system monitor cannot be completed. Lack of completion may be due to ambient temperature conditions that are encountered outside the 40°F to 100°F and BARO range at speeds above

40 mph during a driving cycle in which all continuous and non-continuous monitors were evaluated. In the 1997 through 1999 model years, the bypass logic will consider the evaporative system monitor as completed. If the above conditions are repeated during a second driving cycle, the I/M readiness bit for the evaporative system is set to a "ready" condition.

For 2000 and newer vehicles, a timer is incremented if the evaporative system monitor conditions are met (with the exception of the 40°F to 100°F ambient temperatures or BARO range). The timer value is representative of conditions where the EVAP monitor could have run but did not run due to the presence of these conditions. If the timer continuously exceeds 30 seconds during a driving cycle in which all continuous and non-continuous monitors were evaluated, the evaporative system monitor is then considered complete. If the above conditions are repeated during a second driving cycle, the I/M readiness bit for the evaporative system is set to a "ready" condition.

Another method Ford uses is to measure lambse shift with no purge flow as compared to purge flow. The monitor runs when the engine speed is below 900 rpm, in closed-loop fuel strategy, engine coolant is between 104°F and 215°F, and vehicle speed is below 3.76 mph. When the monitor runs, the lambse is sampled and averaged before fuel vapors are allowed to flow. The EVAP solenoid is then progressively opened in steps beginning with 10 percent to 40 percent duty cycle, then from a 40 percent to 100 percent duty cycle. During this time, the resulting lambse value is compared with the averaged no-flow value. If there is not a sufficient shift in values, the system fails the first execution (Test 1) of the EVAP monitor. Upon completion of Test 1, the monitor will return normal operation of the EVAP system. During this time, the monitor will determine when most of the vapors should have been drawn from the canister. The monitor then takes over and the second execution (Test 2) will run to check for sufficient vapor flow. If this test fails, a fault is set. The MIL is activated after one of the two tests fail on two consecutive drive cycles.

Misfire Monitor.   Cylinder misfire monitoring requires measuring the contribution of each cylinder to engine power. The misfire monitoring system uses a highly accurate crankshaft angle measurement to measure the crankshaft acceleration each time a cylinder fires. A high data rate crankshaft sensor (eighteen position references per crankshaft revolution) is usually used for this function. The PCM monitors the crankshaft acceleration time for each cylinder firing. If a cylinder is contributing normal power, a specific crankshaft acceleration time occurs. When a cylinder misfires, the cylinder does not contribute to engine power and crankshaft acceleration for that cylinder is slowed.

The high data rate input signal is processed using two different algorithms. The first algorithm, called pattern cancellation, is optimized to detect low rates of misfire. The algorithm learns the normal pattern of cylinder accelerations from the mostly good firing events and is then able to accurately detect deviations from that pattern. The second algorithm is optimized to detect one or more continuously misfiring cylinders. This algorithm filters the high-resolution crankshaft velocity signal to remove some of the crankshaft tensional vibrations that degrade signal to noise. This significantly improves detection capability for continuous misfires. Both algorithms produce a deviant cylinder acceleration value that is used in evaluating misfire.

Due to high data processing requirements, the high data rate algorithms are not a function of the PCM's microprocessor. Instead, they are implemented in a separate chip in the PCM called an "AICE" chip. The PCM microprocessor communicates with the AICE chip using a dedicated serial communication link. The output of the AICE chip (the cylinder acceleration values) is sent to the PCM microprocessor for additional processing.

"Profile correction" software is used to "learn" and correct for mechanical inaccuracies in crankshaft tooth spacing under fuel cutoff conditions (requires three 60 to 40 mph no-braking decelerations after KAM has been reset). If KAM has been reset, the PCM microprocessor initiates a special routine that computes correction factors for each of the eighteen position references and sends these correction factors back to the AICE chip to be used for subsequent misfire signal processing. These learned corrections improve the high rpm capability of the monitor. The misfire monitor is not active until a profile has been learned.

In the 2003 model year, a low data rate system (one position reference signal at 10-degree BTDC for each cylinder event) that is capable of meeting the FTP monitoring requirements on most engines and is capable of meeting "full-range" misfire monitoring requirements on 4-cylinder engines was used. The high data rate system continues to be used for full-range misfire monitoring requirements on 6- and 8-cylinder engines. In addition, 2003 software has been modified to allow for detection of any misfires that occur in six engine revolutions after initially cranking the engine. This meets the new OBD-II requirement to identify misfires within two engine revolutions after exceeding the warm drive, idle rpm.

The misfire monitor detects type A and type B engine misfires. When detecting a type A misfire, the monitor checks cylinder misfiring over a 200-rpm period. A type A (catalyst-damaging misfire rate) misfire rate is an rpm/load table ranging from 40 percent at idle to 4 percent at high rpm and loads.

Under this condition, the PCM may shut off the fuel to the misfiring cylinder or cylinders to limit catalytic converter heat. The PCM may turn off up to two injectors at the same time on misfiring cylinders. This fuel shut-off feature is used on many 8-cylinder engines and some 6-cylinder engines. It is never used on a 4-cylinder engine.

If the misfire monitor detects a type A cylinder misfire and the PCM does not shut off the injector or injectors, the MIL light begins flashing. When the misfire monitor detects a type A cylinder misfire and the PCM shuts off an injector or injectors, the MIL light is illuminated continually.

A catalyst temperature model is currently used for entry into the catalyst and oxygen sensor monitors. The catalyst temperature model uses various PCM parameters to infer exhaust/catalyst temperature. The catalyst temperature model has been enhanced and incorporated into the type A misfire monitoring logic. This allows the model to predict catalyst temperature in the presence of misfire. The catalyst damage misfire logic (type A) for MIL illumination has been modified to require that both the catalyst damage misfire rate and the catalyst damage temperature is being exceeded prior to MIL illumination. This is intended to prevent the detection of transit misfire while ensuring that the MIL is properly illuminated for misfires that truly damage the catalyst.

To detect a type B cylinder misfire (emission threshold rate), the misfire monitor checks cylinder misfiring over a 1,000-rpm period. If cylinder misfiring exceeds 1 percent to 2 percent during this period, the monitor considers the misfiring to be excessive. This amount of cylinder misfiring may not overheat the catalytic converter, but it may cause excessive emission levels. When a type B misfire is detected, which can be either a single 1,000 revolutions exceedence from start-up or four subsequent 1,000 revolutions exceedences on a drive cycle after start-up, a pending DTC is set in the PCM memory. If this fault is detected on a second consecutive drive cycle, the MIL light is illuminated. Many 2003 model year vehicles will set a DTC if the type B malfunction threshold is exceeded during the first 1,000 revolutions after engine start-up. This DTC is stored in addition to the normal DTC that indicates the misfiring cylinder(s).

Fuel System Monitoring.   The fuel system monitor checks short-term fuel trim (SFT) and long-term fuel trim (LFT) while the PCM is operating in closed loop. A correction factor is added to the fuel injector pulse width calculation according to the long- and short-term fuel trim as needed to compensate for variations in the fuel system. The fuel system monitor operates continually when the PCM is in closed loop. Once the monitor is activated, it looks for the adaptive tables to reach adaptive clip and lambse to exceed a calibrated limit. When deviation in the parameter lambse is excessive, air-fuel control suffers and emissions increase. When lambse exceeds a calibrated limit and the adaptive fuel table has clipped, the fuel system monitor sets a DTC and illuminates the MIL light if the fault occurs on two consecutive drive cycles.

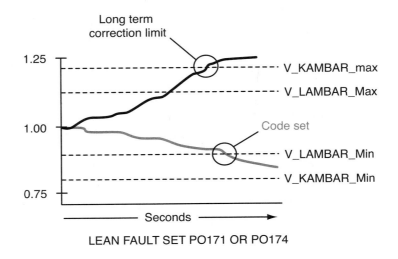

LEAN FAULT SET PO171 OR PO174

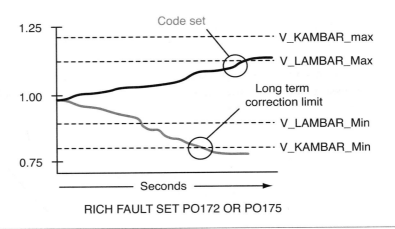

RICH FAULT SET PO172 OR PO175

Figure 8-36 Fuel system monitor failure thresholds.

A fuel system lean fault will be recorded if long-term fuel trim is greater than 25 percent and short-term fuel trim is greater than 5 percent. A rich malfunction is recorded if long-term fuel trim is less than 25 percent and short-term fuel trim is less than 10 percent (Figure 8-36).

Heated Oxygen Sensor Monitor. All HO$_2$S sensors are monitored once per drive cycle, but the heated oxygen sensor monitor provides separate tests for the upstream and downstream sensors. The heated oxygen sensor monitor checks the voltage signal frequency of the upstream HO$_2$S sensors. Excessive time between signal voltage frequency indicates a faulty sensor. At certain times, the heated oxygen sensor monitor varies the fuel delivery and checks for HO$_2$S sensor response. The response rate is evaluated by entering a special 1.5-Hz square wave fuel control routine. This routine drives the air-fuel ratio around stoichiometry at a calibrated frequency and magnitude, thereby producing predictable oxygen sensor signal amplitude. A slow sensor will show reduced amplitude. Oxygen sensor signal amplitude below a minimum threshold indicates a slow sensor malfunction.

A functional test of the rear HO$_2$S sensors is done during normal vehicle operation. The peak rich and lean voltages are continuously monitored. Voltages that exceed the adjustable rich and lean thresholds indicate a functional sensor. If the voltages have not exceeded the thresholds after a long period of vehicle operation, the air-fuel ratio may be forced rich or lean in an attempt to reach the rear sensor to switch. This situation normally occurs only with a catalyst with less than 500 miles. If the sensor does not exceed the rich and lean peak thresholds, a malfunction is indicated.

The fuel system and misfire monitors must have completed successfully before the HO$_2$S monitor is enabled

Oxygen Sensor Heater Monitor. A separate current-monitoring circuit monitors heater current once per driving cycle. The heater current is actually sampled three times. If the current value for two of the three samples falls below an adjustable threshold, the heater is assumed to be degraded or malfunctioning. The malfunction threshold is based on the type of O$_2$ sensor used:

- *NTK*—Heater current is less than 200 ma or greater than 3 amperes.
- *Bosch*—Heater current is less than 400 ma or greater than 3 amperes.
- *NTK Fast Light Off*—Heater current is less than 465 ma or greater than 3 amperes.
- *Bosch Fast Light Off*—Heater current is less than 230 ma or greater than 3 amperes.

EGR System Monitoring. The EGR system contains a **delta pressure feedback EGR (DPFE) sensor** (Figure 8-37). An orifice is located under the EGR valve, and small exhaust pressure hoses are connected from each side of this orifice to the DPFE sensor. During the EGR monitor, the PCM first checks the DPFE signal. If this sensor signal is within the normal range, the monitor proceeds with the tests.

With the engine idling and the EGR valve closed, the PCM checks for pressure difference at the two pressure hoses connected to the DPFE sensor. When the EGR valve is closed and there is no EGR flow, the pressure should be the same at both pipes. If the pressure is different at these two hoses, the EGR valve is stuck open.

The PCM commands the EGR valve to open and then checks the pressure at the two exhaust hoses connected to the DPFE sensor. With the EGR valve open and EGR flow through the orifice, there should be higher pressure at the upstream hose than at the downstream hose.

The PCM checks the EGR flow by checking the DPFE signal value against an expected DPFE value for the engine operating conditions at steady throttle within a specific rpm range. If a fault is detected in any of the EGR monitor tests, a DTC is set in the PCM memory. If the fault occurs during two consecutive drive cycles, the MIL light is illuminated. The EGR monitor operates once per drive cycle.

The **delta pressure feedback EGR (DPFE) sensor** is a ceramic, capacitive-type pressure transducer that monitors the differential pressure across a metering orifice located in the orifice tube assembly. The differential pressure feedback sensor receives this signal through two hoses.

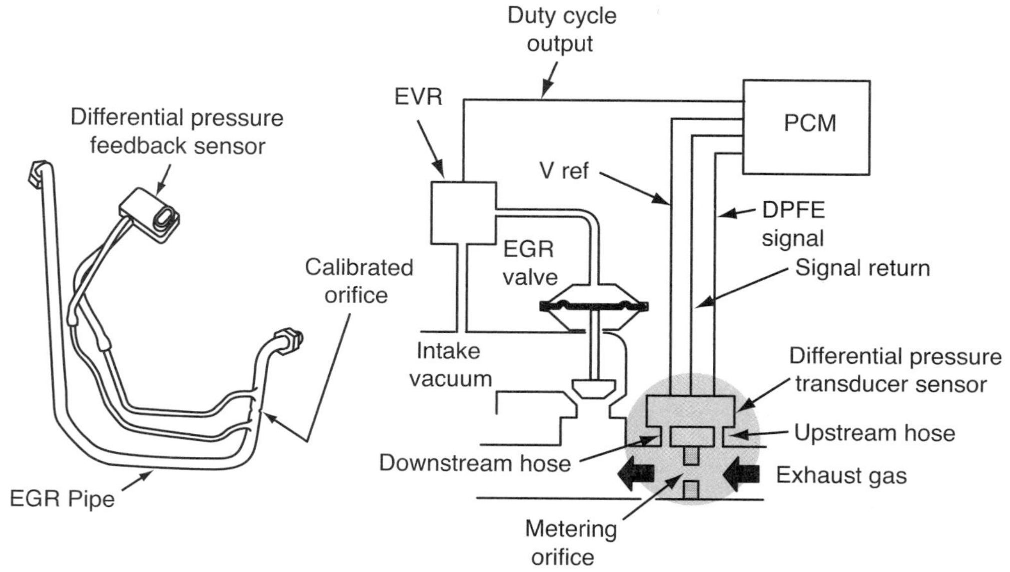

Figure 8-37 The EGR system incorporates a Delta Pressure Feedback EGR (DPFE) sensor.

Since the EGR system normally has water vapors present as the result of the engine combustion process, these vapors may freeze if ambient temperature is too cold. To prevent MIL illumination for temporary freezing, the PCM uses the following logic:

- If an EGR system malfunction is detected above 32°F (0°C), the EGR system and the EGR monitor is disabled for the current driving cycle. A DTC is stored and the MIL is illuminated if the malfunction has been detected on two consecutive driving cycles.
- If an EGR system malfunction is detected below 32°F (0°C), only the EGR system is disabled for the current driving cycle. A DTC is not stored and the I/M readiness status for the EGR monitor will not change. However, the EGR monitor will continue to operate. If the EGR monitor determined that the malfunction is no longer present, the EGR system will be enabled and normal system operation will be restored.

The differential pressure indicated by the DPFE sensor is also checked. The high flow check is performed at idle with zero EGR flow. If the differential pressure exceeds a threshold limit, it indicates a stuck open EGR valve or debris temporarily lodged under the EGR valve seat. This check is performed after the vehicle is started. During vehicle acceleration, the differential pressure is indicated by the DPFE sensor at zero. At this time, EGR flow is checked to ensure that both hoses to the DPFE sensor are connected. Under this condition, the differential pressure should be zero. If the differential pressure exceeds a maximum positive threshold, a downstream DPFE hose malfunction is indicated. If the differential pressure falls below a minimum negative threshold, an upstream DPFE hose malfunction is indicated.

After the vehicle has warmed to allow EGR operation, the low flow check is performed. Since the EGR system is a closed-loop system, the EGR system will deliver the requested EGR flow as long as it has the capacity to do so. If the EVR duty cycle is very high (greater than 80 percent duty cycle), the differential pressure indicated by the DPFE sensor is evaluated to determine the amount of EGR system restriction. If the differential pressure is below an adjustable threshold, a low flow malfunction is indicated.

In the 2002.5 model year, Ford began using a revised DPFE system (Figure 8-38). It functions in the same manner as the conventional DPFE system; however, the system components

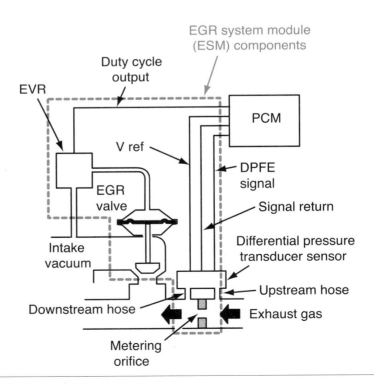

Figure 8-38 Revised DPFE system.

have been combined into a single component called the EGR System Module (ESM). By relocating the EGR orifice from the exhaust to the intake, the downstream pressure signal measures Manifold Absolute Pressure (MAP). The ESM will provide the PCM with a differential DPFE signal, identical to the conventional DPFE system. The DPFE signal is obtained by electrically subtracting the MAP and P1 pressure signals and providing this signal to the DPFE input on the PCM.

The Electric Stepper Motor EGR System (Figure 8-39) uses an electric stepper motor to directly actuate an EGR valve rather than using engine vacuum and a diaphragm on the EGR valve. The EGR valve is controlled by commanding zero to fifty-two steps to get the EGR valve from a fully closed to fully open position. The position of the EGR valve determines the EGR flow. Because there is no EGR valve position feedback, monitoring for proper EGR flow requires the addition of a MAP sensor.

The Stepper Motor EGR Monitor consists of an electrical and functional test that checks the stepper motor and the EGR system for proper flow. The stepper motor electrical test is a continuous check of the four electric stepper motor coils and circuits to the PCM. A malfunction is indicated if an open circuit, short to power, or short to ground has occurred in one or more of the stepper motor coils for a calibrated period of time. If a malfunction has been detected, the EGR system will be disabled and additional monitoring will be suspended for the remainder of the driving cycle until the next engine start-up.

EGR flow is monitored using a MAP sensor. If a malfunction has been detected in the MAP sensor, the EGR monitor will not perform the EGR flow test. The MAP sensor is checked for opens, shorts, or out-of-range values by monitoring the A/D input voltage.

After the vehicle has warmed up and normal EGR rates are being commanded by the PCM, the EGR flow check is performed. The flow test is performed once per drive cycle when a minimum amount of EGR is requested and the remaining entry conditions required to initiate the test are satisfied. If a malfunction is detected, the EGR system, as well as the EGR monitor, is disabled until the next engine start-up.

Observing the behavior of two different values of MAP, the analog MAP sensor reading and **inferred MAP** performs the EGR flow test. During normal, steady-state operating conditions, EGR is intrusively commanded to flow at a calibrated test rate of about 10 percent. At this time, the value of MAP is recorded (EGR-On MAP). The value of inferred MAP EGR-On (IMAP) is also recorded.

Inferred MAP is calculated from the MAF sensor, throttle position, rpm, BARO, etc.

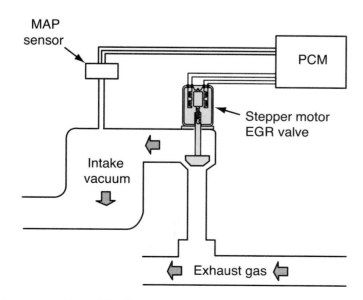

Figure 8-39   Some Ford engines are equipped with stepper motor EGR.

Then, EGR is commanded off. If the EGR system is working properly, there is a significant difference in both the observed and the inferred values of MAP, that is, between the two EGR states. Typically, seven such On/Off samples are taken. After all the samples have been taken, the average EGR-On MAP, EGR-On IMAP, EGR-Off MAP, and EGR-Off IMAP values are stored.

Next, the differences between the EGR-On and EGR-Off values are calculated:

MAP-delta = EGR-On MAP–EGR-Off MAP (analog MAP)

IMAP-delta = EGR-On IMAP–EGR-Off IMAP (inferred MAP)

If the sum of MAP-delta and IMAP-delta exceeds 8.00 in. Hg or falls below 0.62 in. Hg, a flow malfunction is registered. As an additional check, if the EGR-On MAP is greater than BARO (a calibrated value), a low flow malfunction is registered. This test detects reduced EGR flow on systems where the MAP pickup point is not located in the intake manifold but is located just upstream of the EGR valve in the EGR delivery tube.

Secondary Air Injection (AIR) System Monitor. If equipped on the vehicle, the AIR system is monitored with passive and active tests. During the passive test, the voltage of the pre-catalyst HO_2S is monitored from start-up to closed-loop operation. The AIR pump is normally on during this time. Once the HO_2S is warm enough to produce a voltage signal, the voltage should be low if the AIR pump is delivering air to the exhaust manifolds. The secondary AIR monitor will indicate a pass if the HO_2S voltage is low at this time. The passive test also looks for a higher HO_2S voltage when the AIR flow to the exhaust manifolds is turned off by the PCM. When the AIR system passes the passive test, no further testing is done. If the AIR system fails the passive test or the test is inconclusive, the AIR monitor in the PCM proceeds with the active test.

During the active test, the PCM cycles the AIR flow to the exhaust manifolds on and off during closed-loop operation and monitors the pre-catalyst HO_2S voltage and the short-term fuel trim value. When the AIR flow to the exhaust manifolds is turned on, the HO_2S voltage should decrease and the short-term fuel trim should indicate a richer condition. The secondary AIR system monitor illuminates the MIL and stores a DTC in the PCM memory if the AIR system fails the active test on two consecutive trips.

Some Ford vehicles have an electric air pump (EAP) system. In this system, the EAP is controlled by a solid state relay (SSR). The SSR is operated by a signal from the PCM. An air-injection bypass solenoid is also operated by the PCM. This solenoid supplies vacuum to dual air diverter valves (Figure 8-40).

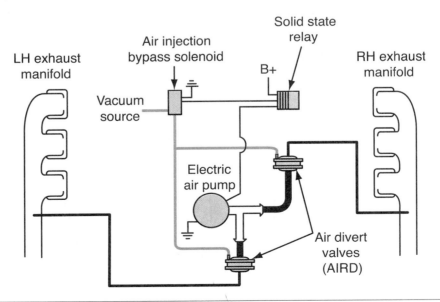

Figure 8-40 Some Ford vehicles are equipped with an electric air pump (EAP) system.

When the engine is started, the PCM signals the SSR to start the EAP, and this module supplies the high current required by the EAP. The EAP may provide a 10-second delay in pump operation after the engine is started. The PCM also energizes the air-injection bypass solenoid. When this solenoid is energized, it supplies vacuum to the dual air diverter valves. This action opens the normally closed air diverter valves, and air from the pump is now delivered to the exhaust manifolds. After the O_2 sensors warm up, the secondary air pump continues to be energized while the fuel system goes into closed-loop fuel. The secondary air system continues to run in closed-loop fuel until the air pump is de-energized. The purpose of the EAP system is to oxidize HC and CO in the exhaust manifolds for 20 to 120 seconds after the engine is started until the catalytic converter begins to function properly. The length of EAP operation after the engine is started depends on engine temperature. Cold engine temperatures provide longer EAP operation. Once the catalytic converter is operating properly, the PCM signals the SSR to shut off the EAP. The PCM also de-energizes the air-injection bypass solenoid, which allows the air diverter valves to close.

The PCM monitors the SSR and the EAP to determine if secondary air is present. The AIR pump flow check monitors the HO_2S signal at idle to determine if secondary air is being delivered into the exhaust system. The air-fuel ratio is commanded open-loop rich and the AIR pump is turned on. The time required for the HO_2S signal to go lean is monitored. If the HO_2S signal does not go lean within the allowable time limit, a low/no flow malfunction is indicated.

This PCM monitor for the EAP system functions once per drive cycle. When a malfunction occurs in the EAP system on two consecutive drive cycles, a DTC is stored in the PCM memory and the MIL is turned on. If this malfunction corrects itself, the MIL is turned off after three consecutive drive cycles in which the fault is not present.

Comprehensive Monitor. The comprehensive component monitor uses two strategies to monitor inputs and two strategies to monitor outputs. One strategy for monitoring inputs involves checking certain inputs for electrical defects and out-of-range values by checking the input signals at the A/D converter. The input signals monitored in this way are:

1. Rear HO_2S inputs
2. HO_2S inputs
3. MAF sensor
4. Manual lever position (MLP) sensor
5. Throttle position (TP) sensor
6. ECT sensor
7. IAT sensor

The comprehensive component monitor (CCM) checks frequency signal inputs by performing rationality checks. During a rationality check, the monitor uses other sensor readings and calculations to determine if a sensor reading is proper for the present conditions. The CCM checks these inputs with rationality checks:

1. Profile ignition pickup (PIP)
2. Output shaft speed (OSS) sensor
3. Ignition diagnostic monitor (IDM)
4. Cylinder identification (CID) sensor
5. Vehicle speed sensor (VSS)

The PCM output that controls the idle air control (IAC) motor is monitored by checking the idle speed demanded by the inputs against the closed-loop idle speed correction supplied by the PCM to the IAC motor.

The output state monitor in the CCM checks most of the outputs by monitoring the voltage of each output solenoid, relay, or actuator at the output driver in the PCM. If the output is off, this voltage should be high. This voltage is pulled low when the output is on.

Monitored outputs include:

1. Wide-open throttle A/C cutoff (WAC)
2. Shift solenoid 1 (SS1)
3. Shift solenoid 2 (SS2)
4. TCC solenoid
5. HO$_2$S heaters
6. High fan control (HFC)
7. Fan control (FC)
8. Electronic pressure control (EPC) solenoid

Import OBD-II

 AUTHOR'S NOTE: It is out of the scope of this book to describe every import OBD-II system. The following Toyota system is offered as an example.

This section provides an example of how import manufacturers have approached OBD-II requirements. Many European countries also have a similar OBD-II system to the United States. Their version is called Europe On-Board Diagnostics (EOBD).

Evaporative System Monitor. Early EVAP systems had to detect 0.040-inch (1 mm) diameter leaks and greater. The systems developed to perform this detection are often called the early type or non-intrusive system. Beginning with the 2,000 model year, a new EVAP monitor system was implemented to meet the new, mandated standard of detecting 0.020-inch (0.5 mm) diameter leaks. This new system is referred to as the late type or intrusive type. In addition, the EVAP monitor has to monitor vapor purge and component performance. All EVAP monitor DTCs require two trips.

In the early type (non-intrusive) pressure detection, the PCM uses a signal from the **vapor pressure sensor (VPS)** to measure the pressures in the EVAP system and in the purge side of the charcoal canister (Figure 8-41). This sensor uses a silicon chip with a calibrated reference

Shop Manual
Chapter 8,
page 342

The **vapor pressure sensor (VPS)** measures the vapor pressure in the evaporative emission control system.

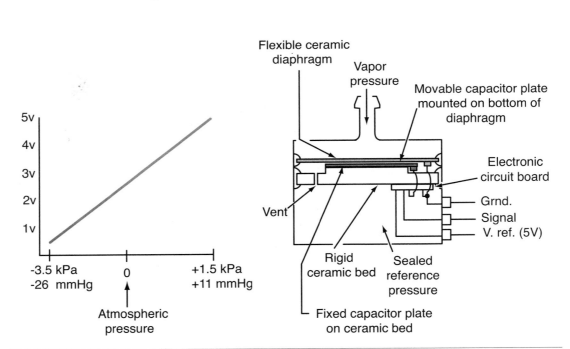

Figure 8-41 Vapor pressure sensor.

pressure on one side of the chip; the other side of the chip is exposed to vapor pressure. Changes in vapor pressure cause the chip to flex and vary the voltage signal to the PCM. The voltage signal depends on the difference between atmospheric pressure and vapor pressure. As vapor pressure increases, the voltage signal increases. The changes in the pressures being measured can be less than 15.5 mm Hg (0.3 psi).

To separate the two pressures, a **three-way vacuum solenoid valve (VSV)** is connected to the VPS, fuel tank and lines, and charcoal canister (Figure 8-42). When the three-way VSV is off, the VPS measures the pressure in the canister purge system. When the PCM activates the VSV, the VPS measures the pressure in the fuel tank.

When the VPS-sensed pressure is higher or lower than atmospheric pressure, the PCM concludes that there are no leaks in that portion of the system being tested. If either the tank or canister purge side is at atmospheric pressure under specific conditions, the PCM determines there is a leak. If a leak is detected on two consecutive trips (providing the monitor ran and completed on each trip), the MIL is illuminated and a DTC is stored. The PCM may take 20 minutes or more to complete testing the fuel tank side.

There are separate DTCs based on the location of the leak. One DTC is for leaks on the fuel tank side of the EVAP system and another for the canister side. The fuel tank side of the test

The **three-way vacuum solenoid valve (VSV)** allows the VP sensor to detect either canister or tank pressure.

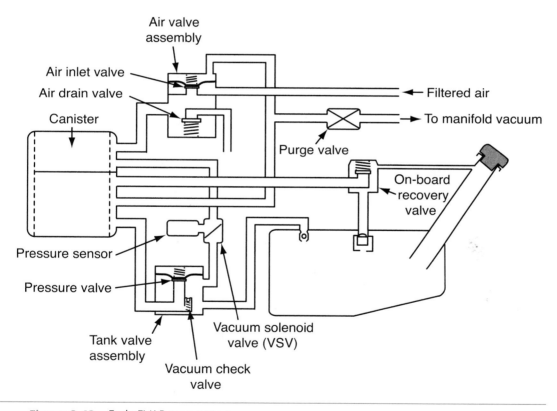

Figure 8-42   Early EVAP-type system.

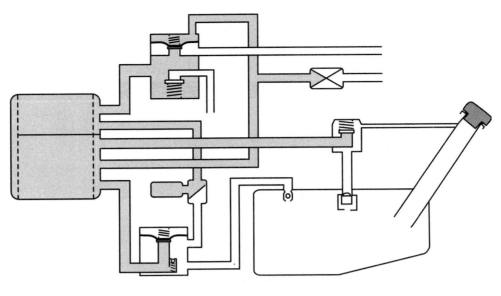

LEAK DETECTION

If a leak is suspected in the canister, check the shaded areas

Figure 8-43 Leak detection in the early type system.

includes the lines between the fuel tank and *part* of the canister. Figure 8-43 illustrates possible locations for a leak on the canister side.

The EVAP purge flow monitor portion of the test is designed to detect:

- Restricted vapor purge flow when the purge VSV is open.
- Inappropriate vapor purge flow when the purge VSV is closed.
- Under normal purge conditions, pressure pulsations generated by the cycling of the purge VSV are present in the canister and are detected by the VPS.

To test for restricted vapor purge flow, the PCM uses the VPS to monitor pressure pulsations. When the EVAP system is being purged under normal conditions, the cycling of the purge VSV generates pressure pulsations as the canister pressure drops. If the VPS does not detect these pulsations and pressure drop in the canister, the PCM determines the EVAP system is not working.

Inappropriate vapor purge flow is tested during starting. With the engine just started, the canister internal pressure should be at atmospheric pressure. If the canister internal pressure drops close to intake manifold pressure (vacuum), this would indicate that the purge VSV is open when it should be closed.

The three-way VSV is monitored for component performance using a test with two modes. The three-way VSV is considered to be functioning properly if there is a pressure difference between the tank and canister when the three-way VSV is switched. If the pressures are equal, then the PCM will look for the following conditions:

- If during purging, pressure pulsations generated by the purge VSV are not present in the canister as detected by VP sensor, the three-way VSV is judged to be defective.
- If there are pressure pulsations detected by the VPS present in the fuel tank, the three-way VSV is judged to be defective.

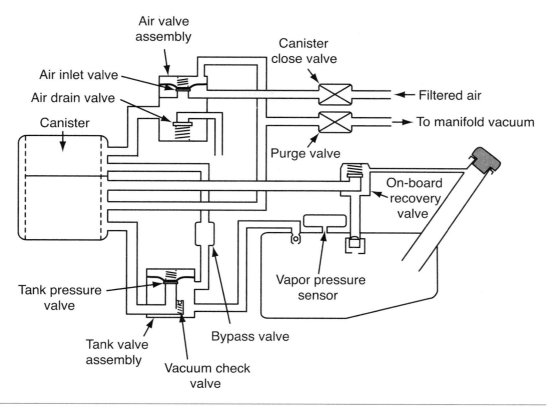

Figure 8-44 Late type EVAP system.

The logic of this test is that during purging, the VPS is supposed to be monitoring pressure pulsations in the canister, not the tank. If the VPS did not measure pulsations in the canister side but measured pulsations in the tank while purging was occurring, the PCM determines that the three-way VSV did not switch.

These two DTCs indicate a faulty VPS or a fault in the VPS circuit. These DTCs are not set instantly. The PCM measures the VPS signal under a variety of conditions and may require the EVAP monitor to complete before determining a fault. Both DTCs require two trips to turn on the MIL.

The later type (intrusive) monitor system was developed to meet the mandated standard of detecting a 0.020-inch (0.5 mm) diameter leak (Figure 8-44). This system uses many of the same components as the early type. However, the following items have been changed:

Shop Manual
Chapter 8,
page 343

- The vapor pressure sensor is connected to the tank and is not switched with the canister.
- The three-way VSV has been replaced with a bypass VSV that connects the canister and tank during monitor operation.
- A closed canister valve (CCV) has been added on the air inlet line allowing the system to be sealed.
- The monitoring for leak detection is different. This system applies a very small vacuum to the EVAP system. The PCM determines if there is a problem in the system based on the vapor pressure sensor signal. All EVAP DTCs require two trips.

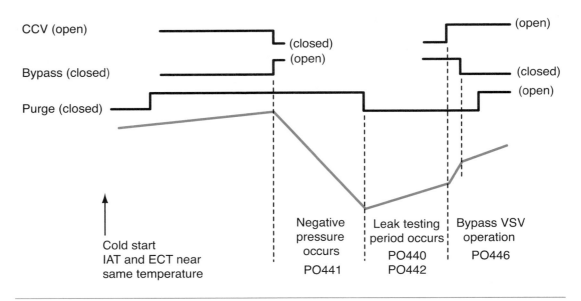

Figure 8-45 EVAP sequence of tests.

The monitor sequence begins with a cold engine start (Figure 8-45). The IAT and ECT sensors must have approximately the same temperature reading. The PCM is constantly monitoring fuel tank pressure. As the temperature of the fuel increases, pressure slowly rises. The PCM will purge the charcoal canister at the appropriate time. With the bypass VSV closed, pressure will continue to rise in fuel tank (Figure 8-46).

At a predetermined point, the PCM closes the CCV and opens the bypass VSV causing a pressure drop in the entire EVAP system. The PCM continues to operate the purge valve until the pressure is lowered to a specified point, at which time the PCM closes the purge valve. If the pressure did not drop or if the drop in pressure decreased beyond the specified limit, the PCM judges the purge VSV and related components to be faulty.

The rate of pressure increase as detected by the VPS indicates whether or not there is a leak and if it is a large or small leak. After purge valve operation, the purge VSV is turned off sealing the vacuum in the system and the PCM begins to monitor the pressure increase. Some increase is normal, however, a rapid increase in pressure indicates that there is a leak in the EVAP system and a DTC is set. A pressure rise just above normal indicates a very small hole.

To check the CCV and vent (air inlet side) operation, the PCM monitors the VPS for a vapor pressure increase to a specified point. The PCM then opens the CCV to cause a rapid increase in pressure (Figure 8-47). The increase is a result of the atmospheric pressure that is allowed into the system. Too low of an increase in pressure indicates a restriction on the air inlet side.

In the next stage, the PCM closes the bypass VSV to block atmospheric pressure from entering the tank side of the system. The pressure rise is no longer as great. If there was no change in pressure, the PCM will conclude the bypass VSV did not close.

Fuel Trim—System Too Lean. Once the engine reaches operating temperature and the air-fuel ratio feedback is stabilized, the fuel trim reaches its limit of correction to the rich side if a fuel system lean fault is detected. The fuel system lean monitor requires two consecutive failures to turn on the MIL.

System Too Rich. Once the engine reaches operating temperature and the air-fuel ratio feedback is stabilized, the fuel trim reaches its limit of correction to the lean side if a fuel system rich fault is detected. The fuel system rich monitor requires two consecutive failures to turn on the MIL.

After a period of driving,
5 - 20 minutes, the PCM
cycles the purge VSV.

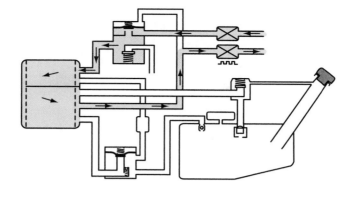

Next, the PCM will close the
CCV and open the bypass VSV
while continuing to operate the
purge VSV. This will lower the
pressure in the EVAP system.

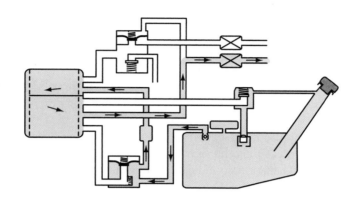

When the pressure reaches a
predetermined point, the purge
VSV is turned off and the valve
is closed. At this point the PCM
will begin to monitor for a leak
by measuring the rate of pressure
increase.

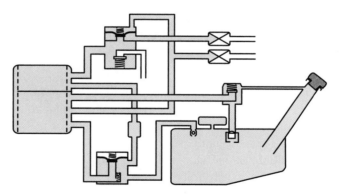

Figure 8-46 Test stages of an EVAP leak monitor.

The CCV is commanded
open by the PCM. The
vapor pressure sensor
will measure a rapid
pressure increase.

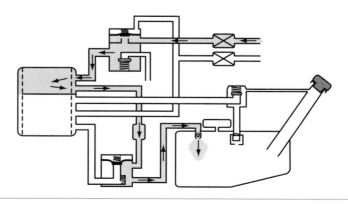

Figure 8-47 Operation of the CCV.

Misfire Monitor. The PCM measures crankshaft speed fluctuations by monitoring the crankshaft position sensor. The crankshaft will accelerate each time a cylinder is on the power stroke. If a cylinder has a misfire, the crankshaft speed slows down. The PCM detects the change in frequency in the crankshaft position sensor signal. The PCM uses the camshaft position sensor to identify the cylinder that has experienced the misfire. Severe rough road conditions may temporarily suspend misfire monitor operation.

The type A misfire is a misfire severe enough to damage the catalytic converter. The MIL will blink if the misfire would result in catalyst temperatures of 1,832°F (1,000°C) or more. The catalyst temperature is calculated by the PCM based on driving conditions and the percentage of misfire. The higher the percentage of misfire and engine load, the more likely the PCM will cause the MIL to blink.

The type B misfire is less severe but will increase emissions above FTP standards. This type of misfire is determined by the percentage of misfire compared to PCM emission specifications. Type B misfire is a two-trip monitor.

Shop Manual
Chapter 8,
page 344

Oxygen Sensor Monitor. The oxygen sensor monitor checks for sensor circuit malfunctions, slow response rate, and for a malfunction of the sensor's heater circuit. There is a DTC for each condition for each sensor. However, the downstream sensor is not monitored for response rate. O_2 sensors are required to be monitored once per trip; however, the PCM continuously monitors O_2 sensor operation.

When the PCM sees the right conditions, the PCM will test the oxygen sensors for performance by measuring the signal response as the fuel injected into the cylinder is varied. The faster the oxygen sensor responds, the better the sensor. A DTC is stored when there is little or no signal response from the O_2 sensor. This DTC is a one-trip code. A fault is determined if the engine is at normal operating temperatures and oxygen sensor output does not indicate rich even when the following conditions are present for at least 1.5 minutes:

- Engine speed: 1,500 rpm or more
- Vehicle speed: 25–62 mph (40–100 km/h)
- Throttle valve does not fully closed
- 140 seconds or more after starting engine

The O_2 sensor's output voltage portion of the monitor determines if the signal to the PCM stays high or low during the test period. The faults are detected if voltage output of the oxygen sensor remains at 0.40 volt or more, or 0.55 volt or less during idling after the engine is warmed up. This is a two-trip fault.

The PCM also monitors the response of the O_2 sensor. This is judged by the time it takes for the O_2 sensor to switch from 350 mV and 550 mV. Response time for the heated oxygen sensor's voltage output to change from rich to lean, or from lean to rich is 1.1 seconds or more during idling after the engine is warmed up. This is also a two-trip failure.

Shop Manual
Chapter 8,
page 344

O_2 Sensor Heater Monitor. The PCM monitors the current required to heat the sensor. If the current is above two amperes or below 200 ma, a DTC will be set. This is a two-trip failure.

Air-Fuel (A/F) Sensor Monitor. The A/F sensor monitor is similar to the O_2 Sensor monitor. Since the A/F sensor has different characteristics than the oxygen sensor, the operating parameters of the monitor are different.

Shop Manual
Chapter 8,
page 344

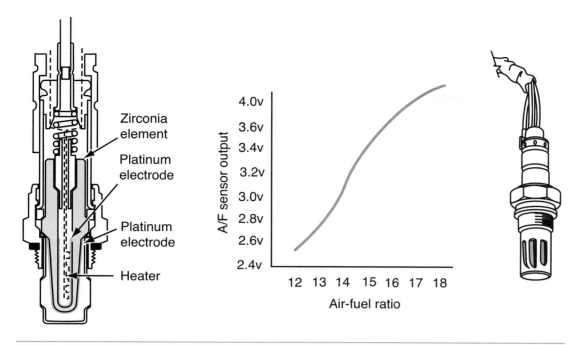

Figure 8-48 Air-fuel ratio sensor.

The A/F sensor (Figure 8-48) operates at approximately 1,200°F (650°C), which is hotter than the typical oxygen sensor that operates at 750°F (400°C). The biggest difference is that the A/F sensor changes its current output in relation to the amount of oxygen in the exhaust stream. A detection circuit in the PCM detects the change and strength of current flow and puts out a voltage signal relatively proportional to exhaust oxygen content (Figure 8-49).

![note icon] **AUTHOR'S NOTE:** This voltage signal can only be measured by using the scan tool. The A/F sensor current output cannot be accurately measured directly. If an OBD-II scan tool is used, refer to the Repair Manual for conversion since the output signal is different.

The A/F sensor is also referred to as a wide range or wide ratio sensor because of its ability to detect air-fuel ratios over a wide range.

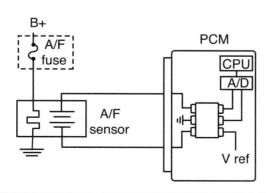

Figure 8-49 Air-fuel ratio sensor circuit.

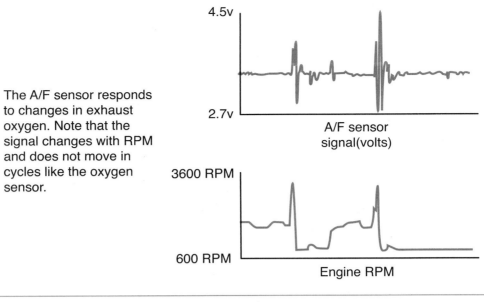

The A/F sensor responds to changes in exhaust oxygen. Note that the signal changes with RPM and does not move in cycles like the oxygen sensor.

Figure 8-50 Air-fuel ratio sensor output signals.

At the stoichiometric air-fuel ratio, there is no current flow (Figure 8-50). The voltage output by the detection circuit at stoichiometric is 3.3 volts. A rich mixture produces a negative current flow. The voltage output by the detection circuit will be below 3.3 volts. A lean mixture produces a positive current flow. The detection circuit will now produce a voltage signal above 3.3 volts.

DTCs are related to voltage output. This DTC will set when a sensor output has very little or no activity. The A/F sensor is monitored for activity (voltage change) when:

- Engine speed is 1,500 rpm or more.
- Vehicle speed is 25 to 62 mph (40 to 100 k/hr).
- TPS does not register idle.
- The condition continues for at least 90 seconds.
- 140 seconds or more must have passed since the engine was started.

OBD-III—A Look Ahead

AUTHOR'S NOTE: It is important to remember that these proposals are not yet made into law. At this time, they are nothing more than proposals. Issues such as vehicle owner privacy and shop equipment requirements must be addressed.

The CARB is currently developing standards for what may eventually become OBD-III technology. The basic concept of OBD-III is OBD-II with wireless communication of inspection/maintenance status. The belief is that many drivers ignore the illumination of the MIL if the engine still appears to be running all right. This belief has led CARB to propose a radio transponder in vehicles to relay OBD status to remote receivers for central data analysis. Owners of vehicles with unsatisfactory reporting data would be contacted to bring their vehicles into a shop for a more detailed inspection. The purpose of OBD-III would be to minimize the delay between the detection of an emissions malfunction by the OBD-II system and the actual repair of the vehicle.

As it is currently envisioned, the driver would have a button for voluntarily reporting the emissions status of his vehicle to authorities. If the OBD-III system reports no emission problems, the vehicle owner is allowed to renew their registration and the vehicle does not have to be tested. However, if the owner does not push the button or does and has an emissions problem, he receives a notice informing him that the vehicle must be brought in for an emissions test.

Options include roadside readers, local station networks, or satellites. The CARB has tested the roadside reader since 1994. It is capable of reading eight lanes of bumper-to-bumper traffic at 100 miles per hour with a data error rate of less than 1 bit in 37 billion bits. Proposed transmitted data includes day and time, full vehicle identification number, and all diagnostic trouble codes and status of monitors. It can be used from a fixed location with portable units or a mobile unit. The local station network has not been tested by CARB, but would allow a location and monitoring service. The satellite system can be used with a cellular phone hookup or location monitoring technology. The vehicle would receive an alert via a cellular phone or the monitoring technology. The location, date, time, VIN, and OBD-II data would be returned to a satellite beacon.

CARB sees a further refinement of the system to offer the capability of owners and their authorized representatives to access data in a secure communication over an established network, such as the Internet. The car owner would benefit from not being tested when the car did not have an emissions problem.

The EPA has not determined if they will implement such a program nationwide at this time. The EPA believes that new technology, such as fuel cells, could eliminate the need for OBD entirely. However, the EPA does see benefits to such a program, such as improved modeling of in-use vehicle emissions and more feedback to manufacturers and regulatory agencies.

The EPA also sees the right OBD-III program as a means of assisting technicians to satisfy their customers. In one possible sequence, a car owner could pre-authorize a particular repair shop to receive notification of a failure at the same time as the owner did. The shop would have access to the vehicle failure data, along with its own in-house service history.

Summary

❏ The MIL is illuminated if vehicle emissions exceed one-and-one-half times the allowable standard for that model year and vehicle certification.

❏ Compared to previous systems, the main difference in an OBD-II system is in the software contained in the PCM.

❏ OBD-II systems must have a 16terminal DLC with specific terminals defined by SAE.

❏ An OBD-II system uses system monitors to check system operation, including catalyst efficiency, engine misfire, fuel system, heated exhaust gas oxygen sensors, EGR, and comprehensive component monitors.

❏ Some manufacturers use Exponentially Weighted Moving Averaging (EWMA) for some of their monitors, such as catalyst and EGR.

❏ The EWMA uses both instantaneous and history data to determine trends to prevent transit conditions from causing the MIL to be illuminated.

❏ To complete an OBD-II drive cycle, the engine must be started and the vehicle driven until the monitors are completed.

❏ A warm-up cycle is defined by vehicle operation after an engine shutdown period. During the vehicle operation, the engine coolant temperature must rise at least 40°F, and the coolant temperature must reach at least 160°F.

❏ Most DTCs are erased after forty warm-up cycles if the problem does not reoccur after the MIL is turned off for that problem.

❏ SAE published J2012 to describe industry-wide standards for a uniform DTC format.

Terms-To-Know

Active

Adaptive numerator

Alternate good trip

Big slope

Calibrated frequency

CARB Readiness Indicator

Catalyst monitor

Comprehensive components

Conflict

Continuous monitor

Delta pressure feedback EGR (DPFE) sensor

Diagnostic executive

Diagnostic management system

Diagnostic switch

Diagnostic weight

Drive cycle

Enhanced EVAP

Excess vacuum test

Exponentially Weighted Moving Average (EWMA)

Fast moving average

Federal test procedure (FTP)

Freeze frame

Fuel system good trip

Fuel system monitor

Gas law principle

Global good trip

Half cycle counts

Idle check

Index ratio

Inferred MAP

Instantaneous switching frequency

Intrusive test

Leak detection pump (LDP)

❑ All OBD-II-monitored systems provide freeze frame data on the vehicle's operating conditions when a maturing code is set.

❑ If a DTC is stored due to misfire or fuel system-related problems, the PCM requires the engine be returned to an SCW to retest for a pass of the monitor.

❑ Similar conditions are defined as within 375 rpm and within 20 percent of the MAP reading at the time of the fault.

❑ Three good trips are required to turn the MIL off.

❑ Forty warm-up cycles are required after the MIL is off to self-erase the fault code.

❑ The CARB Readiness Indicator is used when an emissions I/M test is performed.

❑ The SAE document J1978 describes the minimum requirements for an OBD-II scan tool. This document covers the required capabilities of and conformance criteria for OBD-II scan tools.

❑ Different methods used by manufacturers to comply with the OBD-II requirements.

Review Questions

Short Answer Essay

1. Briefly describe the monitors in an OBD-II system.

2. Describe the main hardware differences between OBD-I and OBD-II systems.

3. Describe an OBD-II warmup cycle.

4. Explain trip and drive cycle in an OBD-II system.

5. Describe how engine misfire is detected in an OBD-II system.

6. What priority does a matured Fuel Control Monitor DTC have in freeze frame?

7. Describe the half-cycle counter.

8. Explain the term "Adaptive Numerator."

9. Define what a rationality test is.

10. What is an intrusive test?

Fill-In-The-Blanks

1. In an OBD-II system, the MIL is illuminated if the emission levels exceed _____ times the standard for that model year.

2. The downstream HO_2S monitors _____ efficiency.

3. Type B engine misfires are excessive if misfiring exceeds _____ to _____ percent in a(n) _____ rpm period.

4. OBD-II vehicles must use a(n) _____ terminal DLC.

5. The _____ _____ detects the catalyst's ability to store and give off oxygen through the use of a(n) _____ _____ oxygen sensor.

6. A _____ _____ is a specific driving method to verify a symptom or the repair for the symptom and to begin and complete a specific OBD-II monitor.

7. The minimum requirement for a(n) _____ is that an ignition switch-off period must precede an OBD-II trip. Following an ignition off period, the engine must be started and the vehicle driven.

8. A _____ _____ is defined by vehicle operation after an engine shutdown period.

9. All OBD-II monitored systems provide _____ _____ data on the vehicle's operating conditions when a maturing code was set.

10. If a DTC is stored due to _____ or _____ system-related problems, the PCM requires the engine be returned to a similar conditions window.

Multiple Choice

1. *Technician A* says OBD-II systems have two heated oxygen sensors downstream from the catalytic converters to monitor converter operation.
   Technician B says the PCM monitors exhaust temperature to determine ignition misfire.
   Who is correct?
   A. A only   **C.** Both A and B
   B. B only   **D.** Neither A nor B

2. *Technician A* says a maturing fault code is a one-trip failure of a two-trip monitor.
   Technician B says a freeze frame is stored when a maturing fault is detected.
   Who is correct?
   A. A only   **C.** Both A and B
   B. B only   **D.** Neither A nor B

3. *Technician A* says in an OBD-II system, the PCM illuminates the MIL if a defect causes emission levels to exceed two-and-one-half times the emission standards for that model year.
   Technician B says if a misfire condition threatens engine or catalyst damage, the PCM flashes the MIL.
   Who is correct?
   A. A only   **C.** Both A and B
   B. B only   **D.** Neither A nor B

4. *Technician A* says if the catalytic converter is not reducing emissions properly, the voltage frequency increases on the downstream HO_2S.
   Technician B says if a fault occurs in the catalyst monitor system on three drive cycles, the MIL is illuminated.
   Who is correct?
   A. A only   **C.** Both A and B
   B. B only   **D.** Neither A nor B

5. *Technician A* says while detecting type A misfires, the monitor checks cylinder misfiring over a 500-rpm period.
   Technician B says while detecting type B misfires, the monitor checks cylinder misfiring over a 1,000-rpm period.
   Who is correct?
   A. A only   **C.** Both A and B
   B. B only   **D.** Neither A nor B

6. While discussing the EGR system and monitor:
   Technician A says two hoses are connected from the delta pressure feedback EGR (DPFE) sensor to an orifice under the EGR valve.
   Technician B says when the EGR valve is open, the pressure should be the same on both sides of the orifice under the EGR valve.
   Who is correct?
   A. A only   **C.** Both A and B
   B. B only   **D.** Neither A nor B

7. A technician is using a DSO to diagnose an OBD-II system. The downstream HO_2S has the same voltage waveform as the upstream HO_2S. *Technician A* says the catalytic converter is defective.
   Technician B says the air-fuel ratio is too rich.
   Who is correct?
   A. A only   **C.** Both A and B
   B. B only   **D.** Neither A nor B

8. *Technician A* says in an enhanced EVAP system there is no leak detection pump.
   Technician B says in a non-enhanced EVAP system there may be a leak detection pump.
   Who is correct?
   A. A only   **C.** Both A and B
   B. B only   **D.** Neither A nor B

9. *Technician A* says a drive cycle is a specific driving method to verify a symptom or the repair for the symptom.
Technician B says it takes three good trips within one drive cycle to extinguish the MIL.
Who is correct?
   A. A only
   B. B only
   C. Both A and B
   D. Neither A nor B

10. *Technician A* says the purpose of the warm-up cycle is to turn off the MIL.
Technician B says the purpose of a trip is to erase the fault code.
Who is correct?
   A. A only
   B. B only
   C. Both A and B
   D. Neither A nor B

I/M 240 Testing Programs

Upon completion and review of this chapter, you should be able to:

- ❑ Describe the purpose of the I/M program.

- ❑ Define what the I/M 240 drive cycle is.

- ❑ Explain the difference between basic and enhanced I/M testing.

- ❑ Explain the differences between the two-speed idle test and the loaded mode test.

- ❑ Describe the equipment required to perform an I/M 240 test.

- ❑ Explain the purpose of the EVAP purge flow test and how it is conducted.

- ❑ Explain the purpose of the EVAP pressure test and how it is conducted.

- ❑ Describe the Acceleration Test Mode (ASM).

- ❑ Explain the features of the California model.

Introduction

In 1990, Congress amended the Clean Air Act with revisions requiring areas that did not meet national ambient air quality standards (NAAQS) to implement either basic or "enhanced" vehicle I/M emissions testing programs depending upon the severity of the area's air quality problem. The act also required that metropolitan areas with populations of more than 100,000 implement enhanced I/M emissions testing regardless of their air quality designation. The EPA, in turn, was required to develop standards and procedures for emissions testing.

In response to the Clean Air Act of 1990, the EPA mandated states to adopt **I/M 240** testing of automotive emissions. Though many states were using a simple idle test to check emissions, those states that had non-attainment areas that did not meet air quality standards were required by the EPA to adopt some type of enhanced emissions testing to catch loaded mode emission problems (Figure 9-1). Although emissions testing is not mandated nationwide, states that elect to implement I/M 240 receive EPA credit in the form of additional funding for their highways.

The I/M 240 requirement specified loaded mode testing for measuring transient emissions on a special dynamometer, as well as checking NO_x emissions and doing an evaporative system purge and pressure test. The I/M 240 test is based on procedures the EPA had developed for certifying new vehicle emission compliance.

Controversy arose around I/M 240 since it specified centralized testing. The EPA believed that the use of "test only" facilities that were administered by independent contractors would eliminate any conflicts of interest (fraud) in shops that both test and repair vehicles. However, in 1995, due to objections from some states, the National Highway System Designation Act was passed. The act included provisions that specifically barred the EPA from mandating I/M 240 exclusively for enhanced emissions testing. Therefore, the EPA was forced to adopt a more flexible posture toward alternative I/M test programs. States are still required to meet air quality standards, but now have a much wider range of options for meeting those standards. The EPA decided it would allow states to choose whatever enhanced test procedure works best for their situation, be it I/M 240, **Acceleration Simulation Mode (ASM)**, or some other loaded test procedure that also includes NO_x and evaporative emissions. As a result of these changes in the requirements, there are different emission testing procedures done in different states. These include basic and enhanced I/M 240, ASM, and **NYTEST239**. This chapter will cover the I/M 240 test procedure since the others are similar.

I/M 240 is an inspection/maintenance vehicle emissions testing program for improving air quality in areas that fail to meet the federal government's ambient air quality standards. The test is similar to the Federal Test Procedure (FTP) that automakers use to certify new vehicle emissions.

Acceleration Simulation Mode (ASM) test procedure usually applies a 50 percent load at 15 mph and a 25 percent load at 25 mph and uses a less expensive emissions analyzer.

NYTEST239 test procedure is a hybrid version of ASM and I/M 240 testing that includes transient mode testing on a dynamometer but uses a less costly exhaust sampling system.

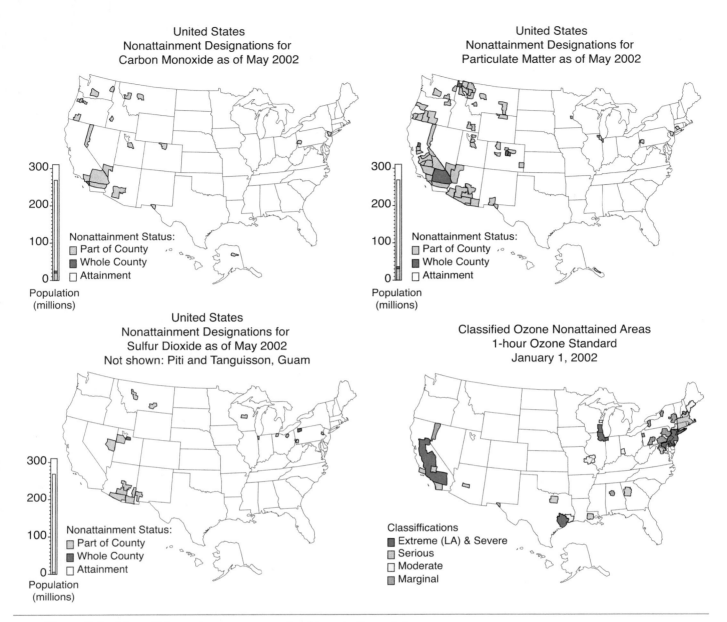

United States Nonattainment Designations for Carbon Monoxide as of May 2002

Nonattainment Status:
☐ Part of County
■ Whole County
☐ Attainment

Population (millions)

United States Nonattainment Designations for Particulate Matter as of May 2002

Nonattainment Status:
☐ Part of County
■ Whole County
☐ Attainment

Population (millions)

United States Nonattainment Designations for Sulfur Dioxide as of May 2002
Not shown: Piti and Tanguisson, Guam

Nonattainment Status:
☐ Part of County
■ Whole County
☐ Attainment

Population (millions)

Classified Ozone Nonattained Areas 1-hour Ozone Standard January 1, 2002

Classifications
■ Extreme (LA) & Severe
☐ Serious
☐ Moderate
▨ Marginal

Figure 9-1   Non-attainment areas cover parts of twenty-three states.

BIT OF HISTORY

Chicago and Cincinnati became the first metropolitan American cities to pass laws that declared excessive smoke emissions a public nuisance. The year was 1881.

Classifications of I/M Programs

The EPA provides each test area with a basic guideline to assist them in developing their particular I/M programs. Each area can adopt the test(s) that are the most appropriate for their level of attainment. Four different types of program can be used:

1. Visual tampering inspections
2. Basic I/M tests.
   a. Two-speed idle test
   b. Loaded mode test
   c. May include evaporative (pressure and purge) tests
3. Enhanced I/M tests
   a. IM 240 transient test
   b. Evaporative (pressure and purge) system check
4. Other Tests
   a. Remote sensing
   b. Acceleration Simulation Mode (ASM). Random roadside visual inspection

Visual Tampering Inspections

Many states may require test vehicles to undergo a visual inspection prior to being emissions tested. Some states will not require a visual inspection if an I/M 240 program is being enforced and includes functional testing of emissions control components.

In a visual inspection, certain systems and components are checked to assure that they are not missing, disconnected, or otherwise tampered with. Visual inspections involve ensuring that "obvious" vacuum and electrical components are connected. If any system or component fails the visual inspection, the vehicle will not receive an emissions test until all failing conditions are corrected.

Some of the items that states include in their tampering inspections include:

- The fuel filler restrictor
- The gas cap
- The PCV valve
- The secondary air system
- The EGR system
- The evaporative emissions system
- The exhaust system
- The MIL indicator
- Quick test for leaded fuel
- Vacuum lines

If any tampering has occurred, the tampered condition must be corrected and waiver amounts do not apply.

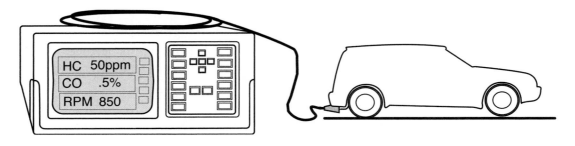

Figure 9-2 The basic I/M test uses a two-gas or four-gas emission analyzer.

Basic I/M Tests

The **basic program** is usually conducted at repair shops rather than through a state-run program. These "decentralized" sites are able to correct any failures that are uncovered during testing. This I/M exhaust test reads HC and CO emissions concentrations of vehicles at two idle speeds (Figure 9-2). The test is performed by holding the engine speed at 2,500 rpm and taking the emissions readings for 30 to 90 seconds. The engine is then allowed to return to idle (must be below 1,100 rpm) for 30 to 90 seconds. At this time, a second set of readings is taken at idle.

Some test areas only take emissions readings during the idle portion of the test. The 2,500-rpm portion of the procedure is used to warm up (or precondition) the vehicle for the idle test.

To pass, the emissions concentration must be below the standard set by the state for that particular test area. The pass/fail standards can vary depending on such factors as the age and technology of the vehicle, whether it is a car or a truck, and so on. When a customer has an emissions test, the passing standards will be printed on their emissions result slip.

Another version of the basic I/M test is the **loaded mode test**. Emissions readings are taken while the vehicle is driven at a constant speed in a "loaded" condition on the dynamometer (Figure 9-3). The load varies based upon vehicle speed. In order to produce an equivalent amount of power to overcome the dynameter's load, vehicles with smaller engines are driven slower on the

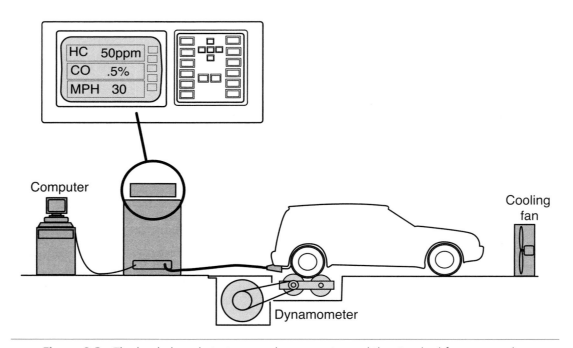

Figure 9-3 The loaded mode test uses a dynamometer and the standard four-gas analyzer.

TABLE 9-1   LOADED MODE TEST SPEED VERSUS ENGINE CYLINDERS

Cylinders	Speed on Dyno (±1 mph)
4 or fewer	24 mph
5 or 6	30 mph
7 or more	34 mph

dynamometer, while vehicles equipped with larger engines are driven faster (Table 9-1). Then, a second set of readings is taken when the vehicle is allowed to return to idle.

Some areas only require the vehicle to have passing values at idle. The load caused by the dynamometer is used only as a means to precondition the vehicle for the idle test.

Enhanced I/M Tests (I/M 240 Test)

An **enhanced test** program requires a more in-depth inspection of the vehicle's emissions. The vehicle is connected to a dynamometer that simulates driving conditions so that the evaporative emissions system and fuel cap may be accurately tested. The test also measures HC, CO, CO_2, and NO_x exhaust emissions. In addition, the evaporative purge system is monitored and a pressure check of the gas cap is conducted. Enhanced testing takes place through a "centralized" network of state-run facilities. These sites are not equipped for vehicle repair following the test.

The EPA's I/M 240 exhaust test actually requires the vehicle to go through a **transient** operation. The driving cycle takes 240 seconds (4 minutes) to complete, thus the name of this test. The actual driving cycle is similar to the first two hills of the FTP used by vehicle manufacturers (Figure 9-4).

One important aspect of the I/M 240 test is that it does more than simply read emissions concentrations. The I/M 240 test procedure takes into account the *volume* of the exhaust being produced and then calculates the *mass* (in grams per mile) of emissions being produced. This measurement is needed since different engine sizes will produce different volumes of exhaust. A large engine produces more volume of exhaust per revolution than a smaller engine (Table 9-2).

Shop Manual
Chapter 9,
page 361

The **enhanced test** uses a transient drive cycle to test for HC, CO, CO_2, and HC, as well as EVAP performance.

Transient means that the vehicle will undergo modes where it accelerates and decelerates following a specific driving cycle.

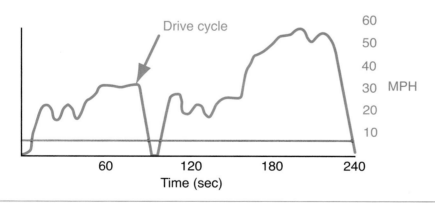

Figure 9-4   The I/M drive cycle test.

TABLE 9-2   ENGINE SIZE DETERMINES THE EXHAUST GAS VOLUME

Engine Displacement	Exhaust Gas Concentration	Exhaust Gas Volume (cu. in. per engine rev.)
2.0L (122 cu. in.)	1.0%	0.6
5.9L (360 cu. in.)	1.0%	1.8
8.0L (488 cu. in.)	1.0%	2.4

Even if their tailpipe readings are the same, the large engine will be producing a greater amount of pollution because it produces a greater volume of exhaust than does the small car. Therefore, it yields a higher emissions reading if we calculate the actual mass of the exhaust being produced.

Shop Manual
Chapter 9,
page 363

Understanding Drive Trace Analysis.   Referring to Figure 9-4 of the 240-second drive trace pattern, the pattern is broken down as follows:

1. First acceleration (0 to 23 mph)
2. Rolling deceleration and acceleration (23 to 17 mph)
3. Rolling light acceleration (17 to 32 mph)
4. Part throttle cruise (32 mph)
5. First deceleration (32 to 0 mph)
6. Second acceleration (0 to 26 mph)
7. Rolling deceleration and acceleration (26 to 23 mph)
8. Part throttle cruise (23 mph)
9. Rolling moderate acceleration (23 to 48 mph)
10. Rolling high-speed acceleration (48 to 53 mph)
11. High-speed deceleration and acceleration (53 to 50 mph)
12. Final deceleration (50 to 0 mph)

A key feature of the I/M 240 test procedure is the use of a special inertia dynamometer (Figure 9-5) to simulate vehicle loads at various speeds during a 240-second drive cycle that includes acceleration, deceleration, and cruise modes. The dynamometer used in an I/M 240 test must be capable of producing a very specific load on each test vehicle. It would not be a fair test to force a small car to drive with the same load placed on a large truck.

The I/M 240 test also measures (NO_x) in the exhaust and uses a special constant volume sampling collection system (Figure 9-6) rather than a partial gas sampling system to analyze the exhaust. The equipment (Figure 9-7) monitors the vehicle's emissions for the duration of the test, then calculates the average emissions and displays the results in grams per mile for NO_x, CO, and HC.

Figure 9-5   Inertia dynamometers are used to simulate actual driving conditions.

Figure 9-6   The constant volume sampling (CVS) hose will cover the entire tailpipe opening to capture all of the exhaust gases. (Courtesy of Envirotest Systems, Corp.)

Figure 9-7 The constant volume sampling (CVS) system determines the volume of exhaust gas then measures the pollutants in grams per mile. (Courtesy of Envirotest Systems, Corp.)

Evaporative System Canister Purge Test. The I/M 240 test also requires an **EVAP purge flow test** while the engine is running to measure the flow rate of the charcoal canister purge valve and an engine off pressure test of the evaporative emission control system to check for fuel vapor leaks.

The purge test verifies that the EVAP system is operating properly. During a purge test, the purge flow rate from the vehicle's evaporative canister is measured (Figure 9-8) using a comput-

Shop Manual
Chapter 9,
page 369

The **EVAP purge flow test** measures the performance of the EVAP to move vapors into the intake manifold.

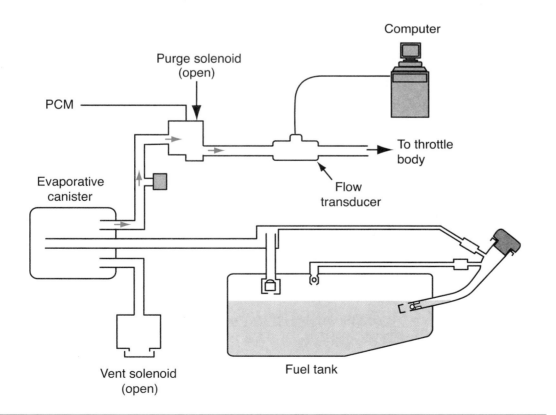

Figure 9-8 The EVAP purge flow is measured by a computer during the drive cyle test.

erized flow meter installed into the EVAP purge line. For many vehicles, the greatest amount of canister purge operation occurs when the vehicle is being driven under a transient load. Therefore, purge tests are usually conducted during an I/M 240 test.

During the I/M 240 driving cycle, the flow transducer measures the amount of purge flow from the canister and sends this information to the computer. The present standard for all test vehicles is that it must flow at least 1 liter during the 4-minute I/M 240 test. Any properly operating vehicle will easily flow many times this amount.

Evaporative System Pressure Test. The **EVAP pressure test** can be run either before or after the exhaust emissions test. In this test (Figure 9-9), the evaporative system is pressurized by a gas (such as nitrogen) until it reaches a pressure of 14 inches water column. Then, the pressure is monitored for 2 minutes. If it bleeds down to less than 8 inches water column, the vehicle fails.

The Constant Volume Sampler (CVS)

A very sophisticated device is required to measure the exhaust volume. This specialized device is called a Constant Volume Sampler (CVS). The CVS dilutes the exhaust with fresh air and measures the flow rate of the mixture. Mass emissions (for each second) are calculated by multiplying this flow rate by the measured concentration of pollutants in the mixture. To arrive at the official test value in grams per mile, the mass emissions for each second are added together; this sum is then divided by the distance (number of miles) traveled over the 240-second test cycle.

The fresh air dilution is vital because it preserves the integrity of the sample and because it protects the emission analyzers from high concentrations of water vapor produced by the vehicle. The dilution process also allows the measurement system to accommodate the differences in exhaust flow between small engines and large engines while measuring the true amount of emissions from each type of engine.

The computer used for an I/M 240 test must generate the driver's trace and project it on to a video screen for the driver to follow. It must take all of the exhaust gas information (read on

Shop Manual
Chapter 9,
page 371

The purpose of the **EVAP pressure test** is to be sure that there are no leaks from the fuel tank to the evaporative canister where fuel vapors could escape into the atmosphere.

Shop Manual
Chapter 9,
page 365

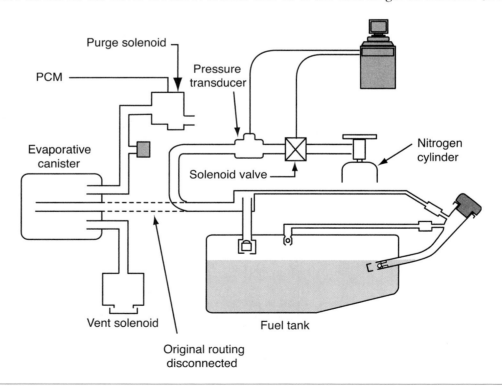

Figure 9-9 Pressure testing the EVAP system for leaks.

high-tech analyzers) and gas volume information from the CVS and continuously calculate the mass emissions. Like the computer in the Basic I/M program, it must determine if the vehicle passes or fails and print this information out for the vehicle owner. Finally, it must store this test information for further analysis.

Other Tests

Other tests that can be used include remote sensing, acceleration simulation mode, random roadside visual inspection, and the California method.

Remote Sensing.   Remote sensing (Figure 9-10) is presently used in some states. For example, Colorado began using this system in 2003. Remote sensing utilizes advanced sensing techniques to measure the pollutant levels of a moving vehicle. Roadside equipment shoots a beam of infrared light through the exhaust stream of a passing vehicle. The reflected beam is analyzed to determine the pollutant levels while the vehicle's license plate is videotaped. This technology allows a high level of inspection and may be useful in identifying heavy polluters.

Acceleration Simulation Mode (ASM).   The ASM test utilizes a dynamometer to apply a very heavy load to the test vehicle during a steady state driving cycle. HC, CO, and NO_x emissions are sampled with a "garage type" analyzer. The emissions readings are in concentrations.

The load that is applied to the test vehicle is based on the load that is applied during the acceleration phase of the second hill of an FTP driving cycle. The acceleration rate is 3.3 mph/second.

For example, in one version of ASM known as ASM 25/25, the vehicle is driven at 25 mph against a load of 25 percent of the horsepower required to simulate FTP acceleration. This method produces relatively high HC and CO emissions. In ASM 50/15, the vehicle is driven at 15 mph against a load of 50 percent of the horsepower required to simulate FTP acceleration. This method produces relatively high levels of NO_x.

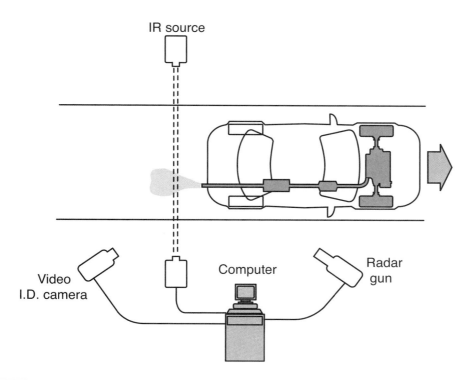

Figure 9-10   Remote sensing for emissions.

The advantage of ASM is that it produces a rigorous test environment without the need for costly CVS. Its chief disadvantage is that the results do not correlate well with the FTP.

Random Roadside Visual Inspection.   This is a spot check of vehicles at roadside inspection lanes. At a roadside inspection, randomly selected vehicles are visually checked to see that all emissions control devices are in place and have not been tampered with.

The California Model

AUTHOR'S NOTE: This section is provided since some states have adopted a hybrid emissions program. This example of the program used in California provides a case study of the hybrid programs.

The Federal Clean Air Act amendments of 1990 originally required that enhanced vehicle emissions testing programs (I/M 240) be performed at a centralized, state-run facility. This mandate would have eliminated all privately-owned "Smog Check" stations. In California, the Smog Check industry produces over 480 million dollars per year. Consequently, California negotiated an alternative plan with the EPA.

The new emissions testing program enhancements were enacted into law in 1994 and approved by the EPA on September 26, 1996. The Smog Check II program has been further refined by subsequent legislation. Those urbanized regions that failed to meet federal or state air quality standards for ozone and CO as of June 8, 1998, were required to have enhanced emissions testing on a dynamometer. The tests included oxides of nitrogen (NO_x) for the first time.

Areas that require enhanced testing use equipment meeting BAR97 specifications, a dynamometer, and a digital storage oscilloscope (DSO). Emission inspection technicians are also required to have an "Advanced Emissions Specialist" (EA) license.

Dynamometer use is limited with some vehicle types. For example, some all-wheel drive and four-wheel drive vehicles can not be driven on a two-wheel dynamometer. Also, some vehicles with traction control systems cannot be driven on a dynamometer. If a vehicle cannot be tested on the dynamometer, it is exempt from the enhanced emissions test. However, the vehicle is required to pass a basic two-speed idle test.

In areas that do not have high levels of ozone and CO, enhanced emissions testing are not required. These "basic areas" use the existing biennial two-speed idle test at licensed test-and-repair stations. In addition, a basic emissions test is required for vehicles that are being sold or are being registered in California for the first time.

Emission test results are transmitted electronically to the California Department of Motor Vehicles. Vehicles that fail the emission test must be repaired. All repairs must be verified and emissions certified at a licensed **Test Only station**.

If a vehicle fails a Smog Check inspection, but the owner does not have the financial resources to make the needed repairs, the owner may be granted a "repair cost waiver." The owner must pay at least $450 in emissions-related repairs at a licensed repair station before the waiver is granted. This waiver is good for two years, and only one waiver will be issued while the motorist owns the vehicle. There is also an "economic hardship extension" for qualified low-income vehicle owners. To obtain an extension, motorists must spend $250 on emissions-related repairs at a licensed Smog Check station, or have an estimate from a licensed station indicating that the repairs would cost more than $250. Neither of the waivers can be issued if the emission system has been tampered with, if the vehicle is being registered for the first time in California, if the vehicle is being sold, or if the vehicle owner was issued a waiver or extension in the previous Smog Check inspection.

A **Test Only station** does not repair the vehicle, it only tests the vehicle.

Summary

❏ In 1990, Congress amended the Clean Air Act with revisions requiring areas that did not meet national ambient air quality standards (NAAQS) to implement either basic or "enhanced" vehicle I/M emissions testing programs depending upon the severity of the area's air quality problem.

❏ In response to the Clean Air Act of 1990, the EPA mandated states to adopt I/M 240 testing of automotive emissions. Though many states were using a simple idle test to check emissions, those states that had non-attainment areas that did not meet air quality standards were required by the EPA to adopt some type of enhanced emissions testing to catch loaded mode emission problems.

❏ Many states may require test vehicles to undergo a visual inspection prior to being emissions tested. Some states will not require a visual inspection if an I/M 240 program is being enforced and includes functional testing of emissions control components.

❏ For the most part, basic programs are conducted at repair shops rather than through a state-run program.

❏ The basic test reads HC and CO emissions concentrations of vehicles at two idle speeds.

❏ Another version of the basic I/M test is the loaded mode test. Emissions readings are taken while the vehicle is driven at a constant speed in a "loaded" condition on the dynamometer.

❏ An enhanced program requires a more in-depth inspection of the vehicle's emissions. The vehicle is connected to a dynamometer that simulates driving conditions so that the evaporative emissions system and fuel cap may be accurately tested. The test also measures HC, CO, CO_2, and NO_x exhaust emissions. In addition, the evaporative purge system is monitored and a pressure check of the gas cap is conducted.

❏ The driving cycle takes 240 seconds (4 minutes) to complete. The actual driving cycle is similar to the first two hills of the Federal Test Procedure (FTP) used by vehicle manufacturers.

❏ The I/M 240 test procedure takes into account the volume of the exhaust being produced and then calculates the mass (in grams per mile) of emissions being produced.

❏ A key feature of the I/M 240 test procedure is the use of a special inertia dynamometer to simulate vehicle loads at various speeds during a 240-second drive cycle that includes acceleration, deceleration, and cruise modes.

❏ The I/M 240 test also requires an evaporative purge flow test while the engine is running to measure the flow rate of the charcoal canister purge valve and an engine off pressure test of the evaporative emission control system to check for fuel vapor leaks.

❏ The ASM test utilizes a dynamometer to apply a very heavy load to the test vehicle during a steady state driving cycle. HC, CO, and NO_x emissions are sampled with a "garage type" analyzer. The emissions readings are in concentrations.

❏ In California, areas that required enhanced testing equipment meeting BAR97 specifications require a dynamometer. Test station technicians are also required to have a Digital Storage Oscilloscope (DSO) to help diagnose emissions system problems and an "Advanced Emissions Specialist" (EA) license.

❏ Repairs can be performed at "Gold Shield Guaranteed Repair" stations that are licensed smog check facilities that meet high performance standards and guarantee the repairs they make on gross polluters.

Terms to Know

Acceleration Simulation Mode (ASM)

Basic program

Enhanced test

EVAP pressure test

EVAP purge flow test

I/M 240

Loaded mode test

NYTEST239

Test Only station

Transient

Review Questions

Short Answer Essay

1. Define what the I/M 240 drive cycle is.
2. Describe the two-speed idle test procedure.
3. Describe the loaded mode test procedure.
4. Explain the enhanced I/M 240 test.
5. Describe the equipment required to perform an I/M 240 test.
6. Explain how the EVAP purge flow test is performed.
7. Describe how the EVAP pressure test is performed.
8. What is the Acceleration Test Mode (ASM)?
9. Why does the I/M 240 test measure pollutants in grams per mile?
10. Besides basic and enhanced I/M testing, what other methods can be used?

Fill In The Blank

1. The idle-speed test measures tailpipe concentrations of _____ and _____.
2. An _____ emissions test is a transient speed test that measures mass emissions, visually inspects tampering, and tests the evaporative emissions control system.
3. A mass emissions test expresses the total weight of contaminants in _____.
4. For the most part, _____ programs are conducted at repair shops rather than through a state-run program.
5. The _____ _____ _____ takes emission readings while the vehicle is driven at a constant speed in a "loaded" condition on the dynamometer.
6. The enhanced I/M test measures _____, _____, _____ and _____ exhaust emissions.
7. The I/M drive cycle is similar to the first two hills of the _____ used by vehicle manufacturers.
8. The I/M 240 test procedure takes into account the _____ of the exhaust being produced and then calculates the _____ of emissions being produced.
9. A key feature of the I/M 240 test procedure is the use of a special _____ dynamometer.
10. The device used to measure the exhaust volume in an I/M 240 test is called a(n) _____ _____ _____.

Multiple Choice

1. The IM 240 emissions test measures which of the following?
   A. Mass emissions in parts per million (ppm)
   B. Emissions concentrations in percent (%)
   C. Mass emissions in grams-per-mile
   D. Emissions concentrations in parts per million (ppm)

2. Which test utilizes a dynamometer to perform a steady state drive cycle to measure HC, CO, and NO_x emissions?
   A. ASM.
   B. Loaded mode test.
   C. Transient test.
   D. Two-speed idle test.

3. During an I/M 240 test, what does a "flow transducer" measure?
   A. Exhaust volume.
   B. Airflow to the catalyst.
   C. Rate of purge from the evaporative.
   D. Data flow to the PCM canister.

4. Oxides of nitrogen (NO_x) are evaluated in which of the following tests?
   A. Basic 11M testing.
   B. Loaded mode testing.
   C. Four-gas testing.
   D. I/M 240 testing.

5. Which of the following would NOT be included in a visual inspection?
   A. The fuel filler restrictor.
   B. The secondary air system.
   C. Proper injector installation.
   D. The EGR system.

6. The basic test checks the emission at what two speeds?
   A. 1,100 and WOT.
   B. 3,500 and 1,100.
   C. 2,500 and 3,500.
   D. 2,500 and idle.

7. *Technician A* says that in a given drive cycle, all engines will produce the same volume of exhaust gas.
   Technician B says the I/M 240 test measures in concentration of pollutants.
   Who is correct?
   A. A only   C. Both A and B
   B. B only   D. Neither A nor B

8. Which of the following would be part of the I/M 240 drive cycle?
   A. High-speed deceleration and acceleration (53 to 50 mph).
   B. Part throttle cruise (65 mph).
   C. Heavy acceleration (from 23 to 65 mph).
   D. Rolling high-speed acceleration (15 to 63 mph).

9. *Technician A* says the device used to measure exhaust volume is a random volume sampler.
   Technician B says the device is called a constant volume sampler.
   Who is correct?
   A. A only   C. Both A and B
   B. B only   D. Neither A nor B

10. *Technician A* says the EVAP pressure test measures the flow of vapors by applying a constant pressure to the fuel tank.
    Technician B says the EVAP pressure test is run during the drive cycle of an I/M 240 test.
    Who is correct?
    A. A only   C. Both A and B
    B. B only   D. Neither A nor B

Alternate Fuel Vehicles

Upon completion and review of this chapter, you should be able to:

- Explain the purpose of alternate fuel vehicles.
- Describe the flow of the fuel delivery system of a liquid propane injection vehicle.
- List and describe the function of the propane fuel system components.
- List and describe the function of typical CNG fuel system components.
- Explain the purpose of propane and CNG tank codes and the regulations concerning these tanks.
- Explain the operation of the flex fuel vehicle.
- Explain the basic operation of electric vehicles.

- Describe the typical operation of a hybrid vehicle.
- Explain the difference between parallel and series hybrids.
- Explain the purpose of regenerative braking.
- Describe how a proton exchange membrane produces electricity in a fuel cell system.
- List and describe the different fuels that can be used in a fuel cell system.
- Describe the purpose of the reformer.
- Explain how different types of reformers operate.

Introduction

Due to the increase in regulations concerning emissions and the public's desire to becoming less dependent on foreign oil, most major automotive manufacturers have developed alternative fuel or alternate power vehicles. This chapter explores many of the most common methods used in these fields. This chapter includes a study on alternate fuel vehicles that run on propane, compressed natural gas, and flex fuel. A study of common hybrid systems is also included. The final section of this chapter covers fuel cell theories and some of the methods that manufacturers are using to approach this alternate power source.

Alternate Gaseous Fuels

Recent federal energy and environmental legislation affects the fuel and vehicle choices of some U.S. fleet operators. The Energy Policy Act of 1992 (EPACT) requires certain federal and state fleet operators and alternative fuel providers to acquire alternative fuel vehicles (AFVs). Some private, municipal, and other fleet operators may also be required to obtain AFVs in the future. The Clean Air Act introduced the Clean Fuel Fleet program, which requires that some fleet operators in cities with the greatest air pollution acquire vehicles that meet special clean fuel vehicle emission standards. State and local requirements also affect the fuel and vehicle choices of fleet operators. The requirements vary from area to area and depend on the specific makeup and location of a fleet.

Original equipment manufacturer (OEM) AFVs are designed specifically to use an alternative fuel, such as propane, CNG, methanol, or ethanol. The engine, emissions, and performance of the vehicle are optimized for that fuel. Such a vehicle features a unique engine, catalyst, and single- or multi-point fuel-injection system that meters fuel more precisely. The on-board computer is designed specifically for the fuel. In addition, the suspension system and shocks may be designed to withstand the added weight of propane or CNG tanks. Often, the fuel system parts are designed and installed specifically for that type of vehicle.

Because the supply of OEM vehicles is limited, some fleet operators may choose to use **converted vehicles** as a way to meet vehicle acquisition requirements. Conversions serve as a transition to the time when more AFVs become available for public sale. Compressed natural gas and propane are the two most common types of fuel for such vehicles.

Propane

Liquefied petroleum gas (LPG), or propane, is a byproduct of natural gas processing and crude oil refining. About 65 percent of the propane supply is extracted from natural gas production, while 35 percent is a byproduct of crude oil refining. The United States is one of the world's largest producers of propane. Almost 92 percent of the U.S. demand for propane gas is supplied domestically.

The automotive propane fuel system is a completely closed system that contains a supply of pressurized LPG. The gas is stored in liquefied form in tanks that are pressurized to around 160 psi. The fuel will remain a liquid as long as it is under pressure.

BIT OF HISTORY

Propane as an engine fuel has been in existence since the 1920s. Today, some 3,500,000 vehicles use propane as their engine fuel worldwide. Approximately 350,000 vehicles in the United States use propane, making it the third most popular fuel behind gasoline and diesel fuel.

Propane gas has the potential to produce less CO and HC than gasoline does. It is clean burning, producing virtually no particulate or sulfur emissions. CARB and U.S. EPA tests have shown that emissions of HC and CO from propane-fueled engines are consistently reduced 50 to 80 percent, and NOₓ emissions are reduced 5 to 30 percent. In addition, since propane is contained in a sealed system, there are no evaporative emissions.

On a per-unit weight basis, liquid propane contains about 5 percent more energy than gasoline; however, the density is 30 percent less. The net result is that propane contains less energy than gasoline for the same quantity. Propane has an octane of about 110. With propane, a greater quantity of fuel is required to travel the same distance. However, because of its gaseous form, the engine burns the fuel more efficiently, which helps offset the extra fuel required. Since propane has about 80 percent of the energy of gasoline on a volumetric basis, fuel consumption is slightly higher with propane as compared to gasoline. However, there are significant maintenance reductions with engines running on propane because of its clean burning characteristics.

Propane is readily available with more than 15,000 refueling outlets in North America, 5,000 refueling stations across Canada, and over 10,000 in the United States of America. Propane, composed primarily of propane and butanes with smaller amounts of propylene and butylenes, is supplied in four grades of different composition. **HD-5** is the only grade appropriate for automotive applications.

LPG Liquid Fuel-Injection Systems. In recent years there has been a decline in propane fuel use. This is partly due to the automobile industry's change from carburetor or throttle body fuel systems to multi-port fuel-injection systems. The propane conversion companies failed to keep up with changing technologies. Today there are a few companies that provide injection systems for converting from gasoline to propane. A few years ago there was a return of the automotive manufacturers to the propane vehicle market. Several automotive manufacturers offered different models with propane fuel systems. Ford offered a number of models through its

Figure 10-1 Liquid propane fuel-injected system tanks.

Qualified Vehicle Modifier (QVM) program with factory-certified conversions. The Chrysler group of DaimlerChrysler Motors offered a completely engineered propane vehicle (not a conversion) including liquid fuel injection and an integrated fuel tank that conforms to the body design. However, even interest by the automotive manufacturers has declined, and the 2004 model year may only see the Ford F150 series trucks being offered with propane.

The liquid fuel propane injection systems operate similar to gasoline fuel injection. The propane electronic fuel injection system uses a special LPG tank (Figure 10-1). The tanks also contain an internal pump (Figure 10-2). The fuel hoses perform two functions

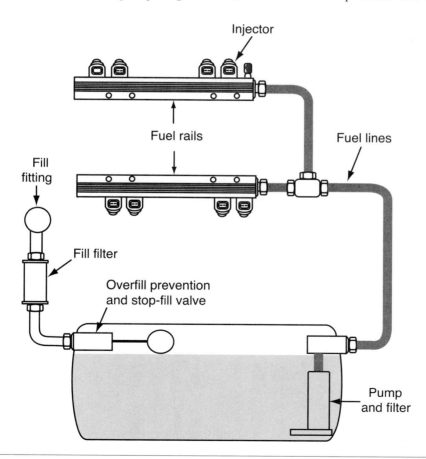

Figure 10-2 Fuel delivery components of the liquid propane injection system.

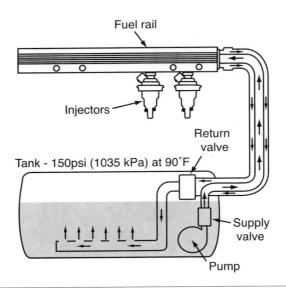

Figure 10-3 A single hose performs the functions of delivery and return.

(Figure 10-3). First, they deliver the liquid fuel to the fuel rail through the center section of the hose. Second, they are the return for vapor and unused fuel back to the tank through the outer section of the hose.

The system is similar enough to gasoline fuel injection that the engine can be designed as a **bi-fuel system** (Figure 10-4). A fuel selector switch allows the driver to change between gasoline and propane fuel. In addition, if the fuel level of the propane tank drops to a predetermined point, the PCM will automatically switch to the gasoline system.

A vehicle equipped with a **bi-fuel system** has two separate fuel systems with the capability to switch from one to the other.

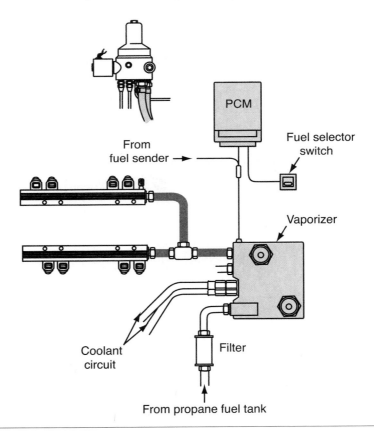

Figure 10-4 LPG bi-fuel system.

Compressed Natural Gas (CNG)

CNG offers several advantages over gasoline and other alternative fuels. Some of these advantages include:

- Lower CO emissions
- Lower HC emissions
- Lower NO_x emissions
- Cleaner burning with fewer deposits
- Higher octane, eliminating engine knock

Shop Manual
Chapter 10,
page 390

Natural gas is composed mainly of methane and has an octane rating between 105 and 130. Unlike LPG, natural gas remains a gas even when compressed to 3,000 to 3,600 psi. Natural gas is lighter than air and when released rises and dissipates into the atmosphere.

CNG also offers the advantage of abundance. It is a plentiful resource, and distribution networks are already in place making it more economical than many other alternative fuels (Figure 10-5).

CNG has a lower British Thermal Unit (Btu) rating per cubic inch than gasoline. Due to its lower energy output, larger storage tanks are required to carry the fuel. As a result, traveling distance per tankfull is somewhat reduced from that of gasoline vehicles. There are a number of natural gas vehicles available now, including the Honda Civic GX, Dodge Caravan, Dodge B-Van, Ford Crown Victoria and Contour, most GM pickups, Chevrolet Cavalier, Toyota Camry, and the Volvo S70 and V70. Many of these models are also available in a bi-fuel configuration, which can run on either compressed natural gas or unleaded gasoline (Figure 10-6).

Natural gas vehicles (NGV) store their fuel in cylinders installed in the rear or under carriage of the vehicle. Inside the cylinders, the gas is under a high pressure between 3,000 and 3,600 psi. The flow of gas is through a fuel regulator located in the engine compartment and is then injected through a specially designed mixer, to achieve the correct air-to-gas mixture.

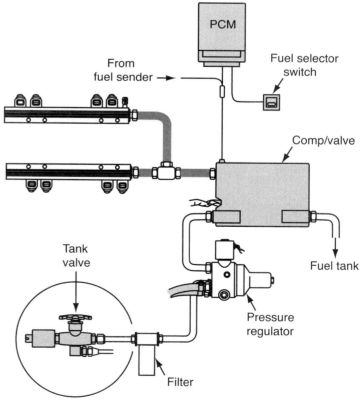

Figure 10-5 Compressed natural gas filling station.

Figure 10-6 CNG bi-fuel system.

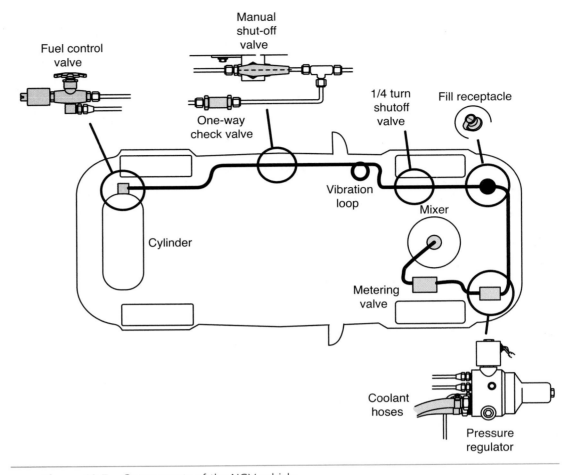

Figure 10-7 Components of the NGV vehicle.

Figure 10-7 shows the typical components of the NGV. The following discusses common components and their function.

Fuel Control Valves With Pressure Relief. Each tank on the CNG vehicle is equipped with its own manually operated fuel control valve (Figure 10-8). Turning the valve handle to the full

Shop Manual
Chapter 10,
page 391

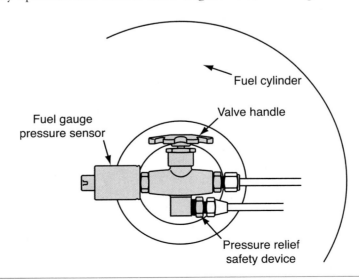

Figure 10-8 Typical fuel control valve.

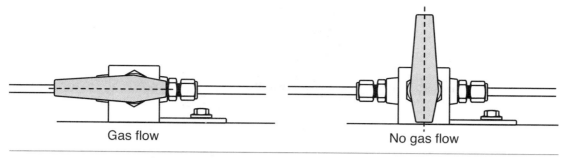

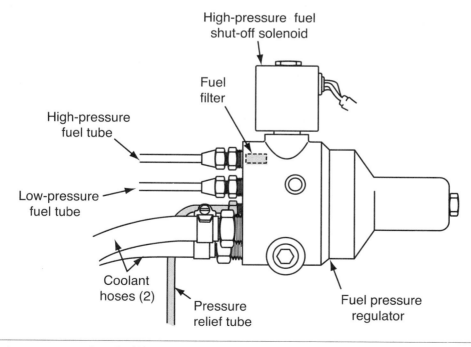

Gas flow

No gas flow

Figure 10-9 Typical manual valve operation.

clockwise position will stop gas flow. Turning the valve to the full counterclockwise position provides gas flow.

Each control valve is equipped with a pressure relief device for safety. The device releases excess cylinder pressure to the atmosphere if the temperature of the cylinder rises above approximately 217°F (103°C). When the relief device opens, the CNG from the entire system (all cylinders) is released into the atmosphere. The relief device is not reusable.

Quarter-Turn Manual Shut-Off Valve. A hand-operated manual shut-off valve is provided to stop fuel flow from all cylinders (Figure 10-9). When closed, the valve will prevent fuel from going to the delivery system. A "Manual Shut-Off Valve" label is affixed to the exterior body panel of the vehicle. It is located directly outboard from where the valve is located.

When the manual shut-off knob is turned parallel to the fuel line, the valve is open. When the valve is turned perpendicular to the fuel line, the valve is closed and fuel cannot reach the delivery system.

Fuel Pressure Regulator. The fuel pressure regulator reduces the fuel cylinder pressure to between 90 and 140 psi. The regulator is located downstream from the manual shut-off valve (Figure 10-10). The regulator contains a built-in pressure relief device and fuel filter. The pressure

Shop Manual
Chapter 10,
page 391

Shop Manual
Chapter 10,
page 397

High-pressure fuel
shut-off solenoid

Fuel
filter

High-pressure
fuel tube

Low-pressure
fuel tube

Coolant
hoses (2)

Pressure
relief tube

Fuel pressure
regulator

Figure 10-10 CNG fuel pressure regulator.

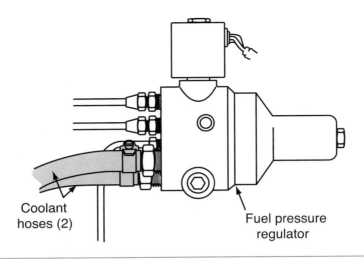

Coolant hoses (2)

Fuel pressure regulator

Figure 10-11 Engine coolant hoses routed to the regulator to prevent freezing.

relief device is a safety item that vents CNG to the atmosphere. It is on the regulator's low-pressure side and vents excess pressure above approximately 225 psi.

Expanding the fuel to a lower pressure has a thermodynamic cooling effect. To counteract this effect, engine coolant is routed to the regulator to prevent freezing (Figure 10-11). The hoses are tied into the vehicle's heater hoses in the engine compartment.

Shop Manual
Chapter 10,
page 401

High-Pressure Fuel Shut-Off Solenoid. The high-pressure fuel shut-off solenoid is an ON/OFF valve that electronically controls high-pressure gas flowing through the fuel pressure regulator. This solenoid is an integral part of the fuel pressure regulator and is controlled by the PCM through the high-pressure fuel shut-off solenoid relay.

Shop Manual
Chapter 10,
page 401

Low-Pressure Fuel Shut-Off Solenoid. The low-pressure fuel shut-off solenoid is used to close off the fuel supply to the fuel injector rail (Figure 10-12). The PCM controls this device through a

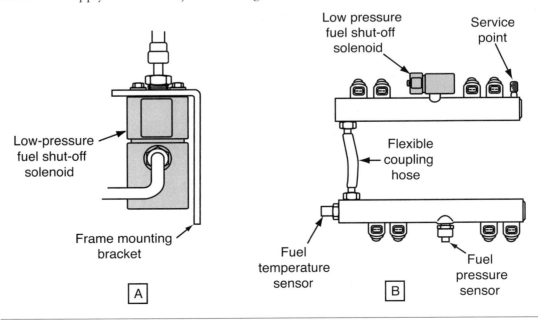

Low-pressure fuel shut-off solenoid

Frame mounting bracket

Low pressure fuel shut-off solenoid

Service point

Flexible coupling hose

Fuel temperature sensor

Fuel pressure sensor

A

B

Figure 10-12 The low-pressure fuel shut-off solenoid can be located in the fuel rail feed line (A), or on the fuel rail (B).

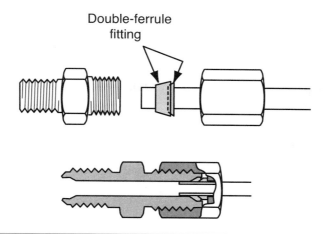

Figure 10-13 CNG fuel line fittings.

relay. When the ignition switch is turned OFF, the PCM de-energizes the relay, turning off the solenoid. When the solenoid is off, the fuel line is closed and CNG cannot enter the fuel rail.

Fuel Lines and Fittings.   High-pressure stainless steel seamless fuel tubes are used in the CNG system. Special stainless steel, double-ferrule, compression-type fittings are used on the high-pressure side of the system (Figure 10-13). All components and fuel tube connections from the inlet side of the pressure regulator to the fuel cylinders must use this type of fitting. Certain components on the high-pressure side require O-rings. Fittings on the low-pressure side of the system (outlet side of the pressure regulator and all downstream connections) use NPT pipe threads and 45-degree flared fittings.

Shop Manual
Chapter 10,
page 394

Filling Receptacle and Check Valves.   The fuel receptacle is mounted behind the fuel filler door (Figure 10-14). The receptacle contains an integral one-way check valve. An additional one-way check valve is installed in the fuel fill line between the fuel pressure regulator and the fuel fill receptacle (Figure 10-15). Both the receptacle check valve and the in-line check valve prevent fuel from escaping through the filling receptacle.

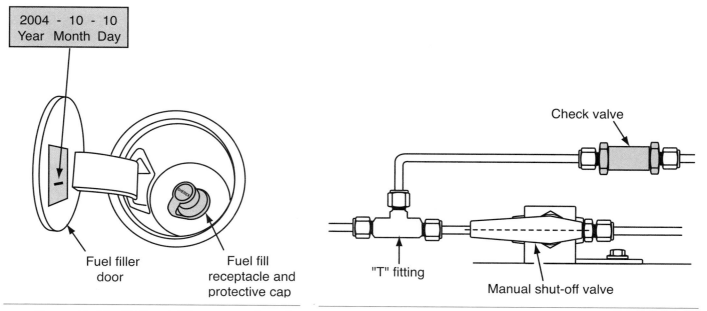

Figure 10-14   CNG tank filling receptacle and cylinder test date sticker.

Figure 10-15   One-way check valve.

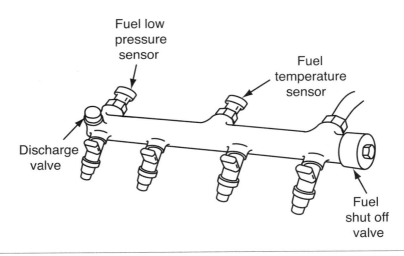

Figure 10-16   Fuel low-pressure sensor and fuel temperature sensor.

Shop Manual
Chapter 10,
page 399

Shop Manual
Chapter 10,
page 400

The Fuel Low-Pressure Sensor.   The fuel low-pressure sensor provides an input to the PCM based on the pressure of the CNG fuel at the fuel rail (Figure 10-16). The PCM uses this input, along with various other sensors, to calculate fuel injector timing.

Fuel Temperature Sensor.   The PCM uses input from the fuel temperature sensor to help calculate fuel injector timing (Figure 10-16).

CNG and Propane Cylinder Codes

CNG is stored in specially approved cylinders at pressures up to 3,600 psi. The cylinders can be made of steel, steel composite, or aluminum composite. Many are constructed of aluminum liners wrapped in fiberglass. The exterior of the cylinders is coated with an epoxy paint and clear polyurethane to provide additional environmental protection.

Government regulations have standardized the testing and use of CNG fuel tanks throughout the automotive industry.

The U.S. Department of Transportation (DOT) requires that each CNG fuel cylinder manufactured on or after March 27, 1995, be removed and reinspected every 3 years in accordance with Federal Motor Vehicle Safety Standard 304. The inspection must be performed by a qualified person in accordance with the cylinder manufacturer's established reinspection criteria and the appropriate Compressed Gas Association, Inc. guideline. Retest markings must be stamped on the cylinder neck or marked on a label securely affixed to the cylinder and must be overcoated with epoxy near the original test date. Reheat treatment or repair of rejected cylinders is not allowed. The fuel cylinder expires and must be removed from service 15 years from the date of manufacture. A label can also be attached on the fuel filler door stating the cylinder retest date and cylinder expiration date.

Canadian requirements state that the cylinder must be reinspected and hydrostatically retested every 3 years in accordance with the Canadian Standards Association (CSA) CAN/CSA-B339, as prescribed for TC-3FCM containers. Retest dates must be stamped on the exposed metallic surface of the cylinder neck or marked on a label securely affixed to the cylinder; and must be overcoated with epoxy near the original test date. Reheat treatment or repair of rejected cylinders is not allowed. The fuel cylinder expires and must be removed from service 15 years from the date of manufacture. A label can also be on the fuel filler door stating the first cylinder retest date and cylinder expiration date.

Because propane is not stored at such high pressures, the testing for propane cylinders is not as rigorous as that for CNG cylinders. Every propane cylinder must be exposed to twice its service pressure, and one out of every 500 is exposed to four times its pressure, or about 960 psi. Two types of containers are authorized for propane: DOT cylinders and American Society of Mechanical Engineers (ASME) tanks.

DOT cylinders are manufactured under the provisions of DOT Hazardous Materials Regulations. They must be requalified for continued use 12 years from the date of manufacture. If the DOT-authorized visual inspection procedure is used, the cylinders must be requalified every 5 years thereafter. They must be checked for physical wear or damage every time they are refilled, and the paint must be kept in good condition. DOT engine fuel cylinders are usually removed from the vehicle and refilled elsewhere, though they may be refilled in place if they are properly installed and equipped for that purpose.

ASME engine fuel tanks are always refilled on the vehicle. These containers are manufactured under the provisions of the ASME Pressure Vessel Code. Periodic requalification is not required, but the tanks should be inspected for unusual wear or physical damage, and the paint must be kept in good condition.

Flex Fuel Vehicles

Most manufacturers produce a version of the **flex fuel vehicle (FFV)**. Initially, most of these vehicles were designed to run on M85 (85 percent methanol). Methanol is typically made from natural gas, though it is possible to produce it by fermenting biomass. M85 has a per-gallon Btu rating between 56,000 and 66,000 with an octane rating of 100. However, methanol is more corrosive than gasoline and requires changes in some of the materials in the fuel-handling systems of both the vehicle and the refueling station.

Due to a lack of availability (due to storage and handling concerns) of M85, all domestic manufacturers now make their flex fuel vehicles to run on E85. Ethanol, or grain alcohol, is produced by fermenting biomass. Ethanol is a renewable resource and contributes nothing in itself to greenhouse gas loading of the atmosphere. As a motor fuel, E85 is blended in a mixture of 85 percent ethanol and 15 percent unleaded gasoline. However, it can be blended with any amount of gasoline in the tank of a flex-fuel vehicle. E85 has a per-gallon Btu rating of about 80,000 and an octane rating of 100.

Most systems will use an alcohol fuel sensor (Figure 10-17) to detect the percentage of ethanol in the mixture. This sensor will provide a signal to the PCM, which optimizes ignition

The flex fuel vehicle (FFV) has a single fuel tank, fuel system, and engine. The vehicle is designed to run on regular unleaded gasoline and an alcohol fuel (either ethanol or methanol) in any mixture.

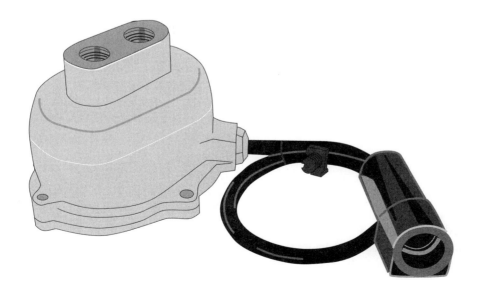

Figure 10-17   Flex fuel sensor.

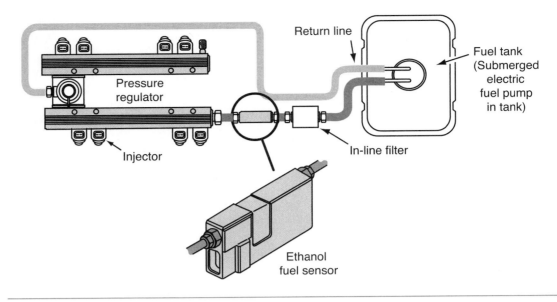

Figure 10-18 The E85 flex fuel system with an ethanol sensor.

timing and air-fuel ratios. The sensor can be located in the tank or in the fuel line between the tank and the injectors (Figure 10-18). Some manufacturers do not use the sensor. Instead, they use the O_2 sensor to indicate that the engine is being run on E85. Used either way, the end result is the PCM will use different strategies based on the detection of gasoline or E85.

Other PCM changes that occur when E85 is detected include how OBD-II monitors will be run. For example, catalyst index ratios for ethanol vary based on the changing concentration of alcohol in the fuel. The malfunction threshold typically increases as the percentage of alcohol increases. A malfunction threshold of 0.5 may be used at E10 (10 percent ethanol) and 0.9 may be used at E85 (85 percent ethanol). The malfunction thresholds are therefore adjusted based on the percentage of alcohol in the fuel.

E85 fuel systems do not look any different from most SFI fuel-injection systems. However, the components of the E85 system are usually different from a standard gasoline system. The components must be able to withstand the actions of ethanol. In most cases, different engine oil is also recommended when running E85 exclusively.

Vehicles with Alternate Power Sources

This section provides a brief description of several alternate vehicle power sources. These power sources are being sold in limited numbers, or they are still in the research and development stage.

Electric Vehicles

The **electric vehicle (EV)** powers its motor from a battery pack.

Since the 1990s, most major automobile manufacturers have developed an **electric vehicle (EV)**. The primary advantage of an EV is a drastic reduction in noise and emission levels. The CARB established a low-emission vehicles/clean fuel program to further reduce mobile source emissions in California during the late 1990s. In this program, emission standards are established for five vehicle types (Figure 10-19): conventional vehicle (CV), transitional low-emission vehicle (TLEV), low-emission vehicle (LEV), ultra low-emission vehicle (ULEV), and zero emission vehicle (ZEV). The electric vehicle meets ZEV standards. The basic components of an electric vehicle are shown in Figure 10-20.

	CV	TLEV	LEV	ULEV	ZEV
NMOG	0.25 *	0.125	0.075	0.040	0.0
CO	3.4	3.4	3.4	1.7	0.0
NOx	0.4	0.4	0.2	0.2	0.0

(*) Emission standards of NMHC

Figure 10-19 California tailpipe emission standards in grams per mile at 50,000 miles.

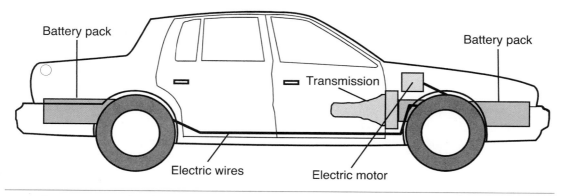

Figure 10-20 The electric vehicle is powered by an electric motor that receives its energy from battery packs.

General Motors introduced the EV1 electric car to the market in 1996. The original battery pack in this car contained twenty-six 12-volt batteries that delivered electrical energy to a three-phase 102-kilowatt (kW) AC electric motor. The electric motor is used to drive the front wheels. The driving range is about 70 miles (113 km) of city driving or 90 miles (145 km) of highway driving. Temperature, vehicle load, and speed affect this range. A 1.2-kW charger in the vehicle's trunk can be used to recharge the batteries. This charger takes about 15 hours to fully charge the batteries. An external Delco Electronics' MAGNE CHARGE 6.6 kW inductive charger operating on 220V/30 amperes can recharge the batteries in 3 to 4 hours (Figure 10-21). The weatherproof plastic paddle is inserted into the charging port located at the front of the vehicle. Power is transferred by magnetic fields.

Figure 10-21 Recharging the EV. (Used with permission from Nissan NA, Inc.)

Figure 10-22 Positioning the EV's battery pack for installation. (Reprinted with permission)

If the batteries are fully charged, the EV1 accelerates from 0 to 60 mph (97 kph) in 9 seconds and has a top speed of 80 mph (130 km/h). The EV1 is equipped with a Galileo electronic brake system that employs a computer and sensors at each wheel to direct power assist braking, regenerative/friction brake blending, four-wheel antilock braking, traction assist, tire pressure monitoring, and system diagnostics. In 1998, nickel/metal/hydride batteries were installed in the EV1 vehicles, which extended the driving range between battery charges to 160 miles (257 km).

The driving range of electric vehicles is their biggest disadvantage. Much research is being done to extend the range and to decrease the required recharging times. Currently, the use of nickel/metal/hydride or lead-acid batteries and permanent magnet motors has extended the operating range. Another disadvantage is the battery pack adds substantial weight to the vehicle (Figure 10-22). Other features, such as **regenerative braking** and highly efficient accessories (such as a heat pump for passenger heating and cooling), are also being installed on electric vehicles. Other disadvantages are the cost of replacement batteries and the danger associated with the high voltage and high frequency of the motors.

BIT OF HISTORY

The electric vehicle has a much longer history than most people realize. According to the book *The Green Car Guide*, the first known electric car was a small model built by Professor Stratingh in the Dutch town of Gröningen in 1835. But the first practical electric road vehicle was probably made either by Thomas Davenport in the United States or by Robert Davidson in Edinburgh in 1842. These pioneers had to use non-rechargeable electric cells. An electric vehicle did not become a viable option until the Frenchmen Gaston Plante and Camille Faure respectively invented (1865) and improved (1881) the storage battery.

Hybrid Vehicles

The first alternative vehicle attempted was the electric vehicle, which had zero emissions and ran primarily on battery power. However, the battery has a limited energy supply and restricts the traveling distance. This limitation was a major stumbling block to most consumers; thus their use is mainly limited to fleets. One method of improving the electric vehicle resulted in the addition of an on-board power generator that is assisted by an internal combustion engine, resulting in the **hybrid electric vehicle (HEV)**. The addition of the internal combustion engine means that the vehicle cannot be classified as a ZEV. However, it did reduce emission levels significantly and increased fuel economy.

A **hybrid electric vehicle (HEV)** has two different power sources. In most HEV's, the power sources are a small displacement gasoline or diesel engine and an electric motor.

BIT OF HISTORY

In 1993, President Clinton signed the Partnership for a New Generation of Vehicles (PNGV). This initiated a cooperative program between the automotive manufacturers and government agencies to develop a new generation of cars by the year 2004. One of the criteria for the new vehicle was fuel economy of 80 mpg of gasoline or gasoline equivalent. In 2002, the Bush administration replaced the PNGV program with a partnership called FreedomCAR (Freedom Cooperative Automobile Research) with DaimlerChrysler, Ford, and General Motors. The new program aims to reduce the United States' dependence on foreign oil while reducing tailpipe pollution. In addition, the program will focus research on the long-term transition from petroleum-based vehicles to the use of clean, renewable hydrogen energy. Overall goals of the program include the development of technology that will enable mass production of affordable hydrogen-powered fuel-cell vehicles and the infrastructure to support them.

Basically, the hybrid electric vehicle relies on power from the electric motor, the engine, or both (Figure 10-23). When the vehicle moves from a stop and has a light load, the electric motor moves the vehicle. Power for the electric motor comes from stored electricity in the battery pack. During normal driving conditions, the engine is the main power source. Engine power is also used to rotate a generator that recharges the storage batteries (Figure 10-24). The output from the generator may also be used to power the electric motor, which is run to provide additional

Figure 10-23 Hybrid power system. (Reprinted with permission)

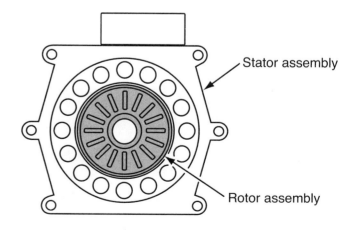

Figure 10-24 Engine power is also used to rotate a generator that recharges the storage batteries and drives the vehicle. The rotor assembly is a very powerful magnet that induces voltage into the stator windings as it is rotated.

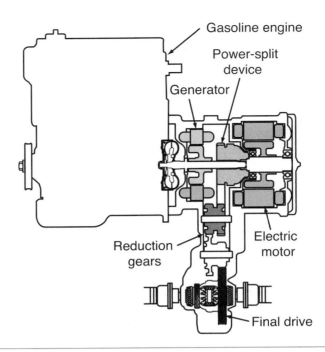

Figure 10-25 Hybrid power system with gasoline engine and electric propulsion motor.

power to the powertrain (Figure 10-25). A computer controls the operation of the electric motor depending on the power needs of the vehicle. During full throttle or heavy load operation, additional electricity from the battery is sent to the motor to increase the output of the power train.

The components of a typical hybrid vehicle include:

- *Batteries:* Some of the batteries that are being used or experimented with now are the lead acid battery, the nickel-metal battery, and the lithium ion battery. In the development of the battery, thermal management must be taken into consideration. The temperature can vary from module to module, so the performance of the battery is dependent on the temperature. An imbalance in temperature will affect the power and capacity of the battery, the charge acceptance during regenerative braking, and vehicle operation. Also, passenger safety is a major topic of concern. The batteries must be kept in sealed containers in order to ensure complete protection. The Toyota Prius (Figure 10-26) seals its non-caustic, nonflammable nickel-metal battery in a carbon composite case positioned in the rear of the vehicle (Figure 10-27).

Figure 10-26 Toyota Prius was the world's first mass-produced hybrid vehicle.

Figure 10-27 Battery pack. (Reprinted with permission)

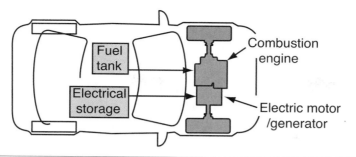

Figure 10-28 Parallel hybrid configuration.

- *Electric Motors:* One of the sources of power is the electric motor. The motor converts electrical energy to mechanical energy. This mechanical energy is what drives the wheels of the vehicle. This motor is designed to allow for maximum torque at low rpm. This gives the electric motor the advantage of having better acceleration than the conventional motor.
- *Regenerative braking:* About 30 percent of the kinetic energy lost during braking is in heat. When decreasing the acceleration of the vehicle, regenerative braking helps to minimize energy loss by recovering the energy used to brake by converting rotational energy into electrical energy through a system of electric motors and generators. When the brakes are applied, the motor becomes a generator by using the kinetic energy of the vehicle to store power in the battery for later use.
- *Ultra capacitors:* The **ultra capacitor** is the primary device in the power supply during hill climbing, acceleration, and the recovery of braking energy. To create a larger storage capacity of the ultra capacitors, the surface area must be increased and, in turn, the voltage is increased. However, because the voltage drops as energy is discharged, additional electronics are required to maintain a constant voltage.

Propulsion. There are two typical ways to arrange the flow of power in a hybrid electric vehicle. If the combustion engine is capable of turning the drive wheels as well as the generator, then the vehicle is referred to as a **parallel hybrid** (Figure 10-28). There is a further distinction between a **mild parallel hybrid** and a **full parallel hybrid**.

Other configurations of the parallel hybrid include the use of an engine to power one axle and an electric motor to power the other (Figure 10-29). Another concept is to use a combination

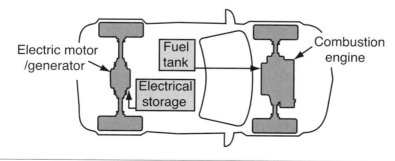

Figure 10-29 Parallel hybrid configuration using an engine to power one axle and an electric motor to power the other.

The **ultra capacitor** is a device that stores energy as electrostatic charge. It is the primary device in the power supply during hill climbing, acceleration, and the recovery of braking energy.

In a **parallel hybrid** configuration, there is a direct mechanical connection between the engine and the wheels. Both the engine and the electric motor can turn the transmission at the same time.

A **mild parallel hybrid** has an electric motor that is large enough to provide regenerative braking, instant engine start-up, and a boost to the combustion engine.

A **full parallel hybrid** uses an electric motor that is powerful enough to propel the vehicle on its own.

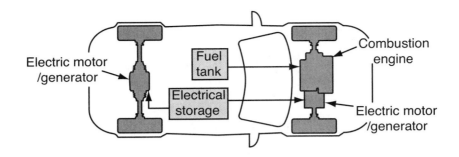

Figure 10-30 Parallel hybrid configuration using a combination where the engine, coupled with an electric motor, powers one axle and another electric motor powers the other axle.

where the engine, coupled with an electric motor, powers one axle, and another electric motor powers the other axle (Figure 10-30).

Most parallel hybrid vehicles use the electric motor to accompany the engine to help drive the wheels. For example, the engine is used for long driving periods, while the electric motor is used for short, low-intensity periods. In other words, the engine is ideal for highway driving, and the electric motor is ideal for a trip around town. The electric motor also provides the vehicle with acceleration. This acceleration, however, is only sustained until the vehicle reaches a certain speed. After this speed is reached, the engine is started and replaces the electric motor. The parallel hybrid combines the alternator, starter, and wheels to create a system that will start the engine, electronically balance it, take power from the engine and turn it into electricity and provide extra power to the drive line when power assist is needed for hill climbing or quick acceleration.

In the **series hybrid** vehicle, there is no mechanical connection between the engine and the wheels. The engine turns a generator, and the generator will either charge the batteries or power the electric motor, which in turn drives the transmission. Therefore, the engine never directly powers the automobile (Figure 10-31).

The power used to give the vehicle motion is transformed from chemical energy into mechanical energy, then into electrical energy, and finally back to mechanical energy to drive the wheels. This configuration is efficient in that it never idles. The automobile turns off completely at rest, such as at a stop sign or traffic light. This feature greatly reduces emissions. There are a variety of options in the configuration and mounting of all of the components. Some series vehicles do not use a transmission.

HEV Examples. The Toyota Prius is considered a Super Ultra Low Emissions Vehicle (SULEV), meaning that it is 90 percent cleaner than an Ultra Low-Emissions Vehicle. This vehicle uses a combined hybrid-electric structure to supply power. This system uses a combination of an inter-

> In a **series hybrid** configuration, propulsion comes directly from the electric motor.

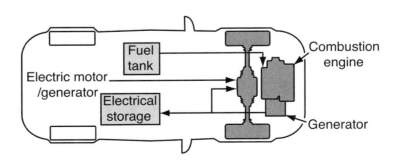

Figure 10-31 Series hybrid configuration.

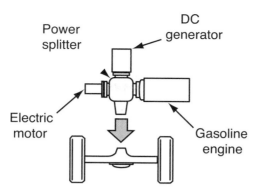

Figure 10-32 The power splitter allows for acceleration using both the engine and the electric motor and to be able to run solely on the electric motor.

nal combustion engine and an electric motor to turn the Electrically Controlled Continuous Variable Transmission (ECVT). However, the Prius is also able to accelerate using both the engine and electric motor and is able to run solely on the electric motor. This is called a power split device (Figure 10-32).

Using a set of planetary gears (Figure 10-33), the vehicle can operate like a parallel vehicle in that the electric motor powers the vehicle, or the gasoline engine powers the vehicle, or both. However, it also can operate as a series hybrid where the engine can operate independently of the vehicle speed either charging the batteries or providing power to the wheels when needed.

AUTHOR'S NOTE: Operating the engine independently of the vehicle speed means that even though the vehicle is traveling at highway speeds, the engine can be close to idle speed since it is only acting as a generator.

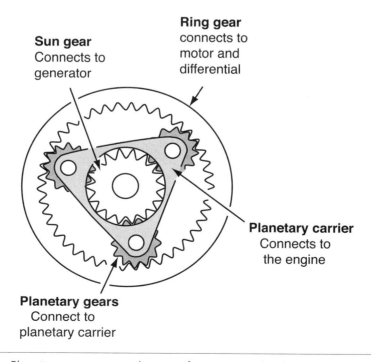

Figure 10-33 Planetary gears are used to transfer power to the drive wheels.

All the propulsion and auxiliary components of the Prius are packaged under the hood. The electric motor is rated at 33 kW and battery power is 21 kW. The nickel-metal hydride battery pack is located between the rear seatback and the trunk.

The Prius is equipped with a variable valve timing four-cylinder engine. The generator also works as the starter of this engine. Gear shifting is not required because of the planetary system that is acting as a continuously variable transmission to keep engine rpms in the range of best efficiency. Instead of a normal transmission, the engine drives the planet carrier of a planetary gear set. The sun gear of that set connects to a motor/generator, and the ring gear drives both the front wheels and a second motor/generator.

The Prius is equipped with a "drive-by-wire" accelerator. The driver inputs to the motor management how much speed is requested. The management system then decides where the necessary power should come from, the engine, the battery, or both. The same accounts for the brake-by-wire system; the driver calls for the appropriate amount of retardation, and the motor management coordinates this between the wheel brakes and the regenerative braking system. The computer also makes sure that the generator runs frequently enough to keep the battery charged.

When the battery is fully charged and the engine temperature is acceptable, the engine can be shut off (if the vehicle speed is low enough). If the drive gently presses the accelerator, the vehicle is moved by battery power. If the driver is requesting a quicker acceleration, the engine is started and powers the vehicle. In normal driving, the Prius maintains the battery state-of-charge within a narrow window. However, driving the vehicle under heavy loads for an extended time may deplete the battery charge.

The Honda Insight is also a parallel-configured hybrid. In this system, the gasoline engine provides the majority of the power. The electric motor is used to assist the gasoline engine to provide additional power during acceleration. The Insight uses regenerative braking technology to capture energy lost during braking. The Honda also has a lightweight engine that uses **lean burn technology** to maximize its efficiency.

The Hybrid Honda Civic uses an Integrated Motor Assist System (IMA) to power the vehicle. The system is comprised of a gasoline and electric motor combination. The electric motor is a source of additional acceleration and functions as a high-speed starter. The electric motor also acts as a generator for the charging systems used during regenerative braking. This way the Civic ensures efficiency by capturing lost energy by using regenerative braking, much like the Prius and Insight.

Lean burn technology, developed in the 1960s, uses high air-fuel ratios to increase fuel efficiency.

BIT OF HISTORY

The 1903 Krieger was a front-wheel drive electric-gasoline hybrid car. A gasoline engine supplemented the battery pack. Between 1890 and 1910, there were many hybrid electric cars and four-wheel drive electric cars. Electric cars were more expensive than gasoline cars, but the electrics were considered more reliable and safer. With the development of the starter motor for gasoline engines and the increased fuel range of gasoline engines, most public interest switched from electrics to gasoline by 1915.

Fuel Cells

Fuel cell powered vehicles have a very good chance of becoming the drives of the future. They combine the reach of conventional internal combustion engines with high efficiency, low fuel consumption, and minimal or no pollutant emission. At the same time, they are extremely quiet. Because they work with regenerative fuel such as hydrogen, they reduce the dependence on crude oil and other fossil fuels.

A **fuel cell** produces current from hydrogen and aerial oxygen.

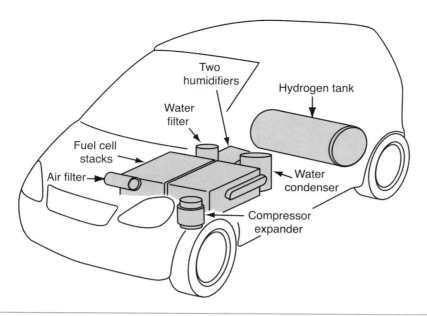

Figure 10-34 Fuel cell vehicle components. Technology has allowed engineers to design fuel cell vehicles without the loss of passengers or cargo space.

A fuel cell powered vehicle (Figure 10-34) is basically an electric vehicle. Like the electric vehicle, it uses an electric motor to supply torque to the drive wheels. The difference is that the fuel cell produces and supplies electric power to the electric motor instead of batteries. Most of the vehicle manufacturers and several independent laboratories are involved in fuel cell research and development programs. A number of prototype fuel cell vehicles have been produced with many being placed in fleets in North America and Europe.

Fuel cells electrochemically combine oxygen from the air with hydrogen to produce electricity. The oxygen and hydrogen are fed to the fuel cell as "fuel" for the electrochemical reaction. There are different types of fuel cells, but the most common type is the **proton exchange membrane (PEM)** fuel cell.

How the Fuel Cell Works

The PEM fuel cell is constructed like a sandwich (Figure 10-35). The electrolyte is situated between two electrodes of graphite paper that is permeable to gas. The electrolyte is a polymer membrane. Hydrogen is applied to the anode side of the PEM and ambient oxygen is applied to the cathode side (Figure 10-36). The membrane keeps the distance between the two gases and provides a controlled chemical reaction.

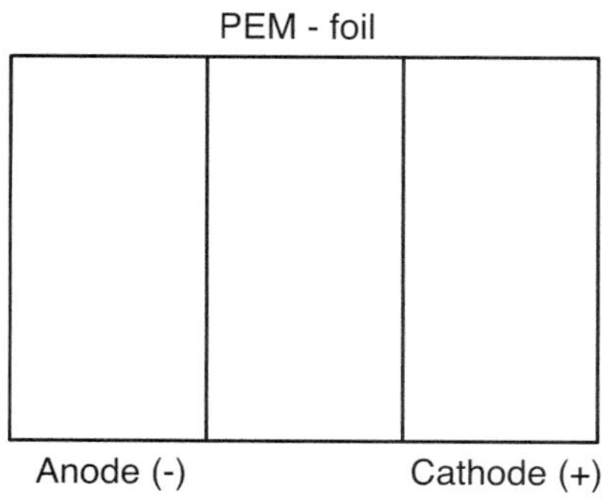

Normally, hydrogen and oxygen bond with a loud bang, but in fuel cells a special **proton exchange membrane (PEM)** impedes the oxyhydrogen gas reaction by ensuring that only protons (H+) and not elemental hydrogen molecules (H_2) react with the oxygen.

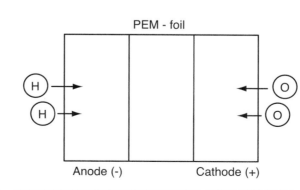

Figure 10-35 The PEM foil.

Figure 10-36 Hydrogen is applied to the anode while oxygen to applied to the cathode.

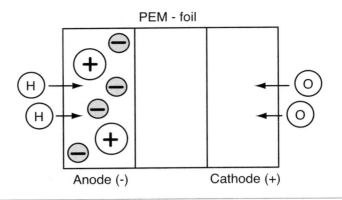

Figure 10-37 The catalyst breaks down the H₂ into protons and electrons.

The anode is the negative post of the fuel cell. It conducts the electrons that are freed from the hydrogen molecules so that they can be used in an external circuit. It has channels etched into it that disperse the hydrogen gas evenly over the surface of the catalyst. The cathode is the positive post of the fuel cell. It also has channels etched into it that distribute the oxygen to the surface of the catalyst. It also conducts the electrons back from the external circuit to the catalyst where they can recombine with the hydrogen ions and oxygen to form water.

A fine coating of platinum is applied to the foil to act as a catalyst. This is used to accelerate the decomposition of the hydrogen atoms into electrons and protons (Figure 10-37). The catalyst is rough and porous so that the maximum surface area of the platinum can be exposed to the hydrogen or oxygen. The platinum-coated side of the catalyst faces the PEM.

The PEM is the electrolyte. This specially treated material (which looks similar to ordinary kitchen plastic wrap) only allows the protons to move across from the anode to the cathode (Figure 10-38). As a result, the anode will have a surplus of electrons and the cathode will have a surplus of protons. If the anode and cathode are connected outside of the cell, current will flow through the conductor (Figure 10-39). The electrons will move through the conductor to the cathode. The electrons will then recombine with the protons and the oxygen to produce water.

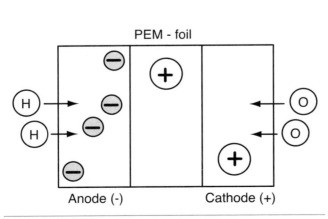

Figure 10-38 The PEM foil only allows the protons to migrate to to the cathode leaving the electrons on the anode.

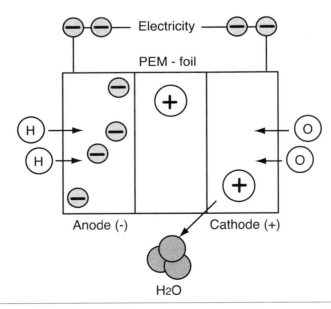

Figure 10-39 With an excess amount of electrons on the anode and excess amount of protons on the cathode, current will flow through an external conductor. The electrons then react with the protons and oxygen to produce water.

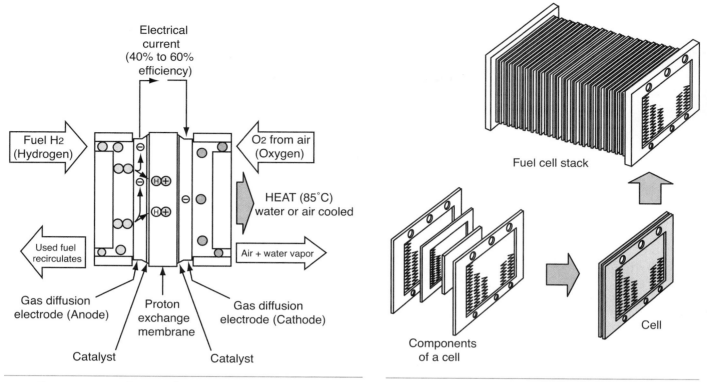

Figure 10-40 The PEM fuel cell.

Figure 10-41 The PEM fuel cell stack.

The entire process is illustrated in Figure 10-40. Pressurized hydrogen gas (H_2) enters the fuel cell on the anode side. This gas is forced through the catalyst by the pressure. When an H_2 molecule comes in contact with the platinum on the catalyst, it splits into two H+ ions and two electrons (e–). The electrons are conducted through the anode where they make their way through the external circuit (doing useful work such as turning a motor) and return to the cathode side of the fuel cell.

Meanwhile, on the cathode side of the fuel cell, oxygen gas (O_2) is being forced through the catalyst where it forms two oxygen atoms. Each of these atoms has a strong negative charge. This negative charge attracts the two H+ ions through the membrane where they combine with an oxygen atom and two of the electrons from the external circuit to form a water molecule (H_2O).

This reaction in a single fuel cell produces only about 0.7 volt. For this voltage to become high enough to be used to move the vehicle, many separate fuel cells must be combined in series to form a fuel-cell stack (Figure 10-41).

Fuels for the Fuel Cell

A fundamental problem with fuel cell technology concerns whether to store hydrogen or convert it from other fuels on board the vehicle. All four principal fuels that automotive manufacturers are considering (hydrogen, methanol, ethanol, and gasoline) pose some challenges.

Hydrogen Fuel. One solution is to store hydrogen on board the vehicle. The ability to use hydrogen directly in a fuel cell provides the highest efficiency and zero tailpipe emissions. However, hydrogen has a low energy density and boiling point, thus on-board storage requires large and heavy tanks. There are three types of hydrogen storage methods under development: compressed hydrogen, liquefied hydrogen, and binding hydrogenate to solids in metal hydrides or carbon compounds.

Compressed hydrogen offers the least expensive method for on-board storage of hydrogen. However, at normal CNG operating pressures of 3500 psi (24 MPa), reasonably sized, commercially available pressure tanks will provide limited range for a fuel cell vehicle (about 190 km, or 120 miles). DaimlerChrysler and Hyundai are now using pressure tanks that are capable of 5000 psi (34 Mpa). Quantum is conducting research of high-performance hydrogen storage systems, looking at pressure tanks that are capable of up to 10,000 psi (69 Mpa), which would permit a 400-mile (645 km) driving range.

Liquefied hydrogen can be stored in large cylinders containing a hydride material (something like steel wool). Liquefied hydrogen does not require the high storage capacity as that of compressed hydrogen for the same amount of driving range. However, the very low boiling point of hydrogen requires that the tanks have an excellent insulation. Maintaining the extreme cold temperature of $-423°F$ ($-253°C$) during refueling and storage is difficult. It is estimated that up to 25 percent of the liquid hydrogen may be boiled off during the refueling process. In addition, about 1 percent is lost per day in on-board storage. Storing liquid hydrogen on a vehicle also involves some safety concerns. As the fuel tank warms, the pressure increases and may activate the pressure relief valve. This action discharges flammable hydrogen into the atmosphere, creating a source of danger and pollution.

A **reformer** is a high-temperature device that converts hydrocarbon fuels to CO and H_2.

Methanol. Several automotive manufacturers are using methanol to power their fuel cells. It is believed that methanol fuel cells could bridge the gap while a hydrogen distribution infrastructure is being built over the next few decades. Using "methanolized" hydrogen as fuel has the advantage that it can be stored on the vehicle similar to gasoline. For the reaction in the fuel cell, a **reformer** on board the vehicle produces hydrogen from the methanol fuel. To produce the hydrogen, the methanol fuel is mixed with water. When it evaporates, it is decomposed into hydrogen and carbon dioxide. Prior to sending the hydrogen and carbon dioxide to the fuel cell, it is purified in additional steps.

Methanolized hydrogen contains more hydrogen atoms and has an energy density that is greater than that of liquid hydrogen. Like hydrogen, methanolized hydrogen is also independent of mineral oil. Compared to hydrogen, vehicles driven by methanol are not completely emission-free, but they produce very little pollutants and much less carbon dioxide than internal combustion engines.

A special type of PEM fuel cell, called the Direct Methanol Air Fuel Cell (DMFC), utilizes methanol combined with water directly as a fuel and ambient air for oxygen. This technology enables use of a liquid fuel without the need for an on-board reformer while still providing a zero-emissions system. However, current research has demonstrated that the power density is lower than other PEM fuel cells.

Ethanol. Ethanol is considered less toxic than either gasoline or methanol. The system will require a reformer be added to the vehicle, similar to that of a methanol system. The fuel cell could use E100, E95, or E85.

Gasoline. Fuel cells can be driven with a special, more pure gasoline. Using reformers for on-board extraction of hydrogen from gasoline is one approach to commercialization of fuel cell vehicles since the gasoline infrastructure is already in place. However, producing hydrogen from gasoline in a vehicle system is much more difficult than from methanol or ethanol. Gasoline reforming requires higher temperatures and more complex systems than the methanol or ethanol reforming. The reformation reactions occur at 1,562°F to 1,823°F (850°C to 1,000°C), making the devices slow to start and the chemistry tempermental. Thus, the drive would work less efficiently and produce more emissions. Moreover, the capabilities for cold starts would be restricted. The size of the reformer is also an issue, making it difficult to fit under the hood of a standard-sized vehicle. Furthermore, there is concern about the sulfur levels in current gasoline and carbon monoxide in the reformer poisoning the fuel cell.

Reformer.   As mentioned, some fuel cell systems may require the use of an on-board reformer to extract hydrogen from liquid fuels such as gasoline, methanol, or ethanol. On-board reformation of a hydrocarbon fuel into hydrogen allows the use of more established infrastructures but adds additional weight and cost and reduces vehicle efficiency. In addition, the reformer does create some emissions.

PEM fuel cell reformers combined fuel and water to produce additional H_2 and convert the CO to CO_2. The CO_2 is then released to the atmosphere. Reformer technologies include steam reforming, partial oxidation, and high-temperature electrolytes reforming.

Steam reforming (SR) uses a catalyst to convert fuel and steam to H_2, CO, and CO_2. The CO is further reformed with steam to form more H_2 and CO_2. A purification step then removes CO, CO_2, and any impurities to achieve a high hydrogen purity level (97 to 99.9 percent). SR of methanol is the most developed and least expensive method for producing hydrogen from a hydrocarbon fuel on a vehicle, resulting in 45 to 70 percent conversion efficiency.

Partial oxidation (POX) reforming is similar to stream reformers since both combine fuel and steam, but this process adds oxygen in an additional step. The process is less efficient than steam reformers, but the heat-releasing nature of the reaction makes it more responsive than steam reformers to variable load. Heavier hydrocarbons can be used in POX, but they have lower carbon-to-hydrogen ratios, which limit hydrogen production.

Hydrogen can also be obtained by **electrolysis** from water. The drawback to this process is it requires a great deal of electrical energy. Recently, the development of high-temperature electrolytes that could operate at temperatures in excess of 212°F (100°C) has shown some positive results. The benefits of high-temperature electrolytes include:

- *Improved CO tolerance.* This allows the manufacturer to reduce or remove the need for an oxidation reactor and for air-bleed. Since these requirements can be reduced, system efficiency is increased by 5 to 10 percent. There is also a considerable reduction in start-up time. The remaining CO will be combusted in a catalytic tail-gas burner to prevent emissions of CO.
- *Facilitated stack cooling.* This reduces the size of the radiator and reduces the fuel cell stack cooling plate requirements.
- *Humidity-independent operation.* Generally, high-temperature membranes require humidifiers and water recovery, whereas this system does not.

Electrolysis is the splitting of water into hydrogen and oxygen.

Solid Oxide Fuel Cells.   Planar Solid Oxide Fuel Cells (SOFCs) operate at high temperatures of 932°F to 1,472°F (500°C to 800°C) and can use CO and H_2 fuel. SOFCs have a good tolerance to fuel impurities and use ceramic as an electrolyte. BMW is currently developing an auxiliary power unit (APU) using an SOFC with Delphi and Global Thermoelectric. Currently, SOFCs use gasoline fuel and require a reformer.

Sodium Borohydride.   DaimlerChrysler introduced a new concept in fuel technologies with a concept vehicle called the Chrysler Town & Country Natrium (Latin for sodium). This research was driven by the lack of a safe, compact way to contain hydrogen. In addition, most fuel cell vehicles are not "on demand," meaning that there is a start-up time required before the reformer can begin producing hydrogen.

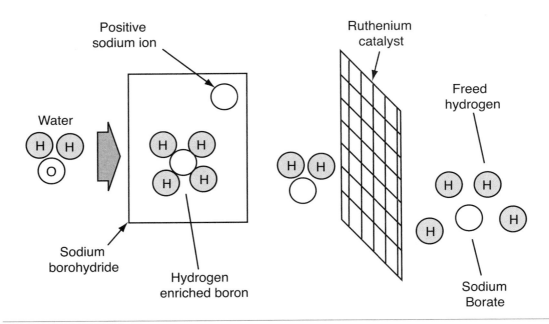

Figure 10-42 The process of extracting hydrogen from sodium borohydride.

A fairly simple chemical process mixes sodium borohydride powder with water. The sodium borohydride powder produces free hydrogen for the fuel cell (Figure 10-42). Unlike the fuel cell systems that use methanol or gasoline, the Natrium produces no pollution and no carbon dioxide.

> **AUTHOR'S NOTE:** The advantage of sodium borohydride is that it can easily be stored and transported in lightweight plastic tanks at normal temperatures and pressures when dissolved in water. Also, the borate solution is neither poisonous nor flammable or explosive.

The sodium borohydride powder holds more hydrogen than the most densely compressed hydrogen tank. The prototype minivan attains a top speed of 80 mph (130 kph) and an operating range of almost 300 miles (500 kilometers). The byproducts of this process are water and sodium borate, which is basically laundry detergent. After use, the spent powder goes into a storage tank where it can be pumped out and reclaimed.

As the name implies, sodium borohydride ($NaBH_4$) is a salt. The salt that powers this system is not ordinary table salt (sodium chloride—NaCl) but is a white salt whose molecules contain a relatively large amount of hydrogen. Through the use of a chemical catalyst, the sodium borohydride reaction with water results in elemental hydrogen. A slurry of sodium perborate ($NaBO_2$) forms during this reaction as well. This compound is chemically related to borax, which is used as a bleaching agent in conventional detergents. The sodium perborate slurry is collected in a special tank and can be effectively recycled in a chemical process. In this reverse reaction, the sodium perborate is reclaimed back to sodium borohydride, which can then be reused as an energy source for the fuel cells.

The catalyst system is called "Hydrogen on Demand TM™" and is a developed Millennium Cell (Figure 10-43). When the driver of the minivan steps on the accelerator, he is actually "throttling" a fuel pump. The sodium borohydride is pumped into the catalyst, which immediately generates hydrogen. This dynamic process slows down or ceases entirely when the fuel pump is throttled or completely shut off. When the driver accelerates once more, hydrogen is immediately produced again, and the fuel cells immediately generate electricity.

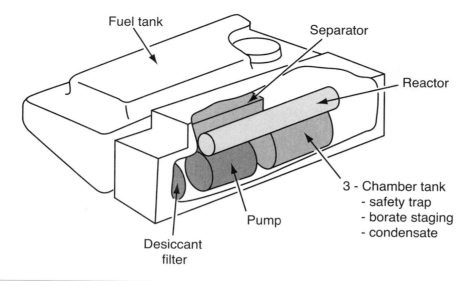

Figure 10-43   The "Hydrogen on Demand™" catalyst system is the heart of the drive train. The hydrogen required for the fuel cell to generate power is extracted from sodium borohydride in a reactor.

An AC motor with an output of 35 kW powers the vehicle. A 55 kW lithium-ion battery serves as a storage unit for electrical energy. The battery is recharged by the fuel cell unit and by regenerative braking. If the fuel cell system should fail, the battery is sufficient to drive the electric motor and move the vehicle. All components of the fuel cell are contained within the frame rail of the vehicle (Figure 10-44).

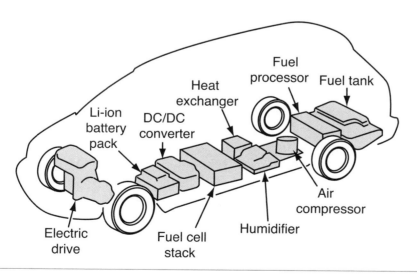

Figure 10-44   Components of the sodium borohydride fuel cell minivan are all located between the frame rails.

Summary

Terms to Know

Bi-fuel system

Converted vehicles

Electric vehicle (EV)

Electrolysis

Flex fuel vehicle (FFV)

Fuel cell

Full parallel hybrid

HD-5

Hybrid electric
vehicle (HEV)

Lean burn
technology

Mild parallel hybrid

Parallel hybrid

Proton exchange
membrane (PEM)

Reformer

Regenerative braking

Series hybrid

Ultra capacitor

❏ Due to the increase in regulations concerning emissions and the public's desire to become less dependent on foreign oil, most major automotive manufacturers have developed alternative fuel or alternate power vehicles.

❏ Original equipment manufacturer (OEM) alternate fuel vehicles are designed specifically to use an alternative fuel, such as propane, compressed natural gas (CNG), methanol, or ethanol.

❏ The automotive propane fuel system is a completely closed system that contains a supply of pressurized liquefied petroleum gas (LPG). The gas is stored in liquefied form in tanks that are pressurized to around 160 pounds per square inch.

❏ The propane tanks are fitted with a pressure relief valve to allow excess vapor to escape outside of the vehicle if the pressure becomes too high.

❏ The liquid fuel propane injection systems operate similar to gasoline fuel injection. The propane electronic fuel-injection system uses a special LPG tank that contains an internal pump.

❏ A bi-fuel system can operate on two different fuel types.

❏ Natural gas vehicles (NGV) store their fuel in cylinders under a high pressure between 3,000 and 3,600 psi.

❏ The CNG gas flows through a fuel regulator and is then injected through a specially designed mixer to achieve the correct air-to-gas mixture.

❏ There are several valves and safety components on a CNG vehicle including: manually operated fuel control valve, a hand-operated manual shut-off valve, a fuel pressure regulator, a high-pressure fuel shut-off solenoid, a low-pressure fuel shut-off solenoid, and one-way check valves.

❏ The CNG system uses high-pressure stainless steel, seamless fuel tubes.

❏ Special stainless steel, double-ferrule, compression-type fittings are used on the high-pressure side of the CNG system.

❏ Sensors used on the CNG system include a low-pressure sensor and a fuel temperature sensor to help calculate fuel-injector timing.

❏ Government regulations have standardized the testing and use of CNG fuel tanks throughout the automotive industry. Tanks must be inspected every 3 years and replaced every 15 years.

❏ Most manufacturers produce a version of the flex fuel vehicle that can run on either unleaded gasoline, M85, or E85.

❏ Most flex fuel systems will use an alcohol fuel sensor to detect the percentage of ethanol in the mixture. Some manufacturers use the O_2 sensor to indicate that the engine is being run on E85.

❏ Electric vehicles powered by an electric motor run off a battery pack.

❏ The hybrid electric vehicle relies on power from the electric motor, the engine, or both.

❏ The hybrid vehicle can be configured as a parallel, series, or combination system.

❏ Regenerative braking recovers the energy used to brake by converting rotational energy into electrical energy through a system of electric motors and generators. When the brakes are applied, the motor becomes a generator by using the kinetic energy of the vehicle to store power in the battery for later use.

❏ A fuel cell powered vehicle is basically an electric vehicle except the fuel cell produces and supplies electric power to the electric motor instead of batteries.

❏ Fuel cells electrochemically combine oxygen from the air with hydrogen to produce electricity.

❏ The most common type of fuel cell is the proton exchange membrane (PEM).

❏ There are several methods being explored for storage of fuel for the fuel cell. These include hydrogen, methanol, ethanol, and gasoline.

❏ Most fuel cell systems require the use of a reformer to extract hydrogen from liquid fuels such as gasoline, methanol, or ethanol.

Review Questions

Short-Answer Essay

1. Explain the purpose of alternate fuel vehicles.
2. Describe the flow of the fuel delivery system of a liquid propane-injection vehicle.
3. Explain the difference between a bi-fuel and a flex fuel system.
4. List typical components of the compressed natural gas (CNG) system and briefly describe their function.
5. Describe the government regulations concerning CNG tanks.
6. Explain the meaning of regenerative braking.
7. Describe the basic operation of a typical hybrid vehicle.
8. Explain the difference between parallel and series hybrids.
9. Briefly describe how the proton exchange membrane (PEM) fuel produces electrical energy.
10. What is the purpose of the reformer?

Fill-in-the-Blanks

1. The automotive propane fuel system is a completely _____ system that contains a supply of pressurized liquefied petroleum gas (LPG).
2. Natural gas vehicles (NGV) store their fuel in cylinders under a high pressure between _____ and _____.
3. Special stainless steel, _____, compression-type fittings are used on the high-pressure side of the CNG system.
4. CNG tanks must be inspected every _____ years and replaced every _____ years.
5. Flex fuel vehicles run on M85, E85, or _____.
6. _____ _____ recovers the heat energy used to brake by converting rotational energy into _____ energy through a system of electric motors and generators.
7. Fuel cells _____ combine oxygen from the air with hydrogen to produce electricity.
8. Most fuel cell systems require the use of a _____ to extract hydrogen from liquid fuels such as gasoline, methanol, or ethanol.
9. The _____ converts the liquid propane to a gaseous vapor form.
10. The _____ has a single fuel tank, fuel system, and engine. The vehicle is designed to run on regular unleaded gasoline and an alcohol fuel (either ethanol or methanol) in any mixture.

Multiple Choice

1. *Technician A* says that propane tanks are pressurized to about 160 psi.
   Technician B says a fuel pump is used in some propane injection systems.
   Who is correct?
   A. A only
   B. B only
   C. Both A and B
   D. Neither A nor B

2. *Technician A* says regenerative braking recovers the energy used to brake by converting rotational energy into electrical energy through a system of electric motors and generators.
   Technician B says when the brakes are applied, the motor becomes a generator by using the kinetic energy of the vehicle to store power in the battery for later use.
   Who is correct?
   A. A only
   B. B only
   C. Both A and B
   D. Neither A nor B

3. The liquid fuel propane injection system tanks contain:
   A. a rollover valve.
   B. a pump.
   C. a regulator/vaporizer.
   D. a float valve.

4. A bi-fuel system:
   A. uses a special reduction hose to change pressures.
   B. uses a special reduction hose to change pressures.
   C. uses a double filler neck to a single tank.
   D. uses separate tanks for each fuel.

5. Natural gas has an octane rating between:
   A. 105 and 130.
   B. 85 and 91.
   C. 120 and 160.
   D. 45 and 65.

6. CNG cylinders must be inspected every:
   A. 15 years.
   B. 6 months.
   C. 3 years.
   D. refill.

7. Flex fuel vehicles can run on gasoline and:
   A. natural gas.
   B. propane.
   C. hydrogen.
   D. E85.

8. Electric vehicles power the motor by:
   A. a generator.
   B. a battery pack.
   C. an engine.
   D. none of the above.

9. *Technician A* says in a parallel hybrid vehicle, propulsion comes directly from the electric motor.
   Technician B says in a series hybrid vehicle, both the engine and the electric motor can turn the transmission at the same time.
   Who is correct?
   A. A only
   B. B only
   C. Both A and B
   D. Neither A nor B

10. *Technician A* says a fuel cell produces electrical energy by breaking down H_2 atoms into electrons and protons.
    Technician B says the compressed hydrogen system requires a reformer to cool the fuel cell.
    Who is correct?
    A. A only
    B. B only
    C. Both A and B
    D. Neither A nor B

GLOSSARY

Acceleration Simulation Mode (ASM) An emissions test procedure that usually applies a 50 percent load at 15 mph and a 25 percent load at 25 mph, and uses a less expensive emissions analyzer than I/M 240.

Modo de Simulación de la Aceleración (ASM) Un método de prueba de emisiones que aplique generalmente una carga de 50 por ciento en 15 mph y una carga de 25 por ciento en 25 mph, y utiliza un analizador menos costoso de las emisiones que I/M 240.

Active Testing that requires the PCM to control the system or components in a specific manner while monitoring the results.

Activo Una prueba que requiere que el PCM controla el sistema o los componentes de una manera específica mientras que supervisa los resultados.

Actuators Perform the actual work commanded by the computer. They can be in the form of a motor, relay, switch, or solenoid.

Actuadores Realizan el trabajo actual ordenado por la computadora. Pueden ser en la forma de un motor, de un relais, de un interruptor, o de un solenoide.

Adaptive memory A function of the PCM to make corrections to the pulse width calculation based on oxygen sensor activity.

Memoria adaptiva Una función del PCM para hacer correcciones a las calculaciones en la anchura del pulso basado en la actividad del sensor del oxígeno.

Adaptive numerator The learned value of the crankshaft tone wheel.

Numerador adaptivo El valor docto de la rueda del tono del cigüeñal.

Advanced timing A spark that is delivered earlier than required.

Chispa Avanzada Una chispa que se entrega antes de que se requerie.

Air-fuel ratio Measurement by weight. The desirable air-fuel mixture is about 15 pounds of air to 1 pound of fuel. A 15:1 air-fuel ratio is 9,000 gallons of air to 1 gallon of fuel.

Cociente del aire-combustible Una medida por el peso. La mezcla deseable del aire-combustible es cerca de 15 libras del aire a 1 libra de combustible. Un cociente 15:1 del aire-combustible es 9.000 galones del aire a 1 galón de combustible.

Air injection Emission system that injects air into the exhaust system to heat the catalysts faster and to supply additional oxygen to the center section of some converters.

Inyección del aire Sistema de emisión que inyecta el aire en el dispositivo de escape para calentar los catalizadores más rápidamente y para proveer oxígeno adicional a la sección central de algunos convertidores.

Air pump switchover valve Valve to prevent exhaust gases from flowing back into the air pump.

Válvula del intercambio de la bomba del aire La válvula para evitar que los gases de escape se reintroduzcan nuevamente dentro de la bomba del aire.

Alternate good trip Used in place of global good trips for comprehensive components and major monitors.

Buen viaje alterno Reemplaza los buenos viajes globales para los componentes comprensivos y los monitores importantes.

Alternative fuels Fuels other than gasoline and diesel fuel. LP gas and alcohol are currently in use in automobiles and trucks. Hydrogen is also another example of an alternative fuel.

Combustible alternativo Los combustibles de otros tipos que no sean de gasolina y de combustible diesel y el alcohol del LP son actualmente en uso en en los automóviles y carros. El hidrógeno es también otro ejemplo de un combustible alternativo.

Amperes Measurement of current. One ampere represents the movement of 6.25 billion billion electrons (or one coulomb) past one point in a conductor in one second.

Amperios Una medida de la corriente. Un amperio representa el movimiento de 6,25 mil millones mil millones electrones (o de un culombio) a través de un punto en un conductor en un segundo.

Anode The positive side of the diode.

Ánodo El lado positivo del diodo.

Aspirated Air injection system that uses the exhaust pulses to bring air into the exhaust system. The word aspirate refers to withdrawing a material with a negative-pressure apparatus.

Aspirado El sistema de inyección del aire que utiliza los pulsos del escape para introducir el aire hacia el sistmea de escape. La palabra aspirar refiere a inducir una material por medio de un aparato de la presión negativa.

ASTM International Formerly known as the American Society for Testing and Materials. Provides a global forum for the development and publication of voluntary consensus standards for materials, products, systems, and services. ASTM standards serve as the basis for manufacturing, procurement, and regulatory activities.

Internacional ASTM Conocido anteriormente como la Sociedad Americana de Pruebas y Materiales. Proporciona un foro global para el desarrollo y la publicación de los estándares voluntarios del consenso para los materiales, los productos, los sistemas, y los servicios. Los estándares de ASTM sirven como la base para la fabricación, la consecución, y las actividades reguladoras.

Atom The smallest part of a chemical element that still has all the characteristics of that element.

Átomo La parte más pequeña de un elemento químico que mantiene todas las características de ese elemento.

Back pressure The resistance of an exhaust system to the flow of gases.

Contrapresión La resistencia de un dispositivo de escape al flujo de gases.

Balanced An atom that has an equal amount of protons and electrons.

Equilibrado Un átomo que tiene una cantidad igual de protones y de electrones.

Banjo fitting A metal connection using a circular filling with a hollow bolt and two steel sealing rings.

Conexión banjo Un conector del metal que usa un relleno circular con un perno hueco y dos anillos de estaquedad de acero.

Base The part of the transistor that is the control.

Base La pieza del transistor que es el control.

Base timing The ignition timing that is set when the engine is at curb idle.

Sincronización basa La sincronización de ignición que se establece cuando el motor está en la marcha lenta parada.

Basic program A one- or two-speed, "no load" emissions test, followed by an evaluation of the emissions control devices themselves.

Programa básico Una prueba de emisión de lavelocidad de uno o dos, "sín carga", seguido por una evaluación de los mismos dispositivos del control de emisiones.

Beaker points Mechanical switches that control current flow to the ignition coil's primary winding.

Puntos beaker Los interruptores mecánicos que controlan el flujo al devanado principal de bobina de la ignición

Bi-fuel system Means that a vehicle has two separate fuel systems with the capability to switch from one to the other.

Sistema dual del combustible Significa que un vehículo tiene dos sistemas de combustible separados con la capacidad de cambiar de la una a la otra.

Big slope The rate of change that an oxygen sensor experiences. It is determined by voltage change over time.

Cuesta grande El índice del cambio que experimenta un sensor del oxígeno.Se determina por medio del cambio del voltaje por un cierto plazo.

Binary code Represented by the numbers 1 and 0. Any number and word can be translated into a combination of 1s and 0s.

Código binario Representado por los números 1 y 0. Cualquier número y palabra se pueden traducir a una combinación de 1s y de 0s.

Bipolar Transistors that have three layers, two of which are the same.

Bipolares Transistores que tienen tres capas, dos de las cuales son iguales.

Bit A one 0 or one 1 of the binary code. Eight bits is called a byte.

Bit Un 0 o 1 del código binario. Ocho pedacitos se llaman un byte.

Block learn Makes long-term corrections to the injector pulse width.

Aprendizaje en bloque Efectú las correcciones de largo plazo a la anchura del pulso del inyector.

Blowby The leakage of fuel and combustion gases into the engine crankcase.

Fuga de gas La salida del combustible y de los gas de combustión en el cárter del motor.

Bottom dead center (BDC) Indicates the piston is at the very bottom of its stroke.

Punto muerto inferior (PMI) Indica que el pistón está la parte más baja de su movimiento.

Brake horsepower The useable power produced by an engine. It is measured on a dynamometer using a brake to load the engine.

Caballos de fuerza de freno La energía usable producido por un motor. Se mide en un dinamómetro usando al freno para aplicar una carga en el motor.

British thermal units (Btus) A measurement of energy. One British thermal unit (Btu) is the amount of heat required to raise one pound of water one Fahrenheit degree.

Unidades Termales Británicas (Btus) Medida de la energía. Una unidad termal británica (BTU) es la cantidad de calor requerida para subir la temperatura de una libra de agua por un grado Fahrenheit.

Bus Refers to the transporting of data from one module to another.

Colectora Refiere al transporte de datos a partir de un módulo a otro.

Calibrated frequency Serves as a high-limit threshold for the test frequency and is determined by the PCM.

Frecuencia calibrada Sirve como umbral del alto-liímite para la frecuencia de la prueba y se determina por el PCM.

Calories (cal) One calorie is the amount of heat required to raise the temperature of one gram (g) of water one Celsius degree.

Calorías (caloría) Una caloría es la cantidad de calor requerida para subir la temperatura de un gramo de agua por un grado centígrado.

Camshaft Tube constructed with lobes that are used to open the intake and exhaust valves.

Árbol de levas Un tubo construido con lóbulos que sirven para abrir las válvulas de la entrada y del escape.

Capacitance The ability of two conducting surfaces to store voltage. The two surfaces must be separated by an insulator.

Capacitancia La capacidad de dos superficies de conducción en almacenar al voltaje. Las dos superficies se deben separar por un aislador.

CARB readiness indicator Identifies whether or not the once-per-trip monitors have been run.

Indicador de preparación CARB Identifica si o no se han funcionado los monitores de una vez por viaje.

Catalyst monitor Tests for the ability of the converter to store oxygen efficiently.

Monitor catalizador Prueba la capacidad del convertidor de almacenar oxígeno eficientemente.

Catalytic converter A catalyst that converts emissions into harmless gases. A catalyst triggers a chemical reaction that changes the exhaust gases that pass through it without itself being changed by the reaction.

Convertidor catalítico Un catalizador que convierte las emisiones a los gases inofensivos. Un catalizador acciona una reacción química que cambie los gases de escape que pasan a través de él sín cambiarse si mismo por la reacción.

Cathode The negative side of the diode.

Cátodo El lado negativo del diodo.

Caustic A material that has the ability to destroy or eat through something. Caustic materials are considered extremely corrosive.

Material cáustico Una material que tiene la capacidad de destruir o de corroer. Los materiales cáusticos se consideran extremadamente corrosivos.

Central gateway (CGW) The manager of the CAN bus system. It is responsible for arbitration of the various bus messages.

Entrada central (CGW) El operador del sistema del collectador CAN. Es responsable por el arbitraje de los varios mensajes del collectador.

Central processing unit (CPU) The brain of the computer.

Unidad central de proceso (CPU) El cerebro de la computadora

Centrifugal advance Uses fly weights to move the breaker plate and advance the timing. Centrifugal force causes the fly weights to move in proportion to distributor shaft speed.

Avance centrífugo Utilisa los pesos móviles para mover la placa del disyuntor y para avanzar la sincronización. La fuerza centrífuga causa que los pesos móviles se muevan en proporción con velocidad del eje de la distribuidor.

Cetane A colorless, liquid hydrocarbon that has excellent ignition qualities. Cetane is rated at 100.

Cetano El hidrocarburo en líquido sín color que tiene calidades excelentes de la ignición. El cetano es clasificado en 100.

Cetane rating Describes how easily diesel fuel will ignite.

Grado del cetano Describe que tal fácilmente se encenderá. el combustible diesel.

Check valve A valve used to prevent the fuel in the lines from returning to the tank after the fuel pump shuts off. It also prevents fuel from entering the fuel lines with the pump off in the event of an accident.

Válvula de retención Una válvula que sirve para prevenir que el combustible en las líneas regresa al tanque después de que se apaga la bomba de gasolina. También evita que el combustible entra en las líneas de combustible cuando la bomba se apaga al ocurrir un accidente.

Circuit Meaning circle. It is the path of electron flow consisting of the voltage source, conductors, load component, and return path to the voltage source.

Circuito Significa circular. Es la trayectoria del flujo del electrón que consiste en la fuente del voltaje, los conductores, el componente de la carga, y la trayectoria regresando a la fuente del voltaje.

Circuit opening relay A dual winding relay used to send current to the fuel pump while the engine is cranking and after it starts.

Relais de circuito abierto Un relai de doble devanado que sirve para enviar la corriente a la bomba de gasolina mientras que el motor está arrancando y después de que arranca.

Clamping diode The function of diode when used to prevent voltage spikes from backfeeding through a circuit.

Diodo de bloqueo La función del diodo cuando se utilisa para evitar que los picos de voltaje retrocesan por un circuito.

Clear flood A PCM mode of operation that is accomplished by pressing the accelerator pedal to the wide-open throttle position while cranking the engine. The combination of these inputs has the PCM turn off the injectors.

Despejo completo Modo de operación del PCM que se efectúa presionando el pedal del acelerador a la posición abierta de la válvula reguladora mientras que se arranca el motor. La combinación de estas entradas hace que el PCM apaga a los inyectores.

Clock A crystal that electrically vibrates when subjected to current at certain voltage levels. As a result, the chip produces a very regular series of voltage pulses.

Registro Un cristal que vibra eléctricamente cuando está sujetado a la corriente en ciertos niveles voltaicos. Consecuentemente, el chip produce una serie de pulsos de voltaje muy regular.

Closed loop Mode of operation where the oxygen sensor is involved in the fuel strategy loop. The voltage value of the oxygen sensor is used as feedback.

Bucle cerrado Modo de operación en la queel sensor del oxígeno está implicado en el bucle estratégico del combustible. El valor del voltaje del sensor del oxígeno se registra como retroalimentación.

Cloud point The low-temperature point at which the waxes present in most diesel fuel tends to form wax crystals that clog the fuel filter. This is called the cloud point because the fuel will appear cloudy when the wax separates out.

Punto de turberidad del aceite El punto de baja temperatura en el cual las ceras presentes en la mayoría del combustible diesel suelen formar los cristales de cera que atascan el filtro de combustible. Esto tambien se llama el punto de nube porque el combustible tiene una aparecencia nublado cuando se separa la cera.

Coil on plug (COP) Ignition system that generally uses one coil pack per spark plug and does not use spark plug cables since the coil is installed directly onto the spark plug.

Bujía de bobina prendida (COP) Sistema de ignición que generalmente utiliza un paquete de bobina por cada bujía y no utiliza los cables de bujía puesto que la bobina está instalada directamente en la bujía de chispa.

Coil pack A coil assembly that contains two or more coils.

Paquete de bobina Una asamblea de la bobina que contiene dos bobinas o más.

Collector The part of the transistor that is connected to the opposite side of the circuit from the emitter.

Colector Una pieza del transistor que está conectado al lado opuesto del circuito del emisor.

Combustion chamber The area where combustion is started. It is above the piston and usually includes some of the cylinder head.

Cámara de combustión El área en donde se comienza la combustión. Queda arriba del pistón e incluye generalmente una parte de la cabeza del cilíndro.

Comprehensive components Any components that may affect emissions. These are generally components that were tested under OBD-I.

Componentes comprensivos Cualquieres de los componentes que puedan afectar las emisiones. Éstos son generalmente los componentes probados bajo el Obd-I.

Compressed natural gas (CNG) A mixture of hydrocarbons, mainly methane (CH_4), and is produced either from gas wells or in conjunction with crude oil production.

Gas natural comprimida (CNG) La mezcla de los hidrocarburos, principalmente del metano (CH_4), producido de pozos de gas o conjuntamente con la producción del petróleo crudo.

Compression fitting A line coupler that compresses the end of a sleeve to provide a seal.

Acoplador de compresión Un acoplador de la línea que comprime la extremidad de una manga para proporcionar un sello.

Compression ignition engine Engines that uses the heat generated by compressing air to ignite the fuel charge. These engines do not use a spark plug to start the combustion process.

Motor del encendido bajo compresión Los motores que utilizan el calor generado por la compresión del aire para encender la carga del combustible. Estos motores no utilizan una bujía de chispa para comenzar el proceso de la combustión.

Compression ratio The ratio of the volume in the cylinder above the piston when the piston is at bottom dead center (BDC) to the volume in the cylinder above the piston when the piston is at top dead center (TDC). It is an indication of how much air-fuel mixture can be brought into a cylinder and how tightly the mixture will be compressed.

Cociente de la compresión El cociente del volumen en el cilindro arriba del pistón cuando el pistón está el centro en el punto muerto inferior (BDC) al volumen en el cilindro arriba del pistón cuando el pistón está el punto muerto superior (TDC). Es una indicación la cantidad de la mezcla del aire-combustible que se puede introducir en un cilindro y que tan firmemente se puede comprimir la mezcla.

Computer An electronic device that stores and processes data. It is capable of controlling other devices.

Computadora Un dispositivo electrónico que almacenes y procesa los datos. Es capaz de controlar otros dispositivos.

Conduction When bias voltage difference base and emitter has increased to where the transmitter is turned on.

Conducción Cuando la base de la diferencia del voltaje polarizado y el emisor ha aumentado al punto de prender al transmisor.

Conductor A material that is capable of supporting the flow of electricity through it.

Conductor Una material capaz de tolerar un flujo de la electricidad que lo atraviesa.

Conflict An OBD-II monitor that is stopped because two monitors have met the enabling conditions to run. One of the monitors will stop running since the testing of the other monitor may cause false failures. A conflict can also occur in some monitors when the A/C compressor is turned on.

Conflicto Un monitor de Obd-II que se para debido a que dos monitores han resuelto las condiciones que permitían funcionar. Uno de los monitores parará de funcionar puesto que la prueba del otro monitor puede causar faltas de avería. Un conflicto puede también ocurrir en algunos monitores cuando se prende el compresor de A/C.

Connecting rod Mechanical connection of the piston to the crankshaft.

Biela Una conexión mecánica del pistón al cigüeñal.

Constant control relay module Module that contains several relays within its housing.

Módulo de control constante del relais Un módulo que contiene a varios relais dentro de su caja.

Continuous monitor An OBD-II monitor that runs all the time to determine the readiness of OBD to set DTCs.

Monitor continuo Un monitor de Obd-II que funciona continuamente para determinar la preparación de OBD para regular los DTCs.

Conventional theory States that proton flow of electricity is from a positive point to a more negative point.

Teoría convencional Indica que el flujo de los protónes de la electricidad es de un punto positivo a un punto más negativo.

Converted vehicles Vehicles that have been changed from the use of gasoline fuel to an alternate fuel source after the vehicle was produced.

Vehículos convertidos Los vehículos que se han convertido del uso del combustible de la gasolina a una combustible alternativo después de que el vehículo fuera producido.

Corporate average fuel economy (CAFE) Standards that are federally imposed regulations requiring vehicle manufacturers to a meet a specific average fleet fuel mileage for all their vehicles.

Economía de combustible corporativos (CAFE) Estándares que son impuestas por el gobierno federal que requieren que los fabricantes de los vehículos cumplen un kilometraje promedio de la flota específico del combustible en todos sus vehículos.

Co-solvents Another substance (usually another alcohol) that is soluble in both methanol and gasoline and is used to reduce the tendency of the liquids to separate.

Co-solventes Otra sustancia (generalmente un alcohol) que es soluble en metanol y gasolina y que reduce la tendencia de los líquidos de separarse.

Covalent bonding Atoms that share electrons with other atoms.

Enlace covalente Átomos que comparten electrones con otros átomos.

Crankcase emissions Engine vapors escaping past the piston rings, out of the engine, and into the atmosphere.

Emisiones del cárter Los vapores de motor que se escapan por los anillos del pistón, del motor, hacia la atmósfera.

Crankshaft A mechanical device that converts the reciprocating motions of the pistons into rotary motion.

Cigüeñal Un dispositivo mecánico que convierte los movimientos del intercambio de los pistones al movimiento giratorio.

Crude oil A mixture of hydrocarbon compounds ranging from gases to heavy tars and waxes. The crude oil can be refined into products such as lubricating oils, greases, asphalts, kerosene, diesel fuel, gasoline, and natural gas.

Petróleo crudo La mezcla de los compuestos de hidrocarburo desde los gases a los alquitranes y a las ceras pesados. El petróleo crudo se puede refinar en productos tales como los aceites lubricantes, grasas, asfaltos, keroseno, combustible diesel, gasolina, y gas natural.

Crystal Term used to describe a material that has a definite atom structure.

Cristalino El término describiendo un material que tiene una estructura atómica definida.

Cutoff Where reverse bias voltage is applied to the base leg of the transistor.

Mecanismo de parada En dónde el voltaje de polarización reversa se aplica al puntal en el base del transistor.

Current The aggregate flow of electrons through a conductor.

Corriente El flujo agregado de electrones a través de un conductor.

Cycle A sequence that is repeated.

Ciclo Una secuencia que se repite.

Cylinder block The main structure of the engine. Most of the other engine components are attached to the block.

Bloque de cilindro La estructura principal del motor. La mayoría de los otros elementos del motor se conectan al bloque.

Cylinder heads Upper portion of the combustion chamber that contains the valves, valve seats, valve guides, and valve springs.

Cabeza del cilíndro La porción superior la cámara de combustión que contiene las válvulas, los asientos de las válvulas, las guías de las válvulas, y las válvulas de resorte.

Darlington pair Used in some amplifier circuits. Two transistors that are connected together to increase current flow.

Par Darlington Usado en algunos circuitos amplificadores. Dos transistores están conectados juntos para aumentar el flujo actual.

Delta pressure feedback EGR (DPFE) sensor A ceramic, capacitive-type pressure transducer that monitors the differential pressure across a metering orifice located in the orifice tube assembly.

Sensor Delta de retroalimentación de presión EGR (DPFE) Un transductor de presión de cerámica, tipo capacitivo que supervisa la diferencial de presión a través de un orificio medidor localizado en la asamblea del tubo del orificio.

Density The relative mass of a matter.

Densidad La masa relativa de una materia.

Density speed Fuel system that is similar to the air density system except the engine will not run if either the rpm or MAP sensor inputs are missing.

Velocidad de la densidad Sistema de combustión similar al sistema de la densidad del aire menos que el motor no funcione si las RPM o las entradas del sensor del MAPA faltan.

Depletion-type FET A special type of FET that cuts current flow.

FET de agotamiento Un tipo especial de FET que corta el flujo del corriente.

Detonation The result of an amplification of pressure waves, such as sound waves, occurring during the combustion process when the piston is near top dead center (TDC). The actual "knocking" or "ringing" sound of detonation is due to these pressure waves pounding against the insides of the combustion chamber and the piston top.

Detonación El resultado de una amplificación de las ondas de presión, tal como las ondas acústicas, ocurriendo durante el proceso de la combustión cuando el pistón está cerca del punto muerto superior (TDC). El sonido o golpeteo del detonación es debido a que las ondas de presión golpean contra los interiores de la cámara de combustión y de la cabeza del pistón.

Diagnostic executive The results of OBD-II testing are stored by the diagnostic management system into a software program.

Ejecutivo diagnóstico Los resultados de la prueba de Obd-II se almacenan por el sistema de gerencia diagnóstico en un programa de la computadora.

Diagnostic management system The software that manages the OBD-II system.

Sistema de gerencia diagnóstico Las programas de la computadora que manejan el sistema de Obd-II.

Diagnostic switch Used to determine correct operation of the EVAP solenoid.

Interruptor diagnóstico Sirve para determinar la operación correcta del solenoide de EVAP.

Diagnostic weight Some conditions have higher precedence, and events that happen in these conditions are weighted.

Peso diagnóstico Algunas condiciones tienen precedencia de importancia, y los acontecimientos que suceden bajo estas condiciones son evaluadas.

Dielectric The insulator in a capacitor. The dielectric can be made of some insulator material such as ceramic, glass, paper, plastic, or even the air between the two plates.

Dieléctrico El aislador en un condensador. El dieléctrico puede ser hecho de del material aislante tal como de cerámica, de cristal, de papel, plástico, o aún del aire entre las dos placas.

Diesel fuel A light oil that is refined as part of the same process that makes gasoline. It has several properties that make it useful as a fuel.

Combustible diesel Un aceite ligero refinado como parte del mismo proceso que produce la gasolina. Tiene varias características que lo hacen servir de combustible.

Diode An electrical one-way check valve that will allow current to flow in one direction only.

Diodo Una válvula eléctrica de retentción unidireccional que permitirá que la corriente fluya en una sóla dirección.

Displacement A measure of engine volume. Theoretically, the larger the displacement, the greater the power output.

Dislocación Una medida del volumen del motor. Teóricamente, cuanto más grande es la dislocación, mayor es la salida de energía.

Distributor Controls the primary circuit and is responsible for distributing the secondary spark to the correct combustion chamber.

Distribuidor Controla el circuito primario y es responsable por distribuir la chispa secundaria a la cámara de combustión correcta.

Distributor cap Fastened to the distributor and has a terminal in the center to receive the secondary voltage from the coil and a terminal tower for each cylinder to send the spark to the spark plug.

Tapa del distribuidor Sujetado al distribuidor y tiene un terminal en el centro para recibir el voltaje secundario de la bobina y un torre terminal para cada cilindro para enviar la chispa a la bujía.

Distributor ignition (DI) In SAE J1930 terminology, this term replaces all previous terms for electronically controlled distributor-type ignition systems.

Encendido por distribuidor (DI) En la terminología del SAE J1930, este término substituye todos los términos anteriores para los sistemas de ignición controlados electrónicamente del tipo de distribuidor.

Diverter valve Directs the air pump's output away from the exhaust system during deceleration.

Válvula desviador Dirige la salida de la bomba del aire en el sentido contrario del dispositivo de escape durante la desaceleración

Double flare fitting A line fitting that uses a procedure that folds the ends of the tube over itself doubling the thickness of the tube end and creating two sealing surfaces.

Guarnición doble abocinado Una guarnición de línea que utiliza un procedimiento que doble los extremos del tubo sobre sí mismo doblando el gruesor del extremo del tubo y creando dos superficies de estanqueidad.

Drive cycle A specific driving method to verify a symptom or the repair for the symptom. A drive cycle is also a specific driving method to begin and complete a specific OBD-II monitor.

Ciclo conducción Un método de conducción específico verificar un síntoma o la reparación para el síntoma. Un ciclo de conducción es también un método específico de conducir para comenzar y para terminar un monitor específico de Obd-II.

Driveability Index (DI) A specification for gasoline used to manage engine performance during cold weather and while the engine is warming up.

Índice de manejo (DI) La especificación de la gasolina utilizada para controlar el funcionamiento del motor durante el tiempo frío y mientras que el motor está calentandose.

Dual plug systems Two spark plugs per cylinder.

Sistemas de doble bujía Dos bujías de chispa por cada cilindro.

Dwell The length of time current flows through the primary windings of the coil.

Reposo La cantidad del tiempo en el cual la corriente atraviesa las bobinas primarias de la bobina.

E coil An ignition coil design that is air-cooled, epoxy-sealed.

Bobina E Un diseño de la bobina de la ignición que es refrigerado por aire y sellado.por epoxy.

Eddy currents Currents that flow in reverse of the main current.

Corrientes de Foucault Las corrientes que fluyen en el sentido revés de la corriente principal.

Efficiency A measure of a device's ability to convert energy into work.

Eficacia Una medida de la capacidad con que los dispositivos convierten la energía al trabajo.

Electric fuel pump Uses a DC motor to drive a pump gear or roller set to draw fuel.

Bomba de gasolina eléctrica Utiliza un motor DC para conducir un sistema del engranaje o del rodillo de la bomba para aspirar al combustible.

Electric vehicle (EV) A vehicle that powers its motor from a battery pack.

Vehículo eléctrico (EV) Un vehículo que impulsa su motor por medio de un paquete de batería.

Electrolysis The splitting of water into hydrogen and oxygen.

Electrólisis La división del agua en el hidrógeno y el oxígeno.

Electromagnetic Interference (EMI) An undesirable creation of electromagnetism whenever current is switched on and off.

Interferencia electromágnetica (EMI) Una creación indeseable del electromagnetismo cuando la corriente se prenda y se apaga.

Electromagnetism A form of magnetism that occurs when current flows through a conductor.

Electromagnetismo Una forma del de magnetismo que ocurre cuando la corriente atraviesa un conductor.

Electromotive force (EMF) Voltage.

Fuerza electromotriz (EMF) Voltaje.

Electron theory Defines electrical movement as from negative to positive. This is the flow of electrons.

Teoría electrónica Define el movimiento eléctrico como de negativo al positivo. Éste es el flujo de electrones.

Electronic fuel injection (EFI) Refers to any fuel injection system that is electronically controlled including TBI, MFI, and SFI systems.

Inyección electrónica del combustible (EFI) Refiere a cualquier sistema de inyección del combustible que se controle electrónicamente incluyendo sistemas de TBI, de MFI, y de SFI.

Electronic ignition (EI) In SAE J1930 terminology, this term replaces all previous terms for distributorless ignition systems.

Ignición electrónica (EI) En terminología del SAE J1930, este término substituye todos los términos anteriores de los sistemas de ignición sín distribuidor.

Electrons Negative charged particles that orbit around the nucleus of an atom.

Electrones Las partículas de carga negativa que se mueven en órbita alrededor del núcleo de un átomo.

Electrostatic field The field that is between the two oppositely charged plates.

Campo electrostático El campo que está entre las dos placas de carga opuesta.

Emitter Part of the transistor that has the same polarity as the side of the circuit it is connected to.

Emisor La parte del transistor que tiene la misma polaridad del lado del circuito con que está conectada.

Enhanced EVAP Evaporative emission systems that use either positive or negative pressures to detect a 0.020-inch (0.5 mm) leak.

EVAP realzado Las sistemas evaporativos de la emisión que utilizan las presiones positivas o negativas para detectar un escape de .020-pulgada (0,5 milímetros).

Enhanced test Uses a transient drive cycle to test for HC, CO, CO₂, and HC, as well as EVAP performance.

Prueba realzada Utiliza un ciclo transitorio de la impulsión para probar para HC, CO, CO₂, y HC, así como el funcionamiento del EVAP.

Enhancement-type FET A special FET that improves current flow.

FET tipo realzada Un tipo especial del FET que mejora un flujo actual del corriente.

Ethyl alcohol An alcohol made from grain.

Alcohol etílico Un alcohol hecho de grano.

EVAP pressure test Used to ensure that there are no leaks from the fuel tank to the evaporative canister where fuel vapors could escape into the atmosphere.

La prueba de presión de EVAP Se utiliza para asegurar que no hay escapes del tanquede gasolina al frasco evaporativo en donde los vapores de combustible podrían escaparse a la atmósfera.

EVAP purge flow test Measures the performance of the EVAP to move vapors into the intake manifold.

La prueba de purga del flujo de EVAP Mide el funcionamiento del EVAP en mover los vapores hacia el múltiple de entrada.

Evaporative canister A charcoal-filled canister used to temporarily store evaporative emissions.

El frasco evaporativo Un frasco lleno de carbón usado para almacenar temporalmente a las emisiones evaporativas.

Evaporative emissions Raw fuel vapors that drift into the atmosphere. This evaporation is a direct result of the property of gasoline to vaporize at low temperatures; they are not the result of the combustion process.

Emisiones evaporativas Los vapores crudos del combustible que se escapan a la atmósfera. Esta evaporación es un resultado directo de una propriedad de la gasolina de vaporizarse en las temperaturas bajas;no son el resultado del proceso de combustión.

Excess vacuum test A passive test designed to detect restrictions in the vent path of the EVAP system.

Pruebade del exceso del vacío Una prueba pasiva diseñado para detectar las restricciones en la trayectoria del respiradero del sistema EVAP.

Exhaust emissions Produced as a result of combustion and escape through the vehicle's exhaust system.

Emisiones de escape Producidas como resultado de la combustión y escapen a través del sistema del escape del vehículo.

Exhaust gas recirculation (EGR) An emission control system that mixes the inert gases of the exhaust into the air-fuel charge to reduce combustion temperatures.

Recirculación del gas de escape (EGR) Un sistema de control de emisión que mezcla los gases inertes del extractor en la carga del aire-combustible para reducir temperaturas de la combustión.

Expansion tank A chamber of the fuel tank that allows for fuel expansion resulting in temperature increases.

Tanque de extensión Un compartimiento del tanque de combustible que permite la extensión del combustible dando por resultado las temperaturas más altas.

Exponentially weighted moving averaging (EWMA) A well-documented statistical data processing technique that is used to reduce the variability on an incoming stream of data.

Promedio exponencial de carga móvil (EWMA) Una técnica bien documentada de la información estadística que se utiliza para reducir la variabilidad en una corriente entrante de datos.

Face shields A shield made of a clear plastic that protects the wearer's entire face.

Protector de cara Un protector hecho de un plástico claro que proteja la cara entera de los que lo lucen.

Fast moving average A combination of instantaneous switch frequencies of the downstream oxygen sensor and stored history. Used to detect a major catalyst failure, such as a severe misfire, that has melted the catalyst substrate.

Promedio móvil Una combinación de las frecuencias instantáneas del interruptor del sensor del oxígeno en sentido descendiente y la historia almacenada. Sirve para detectar una falta importante del catalizador, tal como un fallo de encendido severo, que ha derretido el substrato del catalizador.

Federal test procedure (FTP) A series of test procedures, including drive cycles, that are used to measure a vehicle's emissions.

Procedimiento federal de prueba (ftp) Una serie de métodos de prueba, incluyendo un ciclo de impulsión completo, que se utilizan para medir emisiones de los vehículos.

Feedback signal Data concerning the effects of the computer's commands are fed back to the computer as an input signal.

Señal de retorno Los datos referentes a los efectos de los comandos de la computada que se retroactúan a la computadora como señal de entrada.

Feedback As it relates to fuel systems, means the system uses an oxygen sensor to provide feedback information to the computer concerning combustion quality.

Retorno Relacionado con los sistemas de combustión, significa que el sistema aplica un sensor del oxígeno para proporcionar la información del retorno a la computadora referente a la calidad de la combustión.

Field-effect transistor (FET) A transistor that does not require a constant turn-on voltage.

Transistor del efecto de campo (FET) Un transistor que no requiere un voltaje de abertura constante.

Filler tube The pipe the fuel flows through to fill the tank.

Tubo de relleno La pipa por la cual el combustible atraviesa para llenar el tanque.

Firing order The sequence that the spark plugs are fired in.

Orden de la disparo La secuencia en la cual se disparan los bujías de chispa.

Flammable A substance that is will support combustion.

Inflamable La sustancia que promuebe la combustión.

Flash point The temperature at which the vapors on the surface of the fuel will ignite if exposed to an open flame.

Punto de destello La temperatura en la cual los vapores en la superficie del combustible encenderán si están expuestos a una llama.

Flex fuel vehicle (FFV) A vehicle that uses a single fuel tank, fuel system, and engine. The vehicle is designed to run on regular unleaded gasoline and an alcohol fuel (either ethanol or methanol) in any mixture.

Vehículo del combustible flexible (FFV) Un vehículo que utiliza sólo un tanque de gasolina, sistema de carburante, y motor. El vehículo se diseña para funcionar en gasolina unleaded regular y un combustible del alcohol (etanol o metanol) en cualquier mezcla.

Forward-bias A positive voltage is applied to the P-type material and negative voltage to the N-type material of a diode so it will conduct.

Polarizado delantero Un voltaje positivo se aplica al material del tipo-P y el voltaje negativo al material tipo-N de un diodo para que conducirá.

Freeze frame A snap shot of the vehicle's operating conditions when a fault is first detected.

Instantáneo Las condiciones inmediatas de funcionamiento del vehículo en el momento que se detecta un fallo.

Fuel cap Seals the gas tank, but must also provide for pressure release and fresh air intake.

Casquillo del tanque Sella el tanque combustible, pero debe también prever la descarga de la presión y la entrada del aire fresco.

Fuel cell A battery-like component that produces current from hydrogen and aerial oxygen.

Célula de combustible Un componente parecido a una batería que produce la corriente del hidrógeno y oxígeno aéreo.

Fuel filters Elements made from pleated paper, ceramic, or bronze material used to remove contamination in the fuel delivery system.

Filtros de combustible Los elementos hechos del material plisado del papel, de cerámica, o de bronce que sirven para quitar la contaminación del sistema de entrega del combustible.

Fuel pump The device that draws the fuel from the fuel tank through the fuel lines to the engine's carburetor or injectors.

Bomba de gasolina El dispositivo que toma el combustible del tanque de gasolina a través de las líneas de combustible hacia el carburador o hacia los inyectores del motor.

Fuel pump module An assembly that incorporates the fuel pump, fuel filter, fuel gauge sending unit, and fuel temperature sensor (if used) in a single assembly.

Módulo de la bomba de gasolina Una asamblea que incorpora lal bomba de gasolina, el filtro de combustible, la unidad de la galga de combustible, y el sensor de temperatura de combustible (si se usa) en una sola asamblea.

Fuel rail A manifold system used to supply fuel to the injectors.

Carril de combustible El sistema del múltiple qu se usa para proveer el combustible a los inyectores.

Fuel system good trip Used to turn the MIL off after a fuel system DTC as been set. It requires three good trips.

Buen viaje del sistema de combustión Se usa para apagar el MIL después de que un sistema de combustión DTC esté fijo. Requiere tres viajes buenos.

Fuel system monitor Determines if the running condition of the engine is too rich or too lean to be controlled and brought back to stoichiometric.

Monitor del sistema de combustón Determina si la condición en marcha del motor es demasiado rica o demasiado pobre para ser controlada y regresar de nuevo a stoichiométrico.

Fuel tank Stores the liquid fuel until it is delivered to the engine.

Tanque de gasolina Almacena el combustible líquido hasta que se entrega al motor.

Full parallel hybrid Uses an electric motor that is powerful enough to propel the vehicle on its own.

Híbrido paralelo completo Utiliza un motor eléctrico que sea bastante fuerte para impulsarse si mismo a un vehículo..

Gas law principle States that in a sealed container, a vacuum will be created as the temperature of the gas decreases.

Principio de la ley de combustible Indica que en un envase sellado, un vacío será creado cuando cae la temperatura del gas.

Gasoline A term used to describe a complex mixture of various hydrocarbons refined from crude petroleum oil for use as a fuel in engines. Gasoline contains hydrogen and carbon molecules. The chemical symbol for this liquid is C8H15. Gasoline is a colorless liquid with excellent vaporization capabilities.

Gasolina Un término que describa una mezcla compleja de los varios hidrocarburos refinados del aceite del petróleo para el uso como combustible en los motores. La gasolina contiene las moléculas del hidrógeno y del carbón. El símbolo químico para este líquido es C8H15. La gasolina es un líquido descolorido con capacidades excelentes de la vaporización.

Global good trip All monitors that run once per trip must have run and passed. Usually this means the HO_2S and Catalyst Efficiency monitors must run and pass.

Buen viaje global Todos los monitores que funcionen una vez por viaje deban haber funcionado y haber cumplido. Esto significa generalmente que los monitores de la eficacia de HO_2S y del catalizador deben funcionar y cumplir.

Glow plugs A heating element that is threaded into the combustion chamber and uses electrical current to heat the intake air.

Bujías de resplandor Elemento de calefacción que se enrosca en la cámara de combustión y utiliza la corriente eléctrica para calentar el aire de entrada.

Ground The point of the lowest voltage in an electrical circuit. The ground circuit used in most automotive systems is through the vehicle chassis and/or engine block.

Tierra El punto de la tensión más baja de un circuito eléctrico. El circuito de tierra usado en la mayoría de los sistemas automotores está a través del bloque del chasis y/o del motor del vehículo.

Half cycle counts Indicates the frequency of the O_2 sensor.

Cuenta de medio ciclo Indican la frecuencia del sensor O_2.

Hall-effect switch An input device that operates on the principle that if a current is allowed to flow through thin conducting material that is exposed to a magnetic field, another voltage is produced.

Interruptor de efecto hall Un dispositivo de entrada que funcione bajo un principio de que si una corriente se permite atravesar una material delgada de conducción que se expone a un campo magnético, se produce otro voltaje.

Hazard Communication Standard The basis of the right-to-know laws. It was originally intended for chemical companies and manufacturers that required employees to handle hazardous materials in their work situation.

Estándar de la Comunicación de Peligros La base de los leyes del derecho-a-saber. Fue creado originalmente para las compañías y los fabricantes químicos que requirieron que sus empleados manejaran los materiales peligrosos en su situación de trabajo.

Hazardous materials Materials that can cause illness, injury, or death or pollute water, air, or land.

Materiales peligrosos Los materiales que pueden causar enfermedad, lesión, o muerte o contaminar el agua, el aire, o la tierra.

HD-5 Propane rating designating 95 percent propane and 5 percent butane.

HD-5 Grado del propano que señala el 95 por ciento de propano y 5 por ciento de butano.

Heat range Refers to the heat dissipation properties of the spark plug.

Gama del calor Refiere a las características en la disipación de calor del bujía de chispa.

Heated air inlet Systems that direct warm air into the air cleaner assembly during cold engine operation to help vaporize the fuel.

Entrada de aire caliente Los sistemas de la que dirigen el aire caliente en el montaje del filtro de aire durante la operación en frío del motor para ayudar vaporizar el combustible.

High-side drivers Control the output device by varying the positive (12-volt) side of the circuit.

Conductores del lado alto Controlan el dispositivo de salida variando el lado positivo (12-voltíos) del circuito.

Hole The absence of an electron in an atom. These holes are said to be positively charged since they have a tendency to attract free electrons into the hole.

Agujero La ausencia de un electrón en un átomo. Estos agujeros serían cargados positivamente puesto que tienen una tendencia a atraer electrones libres hacia el agujero.

Horsepower The measure of the rate of work.

Caballomotor La medida del índice del trabajo.

Hybrid electric vehicle (HEV) System that has two different power sources. In most hybrid vehicles (HEV), the power sources are a small displacement gasoline or diesel engine and an electric motor.

Vehículo eléctrico híbrido (HEV) Un sistema que tiene dos diversas fuentes de energía. En la mayoría de los vehículos híbridos (HEV), las fuentes de energía son de un motor de gasolina de dislocación pequeña o motor de diesel y un motor eléctrico.

Hydrocarbons Organic compounds that contain only hydrogen and carbon.

Hidrocarburos Compuestos orgánicos que contienen solamente el hidrógeno y carbón.

I/M 240 An inspection/maintenance vehicle emissions testing program for improving air quality in areas that fail to meet the federal government's ambient air quality standards.

I/M 240 Una programa de prueba de inspección y mantenimiento de las emisiones del vehículo para mejorar la calidad del aire en las áreas que no conformen con los estándares de calidad del medio ambiente del gobierno federal.

Idle check Screens out leak indications in the EVAP system caused by changes in altitude.

Cheque de marcha en vacío Elimina las indicaciones de fuga en el sistema de EVAP causado por los cambios en altitud.

Ignition cables Carry the high voltage from the distributor or the multiple coils (EI systems) to the spark plugs.

Cables de ignición Llevan el alto voltaje del distribuidor o las bobinas múltiples (los sistemas de EI) a los bujías de chispa.

Ignition timing Refers to the precise time a spark is sent to the cylinder relative to the piston position.

Sincronización de ignición Refiere al tiempo exacto en que una chispa se envía al cilindro relativo a la posición del pistón.

Index ratio Method that uses the ratio of signal lengths to determine proper activity.

Cociente índice El método que utiliza el cociente de las longitudes de la señal para determinar actividad apropiada.

Indicated horsepower The amount of horsepower the engine can theoretically produce.

Caballos de fuerza indicados La cantidad de caballos de fuerza que puede producir teóricamente el motor.

Induction The magnetic process of producing a current flow in a wire without any actual contact with the wire.

Inducción El proceso magnético de producir un flujo actual en un alambre sin cualquier contacto actual con el alambre.

Inductive reactance Opposing current as the result of self induction. Inductive reactance is similar to resistance since it resists any increase in current flow in a coil.

Reactancia inductiva Corriente de oposición como resultado del autoinducción. La reactancia inductiva es parecido a la resistencia puesto que resiste cualquier aumento en flujo actual en una bobina.

Inertia switch Used to shut off the fuel pump in the event of an accident that may rupture the fuel line and to cause the engine to shut off after an accident.

Interruptor de la inercia Se usa para apagar la bomba de gasolina en el acontecimiento de un accidente que puede romper la línea de combustible y apagar al motor después de un accidente.

Inferred MAP Calculated from the Mass Air Flow Sensor, throttle position, rpm, BARO, etc.

MAPA deducido Se calcula del sensor de flujo de aire, de la posición de la válvula reguladora, de la RPM, del BARO, etc.

Instantaneous switching frequency The rate at which both the upstream and downstream O_2 sensors cross the rich-to-lean switching points.

Frecuencia instantánea de la conmutación El grado en que los sensores O_2 del sentido ascendente tal como del descendiente cruzan los puntos de conmutación de rico-a-pobre.

Intake manifold Directs the air or air-fuel mixture into the cylinders.

Múltiple de entrada Dirige la mezcla del aire o del aire-combustible hacia los cilindros.

Integrated circuit A complex circuit of thousands of transistors, diodes, resistors, capacitors, and other electronic devices that are formed onto a tiny silicon chip. As many as 30,000 transistors can be placed on a chip that is 0.25 inch (6.35 mm) square.

Circuito integrado Un circuito complejo de millares de transistores, de diodos, de resistores, de condensadores, y de otros dispositivos electrónicos que se forman sobre un chip de silicio minúsculo. Tantos como 30.000 transistores se pueden colocar en un chip que sea de 0,25 pulgadas (6,35 milímetros) cuadrado.

Integrator Represents a short-term correction to the pulse width.

Integrador Representa una corrección a corto plazo a la anchura del pulso.

Interface Has two purposes: protect the computer from excessive voltage levels, and to translate input and output signals.

Interfaz Tiene dos propósitos: proteger la computadora contra niveles voltaicos excesivos, y traducir señales de la entrada y de salida.

Internal combustion engines Engines that burn their fuels within the engine.

Motores de combustión interna Los motores que queman sus combustibles dentro del motor.

Intrusive test A special type of active test in that the action taken by the PCM may affect vehicle performance and emissions.

Prueba intrusa El tipo especial de prueba activa en que la acción tomada por el PCM puede afectar el funcionamiento y emisiones del vehículo.

Ion An atom or group of atoms that has an electrical charge.

Ion Un átomo o un grupo de átomos que tiene una carga eléctrica.

Ionize To electrically charge.

Ionizar Cargar eléctricamente.

Kinetic energy Energy that is in motion. Also called working energy.

Energía cinética La energía que está en movimiento. También llamada energía de trabajo.

Knock sensor A piezoelectric device that generates a voltage based on the pressure caused by detonation.

Sensor de golpeteo Un dispositivo piezoeléctrico que genera un voltaje basado en la presión causada por la detonación.

Lambda The air-fuel mixture is expressed either as the ratio of air to fuel vapor or as a lambda value. The lambda value is derived from the stoichiometric air-fuel ratio. The stoichiometric ratio is 14.7:1 when expressed as an air-fuel ratio, or 1 when expressed as a lambda value.

Lambda La mezcla del aire-combustible se expresa como el cociente del aire al vapor de combustible o como valor de la lambda. El valor de la lambda se deriva del cociente stoichiométrico del aire-combustible. El cociente stoichiometric es 14.7:1 cuando está expresado como cociente del aire-combustible, o 1 cuando está expresado como valor lambda.

Leak detection pump (LDP) A vacuum-driven pump that pressurizes the EVAP system to test for leaks.

Bomba detector de fugas (LDP) Una bomba accionada por vacío que presuriza el sistema de EVAP para probar para los escapes.

Lean An air-fuel mixture in which there is an excess of oxygen mixed with the gasoline.

Pobre Una mezcla del aire-combustible en la cual hay un exceso del oxígeno mezclado con la gasolina.

Lean burn technology Uses lean air-fuel ratios to increase fuel efficiency.

Tecnología de combustión pobre Aplica los cocientes de aire-combustible para aumentarla eficacia del combustible

Lean-burn misfire Results when a given volume of fuel mixture simply does not contain enough gasoline to ignite.

Golpeteo de combustión pobre Resulta cuando un volumen dado de la mezcla del combustible simplemente no contiene bastante gasolina para encenderse.

Light-emitting diode (LED) Similar in operation to the diode, except the LED emits light when it is forward-biased.

Diodo electroluminoso (LED) Similar en su operación al diodo, menos que el LED emita la luz cuando es de polarización delantero.

Linearity Refers to the sensor signal as being constantly proportional to the measured value as possible. It is an expression of the sensor's accuracy.

Linearidad Refiere que la señal del sensor es constantemente proporcional al valor medido como sea posible. Es una expresión de la exactitud de los sensores.

Liquid petroleum (LP) gas The name given to describe a family of light hydrocarbons called "gas liquids." The most prominent members of this family are propane (C3H8) and butane (C4H10).

Gas del petróleo en líquido (LP) El nombre dado para describir una familia de hidrocarburos ligeros llamados "gases en líquido". Los miembros más prominentes de esta familia son el propano (C3H8) y el butano (C4H10).

Loaded canister test A passive test designed to detect leaks in the entire EVAP system.

Prueba de frasco cargado Un prueba pasiva diseñada para detectar las fugas en el sistema entero de EVAP.

Loaded mode test I/M test reads the emissions of vehicles that are driven on a "single weight" dynamometer. This test checks for HC and CO concentrations.

Prueba de modo cargado La prueba I/M lee las emisiones de los vehículos que se conducen en un dinamómetro de un "peso sencillo". Esta prueba comprueba las concentraciones de HC y del CO.

Logic gates Circuits that act as gates to output voltage signals depending on different combinations of input signals.

Puertas lógicas Los circuitos que actúan como puertas para las señales de salida del voltaje dependiendo de diversas combinaciones de las señales de entrada.

Long-term adaptive A fuel strategy memory that is a correction to the pulse width calculation based on short-term activity over a period of time.

Adaptativo a largo plazo Una memoria estratégica del combustible que es una corrección al cálculo de la anchura del pulso basado en la actividad a corto plazo durante un período de tiempo.

Low-side drivers Used to complete the path to ground to turn on an actuator.

Conductores del lado bajo Sirven para completar la trayectoria a tierra para prender un actuador.

Magnetic flux density The concentration of the magnetic lines of force.

Densidad magnética del flujo La concentración de las líneas magnéticas de la fuerza.

Magnetic pulse generators Sensors that use the principle of magnetic induction to produce a voltage signal.

Generadores de pulsos magnéticos Los sensores que utilizan el principio de la inducción magnética para producir una señal del voltaje.

Manifold absolute pressure (MAP) sensor A piezoresistive or capacitance discharge-sensing device used to determine the load on the engine by sensing the difference between atmospheric pressure and engine vacuum.

Sensor de presión absoluta del múltiple (MAPA) Un dispositivo detector de descarga del piezoresistivo o de la capacitancia que determina la carga en el motor detectando la diferencia entre la presión atmosférica y el vacío del motor.

Manifold vacuum Vacuum source coming from a fitting in the intake manifold, downstream of the throttle blade.

Vacío del múltiple Un fuente de vacío que viene de una guarnición en el múltiple de entrada, ubicada descendiente de la lámina de la válvula reguladora.

Mass air flow (MAF) Fuel system that use a MAF sensor that provides a direct input as to the amount of air entering the engine.

Total del flujo de aire (MAF) Un sistema de combustión que utiliza un sensor de MAF que proporcione una entrada directa en cuanto a la cantidad de aire que entra en el motor.

Mass air flow (MAF) sensor Located in the air intake system to directly measure the mass of air entering the engine.

Sensor total del flujo de aire (MAF) Situado en el sistema de la toma de aire para medir la masa del aire que entra en el motor directamente.

Material safety data sheets (MSDS) Contain extensive information and facts about hazardous materials.

Las hojas de datos materiales de seguridad (MSDS) Contienen la información y los hechos extensos sobre los materiales peligrosos.

Mechanical efficiency A comparison of the power actually delivered by the crankshaft to the power developed within the cylinders at the same rpm.

Eficacia mecánica La comparación de la energía entregada actualmente por el cigüeñal a la energía que se convierta dentro de los cilindros en la misma RPM.

Metal detection sensors Operate much like a magnetic pulse generator but use a pick-up coil that is an electromagnet.

Sensores de la detección del metal Funcionan como un generador de pulsos magnéticos pero utilizan una bobina captador que es un electroimán.

Methanol A mixture of gasoline and methyl alcohol.

Metanol Una mezcla del la gasolina y del alcohol metílico.

Mild parallel hybrid Uses an electric motor that is large enough to provide regenerative braking, instant engine startup, and a boost to the combustion engine.

Híbrido paralelo Utiliza un motor eléctrico que es bastante grande para proporcionar el frenado regenerador, el arranque inmediato del motor, y un sobrealimentación al motor de combustión.

Minimum TPS The lowest voltage value the PCM received from the TPS during that key cycle.

Mínimo TPS El valor de la tensión más bajo que recibe el PCM del TPS durante ese ciclo dominante.

Misfire good trip Used to turn the MIL off after a misfire DTC as been set. It requires three good trips.

Fallo del buen viaje Se usa para apagar al MIL después de que se recibe un fallo en DTC. Requiere tres buenos viajes.

Misfire monitor Detects any external torque disturbances of the crankshaft and indicates if the engine is polluting above FTP or if the converter is being damaged.

El monitor de fallos Detecta cualquier disturbio externo del cigüeñal e indica si el motor está contaminando más allá del FTP o si se está dañando al convertidor.

Monolithic A type of construction in which a ceramic honeycomb is coated with a thin layer of the catalyst metals.

Monolítico Un tipo de construcción en el cual un panal de cerámica está cubierto con una capa delgada de los metales del catalizador.

Multiplexing Provides the ability to use a single circuit to distribute and share data between several control modules throughout the vehicle. Because the data is transmitted through a single circuit, bulky wiring harnesses are eliminated.

Multiplexación Proporciona la capacidad de utilizar un solo circuito para distribuir y para compartir datos entre varios módulos de control a través del vehículo. Debido a que los datos se transmiten a través de un solo circuito, se eliminan los arneses de cableado estorbadores.

Multipoint fuel injection (MPI) A fuel system that has one injector per cylinder. The injectors are fired in pairs or in groups of three or four. Usually half of the fuel is delivered on each crankshaft revolution.

Inyección en puntos múltiples (MPI) Un sistema de combustión que tiene un inyector por cada cilindro. Los inyectores se encienden en pares o en grupos de tres o cuatro. La mitad del combustible se entrega generalmente en cada revolución del cigüeñal.

Mutual induction An induction of voltage in an adjacent coil by changing current in a primary coil.

Inducción mutua Una inducción del voltaje en una bobina adyacente cambiando la corriente en una bobina primaria.

MUX The common acronym for multiplexing.

MUX Las siglas comunes para multiplexar.

Negative back pressure EGR EGR valve that opens the bleed valve when exhaust back pressure decreases to stop EGR flow.

EGR negativa de contrapresión Una válvula de EGR que abre la válvula de purga cuando disminuyen los flujos de contrapresión del extractor para parar al EGR.

Negative temperature coefficient (NTC) A thermistor that reduces its resistance as the temperature increases.

Coeficiente de la temperatura negativo (NTC) Un termistor que reduce su resistencia mientras que la temperaturasuba.

Neutrons Particles that have no charge called.

Neutrones Las partículas que no tienen ninguna carga específica.

Non-enhanced EVAP Evaporative emission systems that are certified by the manufacturer to the EPA to have an expected useful life of 100,000 miles.

EVAP No-realizados Los sistemas evaporativos de la emisión certificados por el fabricante al EPA para tener una vida útil prevista de 100.000 millas.

Northeast Trading Region (NTR) Includes Massachusetts, New York, Maine, and Vermont.

Región Negociante Nordestal (NTR) Incluye a Massachusetts, Nueva York, Maine, y Vermont.

N-type material Materials with free electrons. The N means negative and means that it is the negative side of the circuit that pushes electrons through the semiconductor and the positive side that attracts the free electrons.

Materiales tipo-N Los materiales que tienen los electrones libres. La N significa la negativa y significa que es el lado negativo del circuito que empuja electrones a través del semiconductor y es el lado positivo que atrae los electrones libres.

Nucleus Consists of positively charged particles called protons and particles that have no charge called neutrons.

Núcleo Consiste en las partículas cargadas positivamente llamadas los protones y las partículas que no tienen ninguna carga llamada los neutrones.

NYTEST239 An emissions test procedure of a hybrid version of ASM and I/M 240 testing that includes transient mode testing on a dynamometer but uses a less costly exhaust sampling system.

PRUEBANY239 Un método de prueba de las emisiones en de una versión híbrida de las pruebas ASM y I/M 240 que incluye la prueba del modo transitorio utilizando un dinamómetro pero que usa una versión menos costoso del sistema de muestreo del escape.

O_2 feedback systems Use an oxygen sensor to determine if the proper air-fuel mixture was delivered to the combustion chamber.

Los sistemas de la regeneración O_2 Utilizan un sensor del oxígeno para determinarse si la mezcla apropiada del aire-combustible fue entregada a la cámara de combustión.

Occupational safety glasses Glasses that are designed with special high-impact lens and frames and provide for side protection.

Gafas de seguridad Los lentes que se diseñan con la lente y los marcos especiales de alto impacto y preven la protección lateral.

Octane rating A measure of how easily the gasoline can be ignited. The higher the octane, the harder the fuel is to ignite and the slower it burns. Generally, the higher the octane, the greater the number of hydrocarbons containing larger numbers of carbon atoms. More carbon atoms per hydrocarbon means that more oxygen and more heat are needed to burn the fuel. The octane rating provides an indication of its anti-knock qualities. The higher the octane number, the less tendency for knock.

Grado de octano Una medida de que tal facilmente puede ser encendida la gasolina. Cuanto más alto es el octano, más difícil es encenderse el combustible y se quema más lentamente. Generalmente, cuanto más alto es el nivel de octano, mayor es el número de los hidrocarburos que tienen más átomos de carbón. Si tiene más átomos del carbón por hidrocarburo significa que se necesita más oxígeno y más calor para quemarse el combustible. El grado del octano proporciona una indicación de sus calidades antidetonantes. Cuanto más alto el número de octano, menos la tendencia para los golpeteos.

Ohms The measurement of resistance of a conductor such that a constant current of 1 ampere in it produces a voltage of 1 volt between its ends.

Ohmios La medida de la resistencia de un conductor en que un corriente constante que contiene 1 amperio produce un voltaje de 1 voltio entre sus extremidades.

Ohm's Law Defines the relationship between current, voltage, and resistance.

La ley de Ohm Define la relación entre la corriente, el voltaje, y la resistencia.

Oil pressure switch Safety switch that closes when the oil pressure developed by the engine's oil pump overcomes a preset value to allow current flow to the fuel pump.

Interruptor de presión del aceite El interruptor de seguridad que se cierra cuando la presión del aceite creada por la bomba del aceite del motor supera el valor preestablecido para permitir el flujo de corriente a la bomba de gasolina.

On-board diagnostics second generation (OBD-II) A legal regulation that requires vehicle manufacturers to have on-board diagnostic routines to determine if the engine is polluting.

Diagnóstico a bordo de segundo generación (Obd-II) Una regulación legal que requiere que los fabricantes del vehículo tengan las rutinas diagnósticas a bordo para determinar si el motor produce la contaminación.

On-board refueling vapor recovery (ORVR) A fueling system that prevents hydrocarbon vapors from escaping to the atmosphere during refueling by drawing them into the fuel tank.

Recuperación del vapor de reaprovisión a bordo (ORVR) Sistema de combustible que previene que los vapores de hidrocarburo se escapen a la atmósfera durante reaprovisión de combustible reintroduciéndoles al depósito de gasolina.

Once-per-trip monitor An OBD-II monitor that runs once for every driving trip defined by an engine start-stop cycle.

Monitor de una-vez-por-viaje Un monitor de Obd-II que funciona una vez en cada viaje definido por un ciclo de marcha-parada del motor.

One-trip faults OBD-II tests that illuminate the MIL after one failure is recorded. These are also called "A monitors."

Fallo en un sólo-viaje Las pruebas de Obd-II que prenden la MIL después de que se registre una falta. Éstos también se llaman los "monitores A".

Open loop Mode of operation wherein the oxygen sensor is out of the fuel strategy loop. The oxygen sensor is ignored during open-loop operation.

Lazo abierto Modo de operación en el cual el sensor del oxígeno es exterior del lazo de la estrategia del combustible. No se hace caso al sensor del oxígeno durante la operación de lazo abierto.

Organic Refers to a product from a source that was originally living.

Orgánico Refiere a un producto que originalmente tenía vida.

Oxidation A reaction in which oxygen combines chemically with another substance.

Oxidación Una reacción en la cual el oxígeno combina químicamente con otra sustancia.

Oxidation catalysts Catalysts that are designed to change hydrocarbons and carbon monoxide into CO_2 and water vapor.

Catalizadores de la oxidación Los catalizadores que se diseñan para cambiar los hidrocarburos y el monóxido de carbono al del CO_2 y vapor de agua.

Oxidation converters Converters that use oxygen to cause the reaction.

Convertidores de la oxidación Los convertidores que utilizan el oxígeno para causar la reacción.

Oxygenated fuels Mixtures of conventional gasoline and one or more combustible liquids that contain oxygen (oxygenates).

Combustibles oxigenadas Las mezclas de la gasolina convencional y de uno o más líquidos combustibles que contienen el oxígeno (que oxígena).

Parallel hybrid A hybrid vehicle configuration that has a direct mechanical connection between the engine and the wheels. Both the engine and the electric motor can turn the transmission at the same time.

Híbrida paralela Una configuración del vehículo híbrido que tiene una conexión mecánica directa entre el motor y las ruedas. El motor y el motor eléctrico pueden dar vuelta a la transmisión en el mismo tiempo.

Passive test Monitors the system or components during normal operation.

Prueba pasiva Supervisa el sistema o los componentes durante la operación normal.

Pended An OBD-II monitor that stops if the diagnostic routine has found fault in another monitor that will cause this monitor to fail. The monitor is stopped pending repair of the first fault.

Pendiente Un monitor Obd-II que se para si durante la rutina de diagnóstico encuentra una avería en otro monitor que causará una falla en este monitor. El monitor se para hasta que finalice la reparación de la primera avería.

Permeability The term used to indicate the magnetic conductivity of a substance compared with the conductivity of air. The greater the permeability, the greater the magnetic conductivity and the easier a substance can be magnetized.

Permeabilidad Un término que indica la conductividad magnética de una sustancia comparada con la conductividad del aire. Cuanto mayor es la permeabilidad, mayor es la conductividad magnética y más fácilmente puede ser magnetizada una sustancia.

Petroleum A word meaning rock oil.

Petróleo La palabra que significa el aceite que viene de piedra

Photo diode A special diode that allows current to flow in the opposite direction of a standard diode when it receives a specific amount of light.

Fotodiodo Un diodo especial que permite que la corriente fluya en la dirección opuesta de un diodo estándar cuando recibe una cantidad específica de luz.

Photochemical smog A condition that develops when oxides of nitrogen and volatile organic compounds created from fossil fuel combustion interact under the influence of sunlight to produce a mixture of hundreds of different chemicals.

Contaminación fotoquímica Una condición que ocurre cuando los óxidos del nitrógeno y de los compuestos orgánicos volátiles creados por la combustión del combustible fósil coordinan con la influencia de la luz del sol para producir una mezcla de centenares de diversos productos químicos.

Phototransistor A transistor that is sensitive to light.

Fototransistor Transistor que es sensible a la luz.

Pick-up coil Also known as a stator, sensor, or pole piece. It remains stationary while the timing disc rotates in front of it. The changes of magnetic lines of force generate a small voltage signal in the coil.

Bobina captor También conocida como un estator, un sensor, o masa polar. Queda inmóvil mientras que el disco de regulación gira delante de él. Los cambios en las líneas magnéticas de fuerza generan una señal pequeña del voltaje en la bobina.

Pistons Move up and down in the cylinders of the block. The top of the pistons are part of the combustion chamber.

Pistones Se mueven hacia arriba y hacia abajo en los cilindros del bloque. La cabeza de los pistones es parte de la cámara de combustión.

Port fuel injection (PFI) A term that may be applied to any fuel injection system that has the injectors located in the intake ports.

Inyección de combustible por orificio (PFI) Un término que se puede aplicar a cualquier sistema de inyección del combustible que tenga los inyectores localizados en los orificios de entrada.

Ported vacuum A vacuum source above the throttle plates when the throttle needs to open before vacuum is present.

Vacío por orificio Una fuente de vacío del arriba de las placas de la válvula reguladora cuando la válvula reguladora necesita abrirse antes de que esté presente un vacío.

Positive back pressure EGR EGR valve that uses exhaust system back pressure to sense engine load.

EGR positiva de la contrapresión Una válvula EGR que utiliza la contrapresión del dispositivo de escape para detectar la carga en el motor.

Positive crankcase ventilation (PCV) Emission system that picks up crankcase gases and sends them to the intake manifold through a control valve.

Ventilación positiva del cárter del motor (PCV) El sistema de la emisión que toma los gaes del cárter del motor y los envía a través de una válvula de control. al múltiple de producto.

Positive displacement The same volume of fuel is delivered every rotation of the pump, regardless of speed.

Dislocación positiva El mismo volumen de combustible se entrega en cada rotación de la bomba, sin importar la velocidad.

Positive plate The plate of a capacitor that is connected to the positive battery terminal.

Placa positiva La placa de un condensador que está conectado al terminal positivo de la batería.

Positive temperature coefficient (PTC) Thermistors that increase their resistance as the temperature increases.

Coeficiente de temperatura positiva (PTC) Los termistores que aumentan su resistencia al subir la temperatura.

Post-catalyst oxygen sensor An oxygen sensor located after the catalytic converter that is used to measure the O_2 content in the exhaust after the catalyst to determine the converter's storage capacity of O_2.

Sensor de oxígeno pos-catalizador Un sensor del oxígeno situado descendiente del convertidor catalítico que sirve para medir el contenido de O_2 en el escape saliendo del catalizador para determinar la capacidad de memoria de O_2 del convertidor.

Post-combustion control Emission controls that reduce emissions levels after combustion has taken place.

Control pos-combustión Los controles de emisión que reducen los niveles de las emisiones después de que haya ocurrido la combustión.

Potential energy Energy that it is available to be used for a purpose, but is not at this point in time.

Energía potencial La energía que está disponible para ser utilizado para un propósito, pero no en ese momento preciso.

Potentiometer A voltage divider that provides a variable DC voltage reading to the computer.

Potenciómetro Divisor del voltaje que proporciona una lectura del voltaje de C.C. variable para la computadora.

Pour point The lowest temperature at which the diesel fuel is observed to flow.

Punto líquido La temperatura más baja en la cual el combustible diesel se nota fluir.

Power-up vacuum test A passive test used to detect restrictions in the vent path of the EVAP system.

Prueba de ciclo inicial del vacío La prueba pasiva que se usa para detectar las restricciones en la trayectoria del respiradero del sistema de EVAP.

Pre-combustion controls Emission control and engine designs that prevent the formation of harmful emissions prior to the combustion process.

Control de precombustión Los diseños del motor y de control de emisión que previenen la formación de emisiones dañosas antes del proceso de la combustión.

Preignition An explosion in the combustion chamber resulting from the air-fuel mixture igniting prior to the spark being delivered from the ignition system.

Preignición Una explosión en la cámara de combustión como resultado de la mezcla del aire-combustible que enciende antes de que la chispa se entrega del sistema de ignición.

Pressure check mode The PCM will loop back and run in the pump mode again. See pump mode.

Modo de cheque de la presión El PCM comenzará el ciclo y funcionará de nuevo en el modo de la bomba. Véase modo de la bomba.

Pressure regulator Maintains the proper fuel pressure at all times.

Regulador de presión Mantiene constante la presión adecuada de combustión apropiada.

Pressure relief valve A safety valve that opens to prevent damage to the fuel delivery system if fuel pump pressure becomes excessive.

Válvula de descarga de presión Una válvula de seguridad que se abre para prevenir los daños al sistema de la entrega del combustible si la presión de la bomba de combustible llega a ser excesiva.

Primary circuit All of the components that regulate the current in the coil primary windings.

Circuito primario Todos los componentes que regulan la corriente en las bobinas primarias de la bobina.

Primary coil windings The second set of winding in the ignition coil. The primary windings will have about 200 turns to create a magnetic field to induce voltage into the secondary winding.

Devanado de las bobinas primarias Un grupo secundario de bobinas en la bobina de la ignición. Las bobinas primarias tendrán cerca de 200 vueltas para crear un campo magnético para inducir el voltaje en la bobina secundaria.

Program A set of instructions the computer must follow to achieve desired results.

Programa Un serie de instrucciones que la computadora debe seguir para realizar los resultados deseados.

Proton exchange membrane (PEM) Impedes the oxyhydrogen gas reaction in a fuel cell by ensuring that only protons (H+) and not elemental hydrogen molecules (H$_2$) react with the oxygen.

Membrana del intercambio de protón (PEM) Impide la reacción oxhídrica del gas en una célula del combustible asegurándose de que solamente los protones (H+) y las moléculas no elementales del hidrógeno (H$_2$) reaccionan con el oxígeno.

Protons Positively charged particles.

Protones Las partículas de carga positiva.

P-type material Materials with excessive protons. The P means positive and that it is the positive side of the circuit that attracts the free electrons through the semiconductor.

Materiales tipo P Las materiales que contienenl os protones excesivos. El P significa positivo y ése él es el lado positivo del circuito que atrae los electrones libres a través del semiconductor.

Pulse transformer Steps up low-voltage pulses to a higher voltage value by using induction principles.

Transformador del pulso Intensifica los pulsos de baja tensión a un valor más alto del voltaje usando principios de la inducción.

Pulse width The length of time the computer energizes the injector.

Anchura del pulso La cantidad del tiempo en que la computadora energiza el inyector.

Pump mode Cycling of the leak detections pump's on and off periods at a fixed rate to pressurize the system to 7.5 inches of water.

Modo de bomba Un ciclo de los periodos prendidos y apagados del detector de goteo que es de un intervalo fijo para presurizar el sistema a 7,5 pulgadas de agua.

Purge free cells Values placed in adaptive memory cells when the EVAP purge solenoid is off.

Células sín purga Los valores puestos en las células de memoria adaptiva caundo el solenoide de purga de EVAP está apagado.

Purge solenoid leak test An active test designed to detect a leak through the purge solenoid.

Prueba de fuga del solenoide de purga Una prueba activa diseñada para detectar una fuga a través del solenoide de la purga.

Quenching The cooling of gases as a result of compressing them into a thin area.

Temple El enfriamiento de gases resultando de comprimirlos en un área de poco espacio.

Quick-connect Coupler consists of a plastic retainer, two O-rings, and a plastic spacer. The retainer engages and locks into position on a raised bead on the fuel line.

Conexión rápido Un acoplador que consiste de un detenedor plástico, dos anillos en O, y un espaciador plástico. El detenedor engancha y se traba en la posición respecto a una bolita levantado en la línea de combustible.

Rationality Testing whether a particular sensor signal makes sense.

Racionalidad Probar si tiene sentido una señal del sensor en particular.

Reach The distance between the end of the spark plug threads and the seat or sealing surface of the plug. Plug reach determines how far the plug reaches through the cylinder head.

Alcance La distancia entre el extremo de los hilos de rosca del bujía de chispa y del asiento o de la superficie de estanqueidad del bujía. El alcance del bujía determina la distancia a que llega bujía a través de la culata.

Reciprocating engines Receive their name from the up and down or back and forth motion of the piston in the cylinder. This engine type is also referred to as the piston engine.

Motores recíprocos Reciben su nombre del movimiento hacia arriba y hacia abajo o hacia adelante y hacia atrás del pistón en el cilíndro. Este tipo del motor también se refiere como el motor de pistón.

Reed valve A one-way check valve. The reed opens to allow the air-fuel mixture to enter from one direction while closing to prevent movement in the other direction.

Válvula de lámina Una válvula de retención unidireccional. La lamina se abre para permitir que la mezcla del aire-combustible entrea partir de una dirección mientras que se cierra para prevenir el movimiento en la otra dirección.

Refining The process of breaking the crude oil down into different parts, called fractions.

Refinando El proceso de separar el petróleo crudo en sus elementos diversos, llamado fragmentos.

Reformer A high-temperature device that converts hydrocarbon fuels to CO and H$_2$.

Reformador Un dispositivo de alta temperatura que convierte los combustibles de hidrocarburo al CO y H$_2$.

Reformulated gasoline (RFG) A general term for federally mandated gasoline that is specially processed and blended to reduce the emission of pollutants such as hydrocarbons, toxics, and nitrogen oxides.

Gasolina reformulado (RFG) Un término general de la gasolina asignada por mandato federal que se procesa y se mezcla especialmente para reducir la emisión de agentes contaminadores tales como hidrocarburos, elemenos tóxicos, y óxidos del nitrógeno.

Refueling flag Used to prohibit the 0.020-inch (0.5mm) idle test of the EVAP system until the gross leak check is done during cruise conditions.

Seña de reaprovisión Se usa para prohibir la prueba en marcha vacío del 0.020-inch (0.5mm) del sistema de EVAP hasta que se haya comprobado la prueba de fugas completas en condiciones de travesía.

Regenerative braking Braking energy is turned back into electricity instead of heat.

Frenado regenerador La energía del frenado se cambia de nuevo a la electricidad en vez de al calor.

Regulator/vaporizer Converts the liquid propane to a gaseous vapor form.

Regulador/vaporizador Convierte el líquido propano a una forma de vapor gaseosa.

Reid vapor pressure (RVP) The pressure of the vapor above the fuel when the fuel is at 1000°F (380°C).

Presión del vapor de Reid (RVP) La presión del vapor más allá del combustible cuando el combustible está en 1000°F (380°C).

Reluctance The term used to indicate a material's resistance to the passage of flux lines.

Reluctancia El término usado para indicar una resistencia de los materiales al paso de las líneas del flujo.

Reserve voltage The difference between the required voltage and the maximum available voltage is referred to as secondary.

Voltaje de la reserva La diferencia entre el voltaje requerido y el voltaje disponible máximo se refiere como secundario.

Resistance The opposition to current flow. In a circuit, resistance controls the amount of current.

Resistencia La oposición al flujo actual. En un circuito, la resistencia controla la cantidad del corriente.

Resistor plugs Spark plugs with resistor built into the electrode core.

Bujía resistor Los bujías de chispa que tiene incorporado un resistor en la base del electrodo.

Resource Conservation and Recovery Act (RCRA) Stipulates that the users of hazardous materials are responsible for the material from the time they become a waste until they are properly disposed of.

Conservación de los Recursos y el Acto de la Recuperación (RCRA) Estipula que los usuarios de materiales peligrosos son responsables del material a partir del tiempo que se convierten en un deshecho hasta que se disponen de ellos correctamente.

Response time test Monitors the transition times between lean-to-rich and rich-to-lean of the upstream O_2 sensors.

Prueba del tiempo de reacción Supervisa los tiempos de la transición entre pobre-a-rico y rico-a-pobre en los sensores ascendientes de O_2.

Retarded timing A spark that is delivered late as compared to piston travel.

Sincronización retardado Una chispa que se entrega tarde con respecto al recorrido del pistón.

Return-type fuel systems All fuel is routed through the fuel rail, and what is not used by the engine is returned to the fuel tank.

Sistemas de combustión tipo retorno Todo el combustible se encamina a través del carril del combustible, y lo qué no se utiliza por el motor se vuelve al depósito de gasolina.

Returnless-type fuel system Does not have a return line routed from the fuel rail to the fuel tank.

Sistema de combustión tipo sín retorno No tiene una línea de vuelta encaminada del carril del combustible al depósito de gasolina.

Reversed-bias The positive voltage is applied to the N-type material and negative voltage is applied to the P-type material so the diode will not conduct.

Polarización invertido El voltaje positivo se aplica al material tipo N y el voltaje negativo se aplica al material tipo P para que el diodo no conducirá.

Rich An air-fuel ratio that contains excessive fuel in relation to the amount of air entering the cylinders.

Ricos Un cociente del aire-combustible que contiene el combustible excesivo en lo referente a la cantidad de aire que entra en los cilindros.

Right-hand rule Identifies the direction of the lines of force of an electromagnet.

Regla a derechas Identifica la dirección de las líneas de la fuerza de un electroimán.

Right-to-know laws Statutory regulations that protect the worker in the workplace by informing them of what type of hazardous materials are stored there.

Regulaciones del derecho-a-saber Las estatutarias de los leyes que protegen al trabajador en el lugar de trabajo informándoles qué tipo de materiales peligrosos se almacenan.

Rise time According to the SAE, is the amount of time (measured in microseconds) for the output of the coil to rise from 10 percent to 90 percent of its maximum output.

Tiempo de subida Según el SAE, es la cantidad del tiempo (medida en microsegundos) para que la salida de la bobina a subir del 10 por ciento a 90 por ciento de su salida máxima.

Roll-over check valve A safety valve that closes the fuel flow in the event of a roll-over accident. It may also be part of the evaporative emission controls.

Válvula de retención de invertido Una válvula de seguridad que cierra el flujo del combustible en el acontecimiento de un accidente en que se invierte el vehículo. Puede también ser parte de los controles de emisiónes evaporativos.

Rotary valve A valve that rotates to cover and uncover the intake port. It is usually designed as a flat disc that is driven from the crankshaft.

Válvula rotatoria La válvula gira para cubrir y destapar el orificio de admisión. Se diseña generalmente como un disco plano impulsado por el cigüeñal.

Rotational velocity fluctuations Used to determine misfire. As the crankshaft is rotating, every power impulse should increase its speed. If a misfire occurs, the crankshaft slows down.

Irregularidades girtorias de la velocidad Sirven para determinar un fallo del encendido. Mientras que el cigüeñal está girando, cada impulso de la energía debe aumentar su velocidad. Si ocurre una falla del encendido, el cigüeñal retrasa.

Rotor Attached to the top of the distributor shaft, it directs secondary voltage from the coil to the terminal of the distributor cap.

Rotor Conectado a la parte superior del eje de la distribuidor, dirige el voltaje secundario de la bobina al terminal del casquillo de la distribuidor.

Safety goggles Eye protection that provides protection from all sides since they fit against the face and forehead to seal off the eyes from outside elements.

Anteojos o gafas de seguridad La protección para los ojos que proporciona la protección en toda parte puesto que se ajustan contra la cara y la frente para sellar los ojos contra los elementos exteriores.

Saturation The point that the magnetic strength eventually levels off, and where an additional increase of the magnetizing force current no longer increases the magnetic field strength.

Saturación El punto en que la fuerza magnética eventualmente se nivela, y en donde una cantidad adicional de la corriente de la fuerza magnetizante no aumenta más la fuerza magnético del campo.

Schmitt trigger An A/D converter.

Disparador de Schmitt Un convertidor de Análogo a Digital.

Secondary circuit The portion of the ignition system that carries high voltage to the combustion chamber.

Circuito secundario La porción del sistema de ignición que lleva el alto voltaje hacia la cámara de combustión.

Secondary coil windings One of two coils in the ignition coil. This winding has several thousand turns and is where low voltage will be transformed to a high voltage.

Bobinas secundarias Una de dos bobinas en la bobina de la ignición. Esta bobina tiene unas mil vueltas y es en dónde la baja tensión será transformada al alto voltaje.

Self-induction The generation of an electromotive force by a changing current in the same circuit.

Autoinducción La generación de una fuerza electromotriz por una corriente que cambia en el mismo circuito.

Semiconductors Materials that conduct electric current under certain conditions, yet will not conduct under other conditions.

Semiconductores Los materiales que conducen la corriente eléctrica bajo ciertas condiciones, pero que no conducirán bajo otras condiciones.

Sensor voltage test Monitors upstream and downstream HO$_2$S for voltages that are out of range.

Prueba del sensor del voltaje Supervisa HO$_2$S en sentido ascendente y descendente para determinar los voltajes que están fuera de la gama.

Sensors Convert some measurement of vehicle operation into an electrical signal.

Sensores Convierten algúna medida de la operación del vehículo en una señal eléctrica.

Sequential fuel injection (SFI) Fuel system whereby each injector is fired individually in ignition firing order prior to the intake valve opening.

Inyección secuencial del combustible (SFI) Sistema de combustible por el cual cada inyector se encienda individualmente en orden de disparo de la ignición antes de la abertura de la válvula de la entrada.

Sequential logic circuits Flip-flop circuits where the output is determined by the sequence of inputs. A given input affects the output produced by the next input.

Circuitos de lógica secuenciales Los circuitos de apagado-prendido donde la salida es determinada por la secuencia de entradas. Una entrada dada afecta la salida producida por la entrada siguiente.

Sequential sampling The process that the MUX and DEMUX operate on. This means the computer will deal with all of the sensors and actuators one at a time.

Muestreo secuencial El proceso en el cual operan los MUX y los DEMUX. Esto significa que la computadora tratará con todos los sensores y actuadores uno a la vez.

Series hybrid Hybrid configuration where propulsion comes directly from the electric motor.

Híbrida de serie Una configuración híbrida en la cual la propulsión viene directamente del motor eléctrico.

Shell The orbit of the electrons around the nucleus.

Cáspula La órbita de los electrones alrededor del núcleo.

Short-term fuel trim An instantaneous correction to the pulse width calculation. It is driven directly by the oxygen sensor activity.

Ajuste a corto plazo del combustible Una corrección instantánea al cálculo de la anchura del pulso. Es conducido directamente por la actividad del sensor del oxígeno.

Shutter wheel Consists of a series of alternating windows and vanes. It creates a magnetic shunt that changes the strength of the magnetic field from the permanent magnet.

Rueda obturador Consiste en una serie de ventanas y de paletas que se alternan. Crea una desviación magnética que cambie la fuerza del campo magnético del imán permanente.

Similar conditions window (SCW) A frame of stored data that records the operating conditions at the time the fault originated. Used to determine pass/fail of an OBD-II monitor.

Muestra de las condiciones parecidos (SCW) Un cuadro de los datos almacenados que registran las condiciones del funcionamiento en el momento que originó la avería. Determina éxito o reprueba de un monitor de Obd-II.

Slow moving average A combination of instantaneous switch frequencies of the downstream oxygen sensor and stored history. Used to detect a naturally aging high-mileage catalyst.

Promedio móvil retardado Una combinación de las frecuencias instantáneas del interruptor del sensor en sentido descendente del oxígeno y de la historia almacenada. Sirven para detectar el envejecimiento natural de un catalizador del alto-kilometraje.

Spark knock A metallic noise an engine makes, usually during acceleration, resulting from abnormal or uncontrolled combustion inside the cylinder. See detonation.

Golpeteo de la chispa Un ruido metálico que hace un motor hace, generalmente durante la aceleración, resultando de la combustión anormal o incontrolada dentro del cilindro. Véase la detonación.

Spark plug Designed to transfer the high-voltage current to the combustion chamber where the electrical spark that is produced is the motivating force behind establishing combustion of the air-fuel mixture.

Bujía de chispa Diseñado para transferir la corriente de alto voltaje a la cámara de combustión en dónde la chispa eléctrica produce la fuerza motivadora para establecer la combustión de la mezcla del aire-combustible.

Spark scatter Refers to altering the ignition timing rapidly and is used to control idle quality. Since the idle air control motor is not capable of making very fine adjustments, fuel pulse width and spark scatter make fine changes in idle speed.

Dispersión de la chispa Refiere a alterar la sincronización de ignición rápidamente y se utiliza para controlar calidad de marcha en vacío. Puesto que el motor del control del aire de marcha en vacío no es capaz de hacer ajustes muy finos, la anchura del pulso del combustible y la dispersión de la chispa efectúan los cambios precisos en marcha en vacío.

Speed density Refers to the fuel-injection system that uses engine speed and MAP inputs to infer the mass of air entering the combustion chamber and responds based on programmed look-up tables in the PCM's memory chips.

Densidad de la velocidad Refiere al sistema de la inyección de combustión que utiliza la velocidad del motor y las entradas delMAPA para deducir la masa del aire que entra en la cámara de combustión y responde basado en las tablas programadas de referencia en las chips de memoria del PCM.

Static electricity Electricity that is not in motion.

Electricidad estática La electricidad que no está en movimiento.

Stepper motor Contains a permanent magnet armature with two, four, or more field coils.

Motor de pasos Contiene una armadura permanente de imán con dos, cuatro, o más bobinas de campo.

Stoichiometric From Greek words meaning "measured element." Mixing air and fuel at the stoichiometric ratio of 14.7:1 is the single most important technique that is used to control emissions levels. When the air-fuel ratio is at stoichiometric, every fuel molecule combines with every oxygen molecule, with nothing left over (theoretically).

Stoichiometrico Viene de una palabra griega significando "un elemento medido" La mezcla del aire y el combustible en el cociente stoichiometrico de 14.7:1 es la técnica más importante que se utiliza para controlar los niveles de las emisiones. Cuando el cociente del aire-combustible está en stoichiometrico, cada molécula del combustible combina con cada molécula del oxígeno, y no queda nada en exceso (teóricamente).

Stratified Charge Engine Refers to the layering of air-fuel mixture.

Motor de carga estratificada Refiere a acodar la mezcla del aire-combustible.

Stroke The movement of the piston from one end of its travel to the other.

Carrera El movimiento del pistón a partir de un extremo de su recorrido al otro.

Suspended An OBD-II monitor that stops when it has recorded a failure in the first test, and also a failure in the second test. The fault code maturing may be suspended until other components are run again to check their function since they may be what is causing this monitor to fail.

Suspendido Un monitor de Obd-II que para cuando ha registrado una falta en la primera prueba, y también una falta en la segunda prueba. La entregada del código de avería se puede suspender hasta que otros componentes se accionan otra vez para comprobar su función puesto que pueden ser que están causando fallar a este monitor.

Switch points The upper and lower ranges of the O₂ sensor voltage.

Puntos interruptor Las gamas superiores y más bajas del voltaje del sensor O₂.

Switch ratio Calculated by dividing the number of downstream oxygen sensor switches by the number of upstream oxygen sensor switches.

Cociente de interuptor Calculado al dividir el número de los interruptores del sensor del oxígeno en el sentido descendiente por el número de los interruptores del sensor del oxígeno en el sentido ascendiente.

Task manager A software program that directs the functions of the OBD-II system.

Gerente de tareas Una programa del software que dirige las funciones del sistema de Obd-II.

Termination resistors Resistors that are used to control induced voltages. Since voltage is dropped over resistors, the induced voltage is terminated.

Resistores de la terminación Los resistores que se utilizan para controlar los voltajes inducidos. Puesto que el voltaje se cae sobre los resistores, se termina el voltaje inducido.

Test fail actions The responses the PCM has to function when a test fails.

Acciones fallo de prueba Las respuestas que tiene que implementar el PCM cuando una prueba falla.

Test frequency The actual measured output frequency of the downstream HO₂S.

Prueba de frecuencia La medida actual de la frecuencia de la salida del HO₂S en sentido descendiente.

Test mode When only the leak detection pump's on period is cycled at a fixed rate. The pump is operated until the system is pressurized and the pump stalls.

Modo de prueba Cuando solamente las bombas de la detección del escape el período se completan un ciclo en una tarifa fija. Se funciona la bomba hasta que se presuriza el sistema y las paradas de la bomba.

Test-Only station An emissions test location that does not repair the vehicle, it only tests the vehicle.

Estación solo para Prueba Una localización de pruebas de las emisiones prueba que no repara el vehículo, solamente prueba al vehículo.

Thermal efficiency A measurement comparing the amount of energy present in a fuel and the actual energy output of the engine.

Eficacia termal Una medida que ompara la cantidad de energía presente en un combustible y la salida actual de la energía del motor.

Thermistor A solid state variable resistor made from a semi-conductor material that changes resistance in relation to temperature changes.

Termistor Un resistor variable de estado sólido hecho de un material de semiconductor que cambia la resistencia referente a los cambios de temperatura.

Thermodynamics The study of the relationship between heat energy and mechanical energy.

Termodinámica El estudio de la relación entre la energía térmica y la energía mecánica.

Three-way converter Used to help clean up the NO$_x$, HC, and CO emissions in the engine exhaust (hence the term three-way).

Convertidor de tres vías Sirve para limpiar las emisiones de NO$_x$, de HC, y del CO en el escape del motor (por lo tanto el término de tres vías).

Three-way vacuum switch valve (VSV) Allows the vapor pressure sensor to detect either canister or tank pressure.

Válvula interruptor del vacío de tres vías del (VSV) Permite que el sensor de la presión del vapor detecte el frasco o la presión del tanque.

Throttle body injection (TBI) A system that has the injector(s) located in a throttle body assembly and above the throttle plates.

Regulador de la inyección del cuerpo (TBI) El sistema que tiene el (los) inyector(es) situado en un montaje del cuerpo de la válvula reguladora y arriba de las placas de la válvula reguladora.

Throttle body An assembly that contains a pressure regulator, injector (or injectors), TPS, idle speed control motor, and throttle shaft and linkage assembly.

Cuerpo del regulador Una asamblea que contiene un regulador de presión, un inyector (o inyectores), un TPS, un motor del control de la marcha en vacío, y un eje de la válvula reguladora y un montaje del acoplamiento.

Thyristor A semiconductor switching device composed of alternating N and P layers. It can also be used to rectify current from AC to DC.

Tiristor Un dispositivo de la conmutación del semiconductor compuesto de capas alternantes de N y de P. Puede también ser utilizado para rectificar la corriente de la CA a la C.C.

Time to activity test Monitors the heater system by measuring the time the sensor requires to become active.

Prueba de tiempo de la actividad Supervisa el sistema del calentador midiendo el tiempo que requiere el sensor en ser activo.

Timing disc Also known as an armature, reluctor, trigger wheel, pulse wheel, or timing core. It is used to conduct lines of magnetic force.

Disco sincronizador También llamado una armadura, un reluctor, una rueda del disparador, una rueda del pulso, o una base de la sincronización. Se utiliza para conducir las líneas de la fuerza magnética.

Top dead center (TDC) Indicates the piston is at the very top of its stroke.

Punto muerto superior (PMS) Indica que el pistón está en la parte más alta de su movimiento.

Torque A rotating force around a pivot point.

Torsión Una fuerza que gira alrededor de un punto de pivote.

Total advance The total of base timing plus centrifugal and vacuum advance.

Avance total El total de sincronización base más el avance del centrífugo y el vacío.

Total engine displacement The sum of displacements for all cylinders in an engine.

Dislocación total del motor La suma de las dislocaciones de todos los cilindros en un motor.

Transient A part of the emissions tests wherein the vehicle will undergo modes where it accelerates and decelerates following a specific driving cycle.

Transeúnte Una parte de la prueba de emisiones en que el vehículo es sometido a los modos donde acelera y decelera siguiendo un ciclo de conducción específico.

Transistor A three-layer semiconductor. It is used as a very fast switching device.

Transistor Un semiconductor de tres capas. Se utiliza como dispositivo muy rápido de la conmutación.

Trip A set of vehicle operating conditions that must be met for a specific monitor to run. All trips require a key cycle.

Viaje Un grupo de las condiciones de funcionamiento del vehículo que se deben resolver para que un monitor específico funcione. Todos los viajes requieren un ciclo dominante.

Turn-on voltage The voltage required to jump the PN junction of a diode or transistor and allow current to flow.

Voltaje de abertura El voltaje requerido para activar la ensambladura del PN de un diodo o de un transistor y para permitir que fluya la corriente.

Two-trip faults OBD-II tests that require two failures in a row before the MIL is illuminated. This allows the system to doublecheck itself to prevent the MIL from illuminating when a fault really does not exist. These are also called "B monitors."

Fallos de dos-viaje Las pruebas de Obd-II que requieren dos faltas una atrás de la otra antes de que prende la MIL. Esto permite que el sistema se verifica si mismo para evitar que la prende la MIL cuando no existe una avería realmente. Éstos también se llaman los "monitores B".

Ultra capacitor A device that stores energy as electrostatic charge. It is the primary device in the power supply during hill climbing, acceleration, and the recovery of braking energy.

Ultra condensador Un dispositivo almacena energía en forma de carga electrostática. Es el dispositivo primario en la fuente de alimentación al subir una colina, la aceleración, y la recuperación de la energía del frenado.

Vacuum advance System that measures engine load through manifold vacuum and moves the breaker plate to make adjustments.

Vacío anticipado Un sistema que mide la carga del motor por medio del vacío del múltiple y mueve la placa del triturador para efectuar los ajustes.

Valves Devices that control the flow of gases into and out of the engine cylinder.

Válvulas Los dispositivos que controlan el flujo de gases entrando y saliendo del cilíndro del motor.

Vapor lock A result of fuel boiling in the fuel line or tank. Unlike liquid fuel, vapor is compressible. This means that the fuel cannot be pumped to the carburetor or injectors, so the engine stalls. After the fuel line cools sufficiently, the engine will run again. Fuel-injected engines use electric fuel pumps that keep the fuel under higher pressure. Higher pressure raises the fuel's boiling point so vapor lock is less likely to occur.

Tapón de vapor Un resultado de el combustible que hierve en la línea o en el tanque de combustible. Desemejante del combustible líquido, el vapor es compresible. Esto significa que el combustible no se puede bombear al carburador o a los inyectores, así que el motor se cala. Después de que la línea de combustible se enfría lo suficiente, el motor funcionará otra vez. los motores de combustible-inyectados utilizan las bombas de gasolina eléctricas que mantienen el combustible bajo una presión más alta. Una presión más alta sube la temperatura en que que hierven los combustibles,así que el tapón de vapor es menos probable.

Vapor pressure sensor (VPS) Measures the vapor pressure in the evaporative emission control system.

Sensor de la presión del vapor (VPS) Mide la presión del vapor en el sistema de control evaporativo de emisión.

Vapor pressure Pressure exerted by a vapor in equilibrium with its liquid state.

Presión de vapor La presión ejercida por un vapor en equilibrio con su estado líquido.

Viscosity A measure of a liquid's resistance to flow.

Viscosidad Una medida de la resistencia en fluir.de un líquido.

Volatile A substance that easily vaporizes or explodes.

Volátil Una sustancia que se vaporiza o estalla fácilmente.

Volatility A measure of how easily a fuel evaporates (forms a vapor). When fuel does not vaporize easily, said to have low volatility.

Volatilidad Una medida de que tal fácilmente un combustible se evapora (forma un vapor). Cuando el combustible no se vaporiza fácilmente, se dice que tiene volatilidad baja.

Voltage The difference or potential that indicates an excess of electrons at the end of the circuit the farthest from the force. It is the electrical pressure that causes electrons to move through a circuit.

Voltaje La diferencia o el potencial que indica un exceso de electrones en el extremo del circuito el más lejano de la fuerza. Es la presión eléctrica que hace electrones moverse a través de un circuito.

Voltage drop A resistance in the circuit that reduces the electrical pressure available after the resistance. A resistance can be the load component, the conductors, any connections, or unwanted resistance.

Caída de voltaje La resistencia en el circuito que reduce la presión eléctrica disponible después de la resistencia. Una resistencia puede ser el componente de la carga, los conductores, cualquier conexión, o resistencia indeseada.

Volumetric efficiency A measurement of the amount of air-fuel mixture that actually enters the combustion chamber compared to the total amount that could.

Eficacia volumétrica La medida de la cantidad de mezcla del aire-combustible que entra actualmente en la cámara de combustión comparado a la cantidad total que podría entrar.

Warmup cycle Counts are used to erase OBD-II fault codes from the PCM's memory. The number of warm-up cycles does not increase until the MIL is turned off by three good trips. Defined as an increase in engine temperature of 40A1F and crossing over 160A1F.

Ciclo del calentamiento Las cuentas que se utilizan para borrar los códigos de fallos del OBD-II de la memoria del PCM. El número de los ciclos del calentamiento no aumenta hasta que la MIL se apaga por tres buenos viajes. Definido como aumento en la temperatura del motor de 40A1F y de la travesía 160A1F.

Waste spark A spark that occurs during the exhaust stroke of a piston.

Chispa desperdiciada Una chispa que ocurre durante la carrera de escape del pistón

Weak vacuum test An active test designed to detect large leaks in the EVAP system.

Prueba débil del vacío Una prueba activa diseñado para detectar las fugas grandes en el sistema de EVAP.

Wheatstone bridge A series-parallel arrangement of resistors between an input terminal and ground.

Puente wheatstone Un arreglo de resistores en serie y paralelo entre un terminal de entrada y tierra.

Workplace hazardous materials information systems (WHMIS) The Canadian equivalent to the MSDS used to inform workers of hazardous material facts.

Sistemas de información peligrosos de materiales del lugar de trabajo (WHMIS) El equivalente para los canadienses al MSDS que informa a los trabajadores de hechos de materiales peligrosos.

Zener diode A diode that will conduct in a reverse direction once the required amount of voltage is applied.

Diodo Zener Un diodo que conducirá en la dirección contraria una vez que se le aplica la cantidad requerida de voltaje.

Zener voltage The voltage that is reached when the diode conducts in reverse direction.

Voltaje Zener El voltaje que se realiza cuando el diodo conduce en la dirección contraria.

INDEX

U

Ultra capacitors, 349

V

Valve timing, 28
Vapor flow, 152, 153
Vapor lock, 22
Vapor management valve, 295
Vapor pressure, 22
Vapor pressure sensor (VPS), 306–310
Variable pulse width modulation (VPWM) voltage,
    87, 88
Vent solenoid valve assembly, 151
Viscosity, 36–37
Visual tampering inspections, 321
Volatile substances, 10
Volatility, 22–23
Volumetric efficiency, 99, 101

W

Warm-up cycle, 266, 267
Warm-up spark advance, 125
Waste pack, 127, 128
Weak vacuum test, 292
Wheatstone bridges, 69–70
Wide-open throttle (WOT), 187
Wide range sensor, 313
Wide ratio sensor, 313
Winter blend gasoline, 24
Workplace hazardous materials information systems
    (WHMIS), 11

X

XOR gate, 54

Z

Zener diode, 60
Zener voltage, 60